AF443326

01

$102
1992

APPLICATIONS OF ENZYME BIOTECHNOLOGY

INDUSTRY–UNIVERSITY COOPERATIVE CHEMISTRY PROGRAM SYMPOSIA

Published by Texas A&M University Press

ORGANOMETALLIC COMPOUNDS
Edited by Bernard L. Shapiro

HETEROGENEOUS CATALYSIS
Edited by Bernard L. Shapiro

NEW DIRECTIONS IN CHEMICAL ANALYSIS
Edited by Bernard L. Shapiro

APPLICATIONS OF ENZYME BIOTECHNOLOGY
Edited by Jeffery W. Kelly and Thomas O. Baldwin

CHEMICAL ASPECTS OF ENZYME BIOTECHNOLOGY: Fundamentals
Edited by Thomas O. Baldwin, Frank M. Raushel, and A. Ian Scott

DESIGN OF NEW MATERIALS
Edited by D. L. Cocke and A. Clearfield

FUNCTIONAL POLYMERS
Edited by David E. Bergbreiter and Charles R. Martin

METAL–METAL BONDS AND CLUSTERS IN CHEMISTRY
AND CATALYSIS
Edited by John P. Fackler, Jr.

OXYGEN COMPLEXES AND OXYGEN ACTIVATION BY
TRANSITION METALS
Edited by Arthur E. Martell and Donald T. Sawyer

APPLICATIONS OF ENZYME BIOTECHNOLOGY

Edited by

Jeffery W. Kelly and Thomas O. Baldwin

Texas A&M University
College Station, Texas

PLENUM PRESS • NEW YORK AND LONDON

Library of Congress Cataloging-in-Publication Data

Texas A & M University. IUCCP Symposium on Applications of Enzyme
  Biotechnology (9th : 1991)
    Applications of enzyme biotechnology / edited by Jeffery W. Kelly
  and Thomas O. Baldwin.
        p.   cm. -- (Industry-university cooperative chemistry program
  symposia)
    "Proceedings of the Texas A & M University, IUCCP Ninth Annual
  Symposium on Applications of Enzyme Biotechnology, held March 18-21,
  1991, in College Station, Texas"--T.p. verso.
    Includes bibliographical references and index.
    ISBN 0-306-44095-4
    1. Enzymes--Biotechnology--Congresses.   I. Kelly, Jeffery W.
  II. Baldwin, Thomas O.   III. Title.   IV. Series.
  TP248.65.E59T47   1991
  660'.634--dc20   91-41625
                                                                   CIP

Proceedings of the Texas A&M University, IUCCP Ninth Annual Symposium
on Applications of Enzyme Biotechnology, held March 18-21, 1991, in College Station, Texas

ISBN 0-306-44095-4

© 1991 Plenum Press, New York
A Division of Plenum Publishing Corporation
233 Spring Street, New York, N.Y. 10013

Printed in the United States of America

FOREWORD

The Industry-University Cooperative Chemistry Program (IUCCP) has sponsored eight previous international symposia covering a range of topics of interest to industrial and academic chemists. The ninth IUCCP Symposium, held March 18-21, 1991 at Texas A&M University was the second in a two part series focusing on Biotechnology. The title for this Symposium "Applications of Enzyme Biotechnology" was by design a rather all encompassing title, similar in some respects to the discipline. Biotechnology refers to the application of biochemistry for the development of a commercial product. Persons employed in or interested in biotechnology may be chemists, molecular biologists, biophysicists, or physicians. The breadth of biotech research projects requires close collaboration between scientists of a variety of backgrounds, prejudices, and interests.

Biotechnology is a comparatively new discipline closely tied to new developments in the fields of chemistry, biochemistry, molecular biology and medicine. The primary function of Texas A&M University is to educate students who will be appropriately trained to carry out the mission of biotechnology. The IUCCP Symposium serves as an important forum for fostering closer ties between academia and industry and exchanging ideas so important to this evolving area.

The topics that were discussed during this conference, include the oxidation of alkanes by enzymes, protein folding, waste remediation, protein purification techniques, and protein expression systems. These titles represent a smorgasbord of topics of importance to the biotechnology industry. The manuscripts submitted point out not only the tremendous progress made in each one of those areas, but also discuss the challenges still facing the industry as a whole. Many of the problems facing the biotech companies are the same problems that academic biochemists and molecular biologists face on a daily basis. It was clear to all participants that general solutions to thorny problems such as protein expression, waste remediation, and refolding recombinant proteins could form the basis for very successful companies. This pioneering and entrepreneurial spirit is what makes biotechnology so exciting and what attracts some of the brightest people to this area.

We are deeply indebted to the IUCCP sponsoring companies Abbott Labs, Hoechst-Celanese, Monsanto Chemical Company, BF Goodrich, Dow Chemical Company for providing the necessary resources to carry out this endeavor.

The co-chairmen of the conference were Professor Thomas O. Baldwin, and Frank M. Raushel of the Texas A&M University Chemistry Department. The program was developed by an academic steering committee consisting of the co-chairmen and members appointed by the sponsoring chemical companies Dr. James Burrington, BP America; Dr. Robert Durrwater, Hoechst-Celanese; Dr. Barry Haymore, Monsanto Chemical Company; Dr. Mehmet Gencer, BF Goodrich; Dr. Paul Swanson, Dow Chemical Company; and Professor Arthur Martell, Texas A&M IUCCP Coordinator.

In closing, the organizers of the Ninth IUCCP Symposium must recognize the contributions that have been made to the symposium by Mrs. Mary Martell, who dealt with the innumerable details necessary for a successful symposium. Her pleasant nature and efficiency are appreciated. Finally, we wish to thank the Texas A&M graduate students who donated their time to ensure smooth operations.

Thomas O. Baldwin
Jeffery W. Kelly

CONTENTS

RADIOLABELED ANTIBODIES: INTRODUCTION AND METAL

CONJUGATION TECHNIQUES

Sally W. Schwarz **and**   Michael J. Welch

Mallinckrodt Institute of Radiology
Washington University, 510 S. Kingshighway
St. Louis, MO, U.S.A.

INTRODUCTION

The use of radiolabeled antibodies in the detection and treatment of cancer has been in practice since the early 1980's. Radioimmunoimaging is an *in vivo* diagnostic technique where a radiolabeled antibody is taken up or bound to an antigen in a target tissue. This allows for non-invasive imaging of the antigen containing tissue, using a gamma camera or a positron emission tomograph (PET) scanner, for subsequent therapy or resection of the tissue if necessary. Radioimmunotherapy is the delivery of a therapeutic quantity of a radioisotope to the same antigen containing tissue to ablate or reduce a primary or metastatic carcinoma.   This chapter will cover the basic principles of antibodies, subsequent conjugation with bifunctional chelates and radiolabeling for the purpose of radioimmunoimaging or radioimmunotherapy.

Antibodies (Ab) are immunoglobulins produced as a result of the body's immune response.   This reaction is triggered when the body is faced with foreign matter. "Immunogens" or antigens can be bacteria, viruses, fungi, or any foreign proteins.

Each antigen (Ag) has more than one epitope or antigenic determinant (figure 1). These epitopes represent only a small fraction of the Ag molecule.

Once a single Ag is injected into the bloodstream it interacts either by direct associaton with a ß-lymphocyte or the Ag is "processed" by macrophages and "presented" to the ß-lymphocyte by a T-helper cell. After interaction with the ß-lymphocyte the ß-cell is activated, and proliferates. It then differentiates into a plasma cell which secretes the Ab specific for the Ag (figure 2). After Ag injection there is a lag time of approximately one week before production of these Ag specific Ab.

Antibodies are glycoproteins consisting of five different classes: IgG, IgM, IgA, IgE and IgD (figure 3). Each group contains one or more subunits of a Y shape. Each Y unit contains 4 polypeptides, two identical segments called heavy chains and 2 segments called light chains (figure 4). Each class of Ab has a specific type of heavy chain, but there are only 2 types of light chain polypeptides known as kappa (κ) and lambda (λ). These chains are held together with multiple inter and intra disulfide bonds and non-covalent bonds. When the body is faced with an unknown Ag, the IgM class of immunoglobulins is

Applications of Enzyme Biotechnology, Edited by J.W. Kelly and
T.O. Baldwin, Plenum Press, New York, 1991

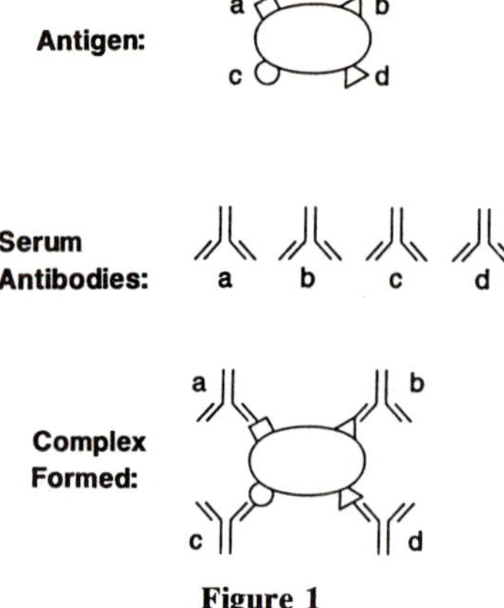

Figure 1

activated. These form memory cells which recognize the Ag in later years. IgG's are primarily produced as the secondary response to reinfection. IgA is found in mucous membranes and secretions. IgD and IgE make up less than 1% of total serum immunoglobulins.

Each Y unit (figure 5) of an intact Ab has 3 protein domains. Two of these are identical, called Fab; the third domain is the Fc region. Each Fab portion consists of a heavy chain and a light chain held together by disulfide bridges. The amino acids of the amino terminal end of these peptide chains determine the Ag binding site, and vary from Ab to Ab making it known as the variable region. A heavy chain and a light chain are required for tight Ag-Ab binding. The carboxyl-terminal regions of the two heavy chains form the Fc domain.

The IgG Ab is susceptible to enzymatic digestion (figure 6)[1]. Using papain the IgG molecule will split into three parts, two Fab fragments and the Fc complement.

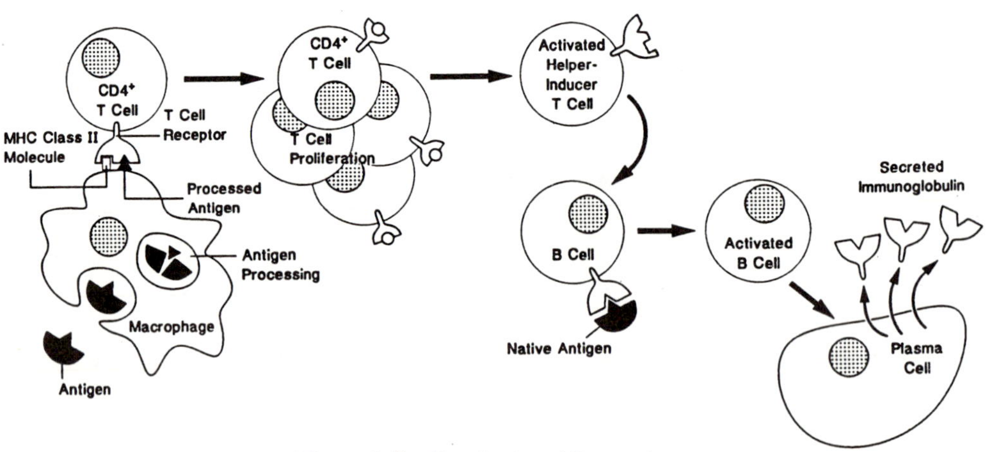

Figure 2. β cell activation; AB secretion.

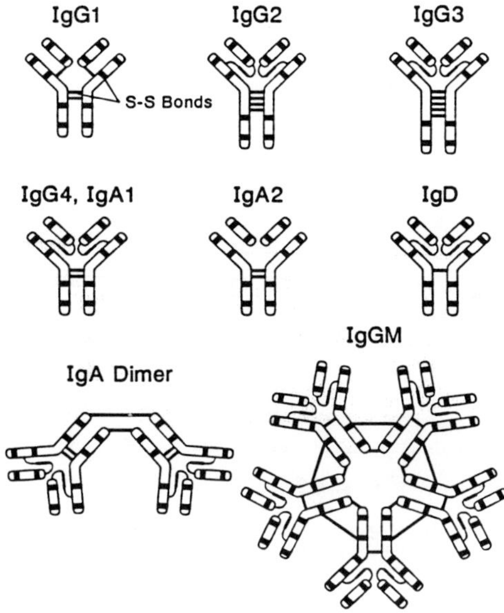

Figure 3. Immunoglobin classes.

Alternatively, pepsin digestion gives two $F(ab')_2$ fragments and a small Fc complement. Further digestion of the $F(ab')_2$ with ß-mercaptoethanol yields two Fab fragments. Since the Fc fragment (fragment which crystallizes) is often the portion of the Ab which causes allergic reaction, removal of this segment of the Ab should cause less immune response. Additionally, the Fc portion tends to bind non-specifically to numerous tissues. Removal of the Fc segment would reduce non-specific tissue binding.

An Ag injected into the bloodstream contains more than one epitope. Antibodies are formed in response to each epitope. The resulting Ab production is termed polyclonal. These Ab can recognize several different epitopes on the Ag and therefore can bind non-specifically to several similar epitopes.

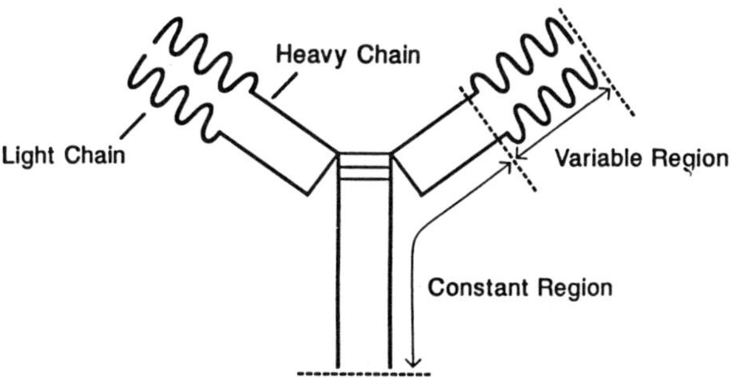

Figure 4. Antibody.

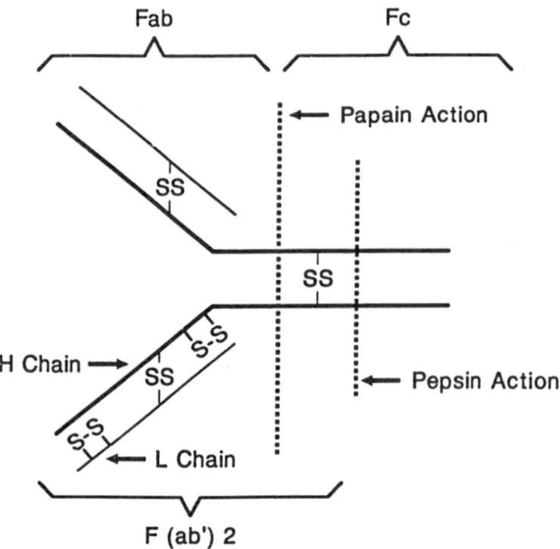

Figure 5. Diagram of the intact monoclonal antibody of the IgG type and its fragments.

Polyclonal Ab can be purified using affinity chromatography (figure 7)[1], a technique used to isolate biomolecules on the basis of biological function, and in this case, yield a more specific Ab. An Ag can be covalently attached to a gel matrix. The polyclonal Ab can then be applied to the column and eluted using a pH gradient. The result is a relatively specific Ab, but cross reactivity of Ab does occur.

Since the plasma cell secretes the Ab, an ideal situation would exist if it was possible to isolate a single plasma cell and clone it in tissue culture. This would lead to the production of a single Ab, with no need for further purification. The problem which exists is the plasma cell will not live in tissue culture.

In 1975 Kohler and Milstein[2] developed hybridoma technology, the fusion of two different types of cells to produce a hybrid-melanoma (hybridoma). This cell line has characteristics of each parent cell. Fusion is achieved by exposing "immortal" myeloma cells (hardy cells capable of producing large quantities of IgG) and splenic ß-cells (cells from a mouse previously immunized with a selected Ag) to polyethylene glycol (figure 8).

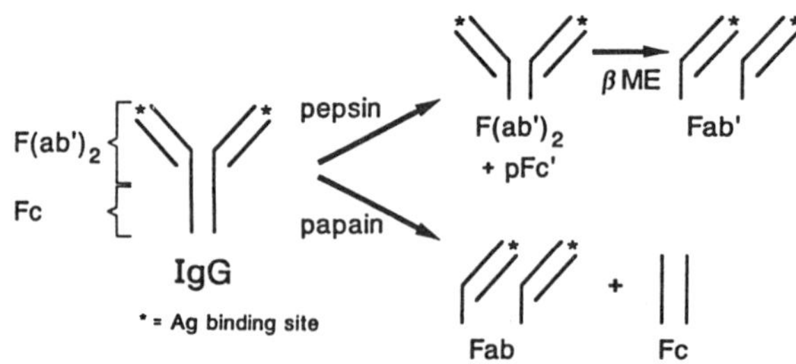

Figure 6. The structure of intact IgG and enzyme-digested fragments.

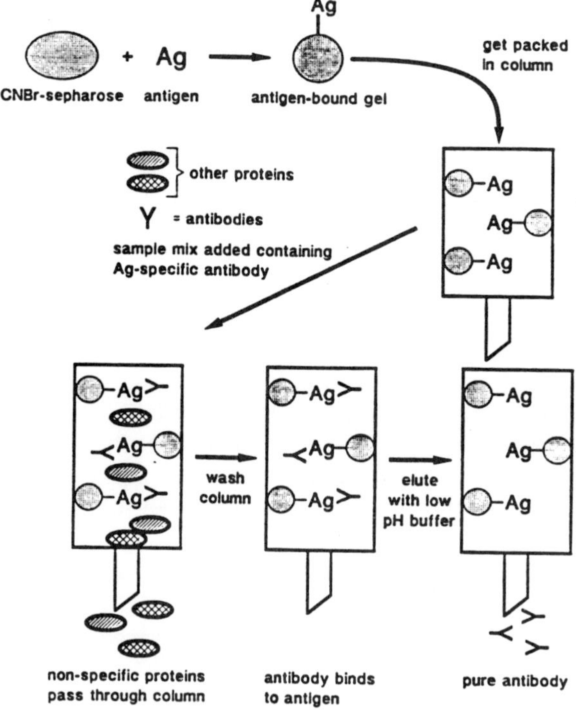

Figure 7. Affinity chromatography.

It is necessary to select a myeloma cell line which lacks the enzyme hypoxanthine-guanine phosphoribosyl transferase (HPRT). This enzyme is essential in the salvage DNA pathway. When myeloma cells and ß-lymphocytes are fused in the selection medium, hypoxanthine/aminopterin/thymidine (HAT), the aminopterin blocks the denovo DNA synthesis of cells in the HAT medium, and forces the cells to use the salvage pathway.

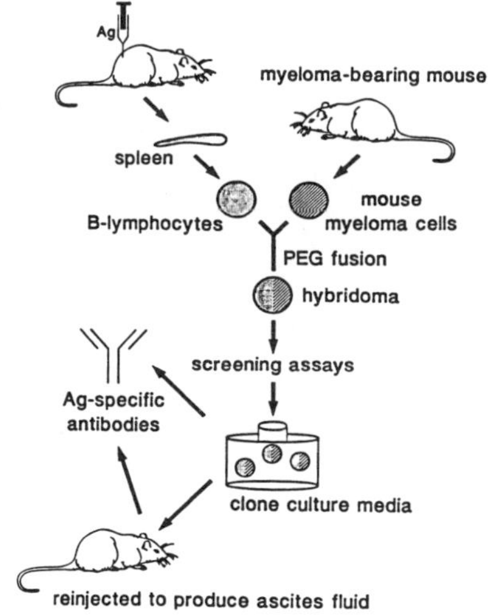

Figure 8. Schematic representation of hydridoma and monoclonal antibody production.

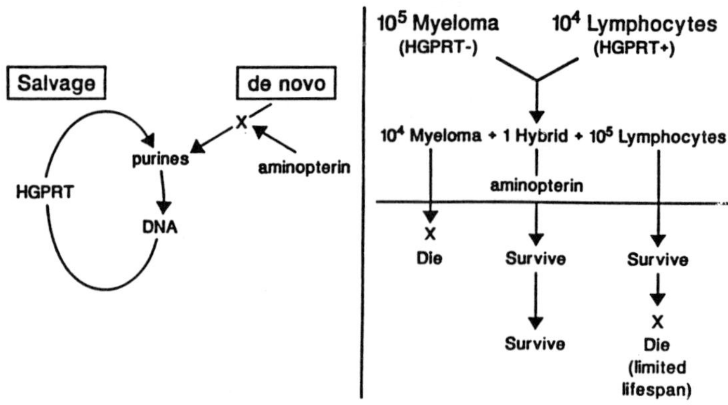

Figure 9. Production of hybrids.

The cells containing the non-functional HPRT enzyme will not survive in these conditions. Thus the myeloma cells are selected against, and the unfused ß-cells die in tissue culture. The only remaining cells in the HAT medium are the hybrid cells (between the myeloma cells, with nonfunctional HPRT, and ß-cells with functional HPRT), or the monoclonal Ab (MAb)(figure 9). Various screening assays, such as the

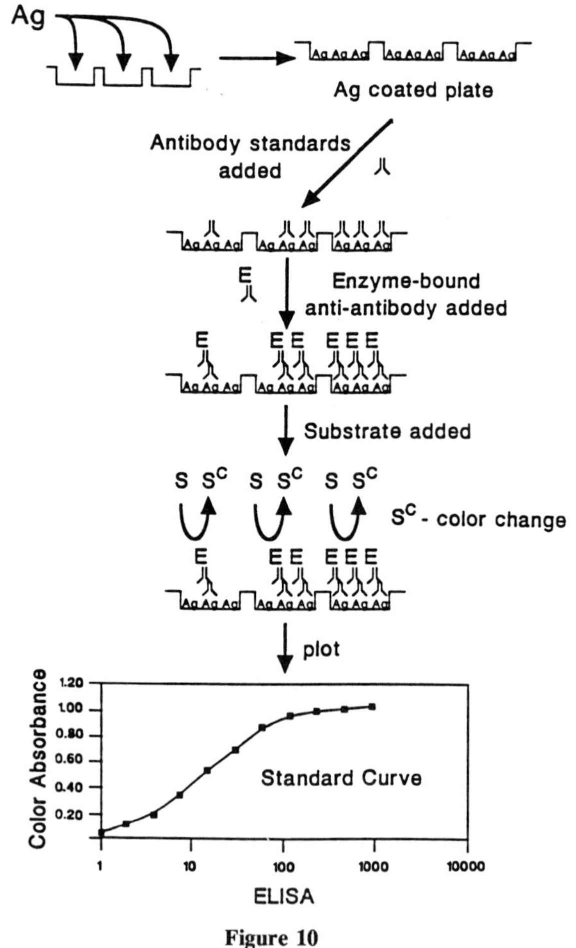

Figure 10

Nuclide	Half-Life $(t_{1/2})$	Main Gamma-Emission(keV)	Beta-Emission
^{131}I	8 days	364	Yes
^{123}I	13.3 hours	159	No
^{111}In	2.8 days	172 247	No
^{99m}Tc	6 hours	140	No
^{62}Cu	9.7 min	511	No

Figure 11. Imaging radionuclides.

enzyme-linked immunosorbent assay (ELISA), are then used to screen the clone against Ag for specific Ag-Ab selectivity (figure 10)[1].

Once the clone is identified, large quantities of the desired monoclonal Ab can be produced either by transplantation of the clone into the peritoneal cavity of a mouse, and harvested from the ascites fluid, or the clone can be vat-produced in culture media. Either method requires purification of the MAb from other proteins present.

SELECTION of RADIONUCLIDE

When developing a radiolabeled Ab for radiodiagnosis or therapy it is necessary to select a radionuclide with the preferred nuclear properties of the atom. If the radionuclide is intended for diagnosis, the match between the gamma ray emitted and the imaging device is of primary importance. The nuclear properties should be such that little or no particle emission occurs, thus minimizing the radiation dose to the patient. The radiation dosimetry limits the quantity of the material that can be administered to the subject, which in turn limits the quality of the image. For therapeutic applications the absorbed radiation dose to the target is important, and should be maximized. The absorbed radiation dose to nontarget (normal) tissues should be minimized. Metal radionuclides that have been used for diagnosis include gallium-67 (half-life = 78 hr) and indium-111 (half-life = 2.8 days). Other radionuclides which have been suggested for labeling Ab utilizing positron emission tomography include zirconium-89 (half-life = 78 hr), gallium-68 (half-life = 68 min) and copper-64 (half-life = 12.7 hr) (figure 11). For therapy, nuclides such as copper-67, rhenium-186 and yttrium-90 have been suggested (figure 12).

SELECTION OF ANTIBODY

Besides the nuclear properties there are biological factors to consider when radiolabeling Ab. Since Ab are biomolecules, the method of labeling must not significantly alter or destroy the immunoreactivity of the Ab. The route of metabolism and excretion should be determined and compared to uptake in the target tissue over time. For successful radioimmuno-imaging sufficient Ab must leave the vascular space and bind Ag for a high target to nontarget ratio. IgG leaves the vascular space with difficulty, while smaller fragments diffuse more readily. Clearance from the blood of IgG and F(ab')$_2$ are similar. Fab clears more quickly (figure 13). Although the blood clearance will improve blood

Nuclide	Half-Life ($t_{1/2}$)	Emission	E(keV)
^{67}Cu	2.58 days	β	395 484 577
^{90}Y	6.41 hours	β	2288
^{109}Pd	13.4 hours	β	1028
^{131}I	8.08 days	β	336 606
^{186}Re	90.6 hours	β	934 1072
^{188}Re	16.9 hours	β	1973 2128
^{212}Bi	60.6 min	β	1550 2250
^{211}At	7.2 hours	α	5866 7450

Figure 12. Radionuclides for therapy.

background, it may be a disadvantage since the Ab fragment may not persist long enough to diffuse into the target and bind it. Intact Ab is cleared mainly through the liver and fragments are cleared to a greater extent through the kidneys.

When use of radiolabeled Ab began, the choice of radionuclides was limited to I-125 and I-131, with I-125 being used for biodistribution studies and I-131 for *in vivo* imaging. With the advent of bifunctional chelates for use with the transition metals, choice of radionuclides was significantly expanded. The bifunctional chelates are covalently attached to the Ab, and the radiometal, most commmonly In-111, is then coupled to the metal binding site of the conjugate.

TECHNIQUES FOR LABELING ANTIBODIES WITH METALS

There have been a variety of approaches used to attach bifunctional metal-chelating groups to biomolecules. Sundberg et al.[3] initially prepared benzenediazonium EDTA using a seven-step synthesis, starting with 1-phenylglycinonitrile. The difficulty of this synthesis lead Yeh et al.[4] to develop a simpler route using readily available amino acid amides as starting materials, including Azo-EDTA. This method eliminated much of the purification of intermediates necessary with Sundbergs method.

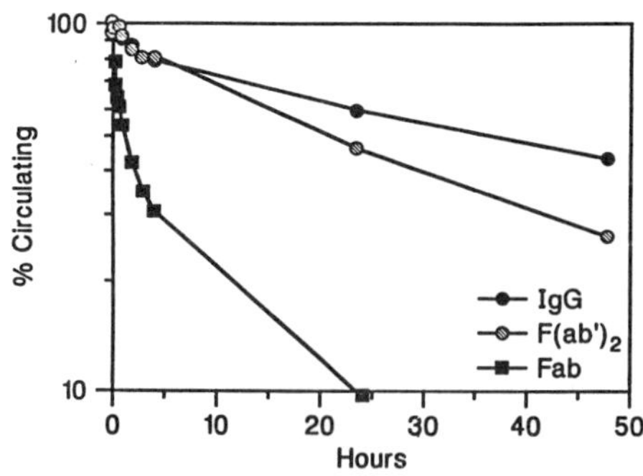

Figure 13. Typical blood clearance of intact I-131 labeled IgG and fragments in a dog.

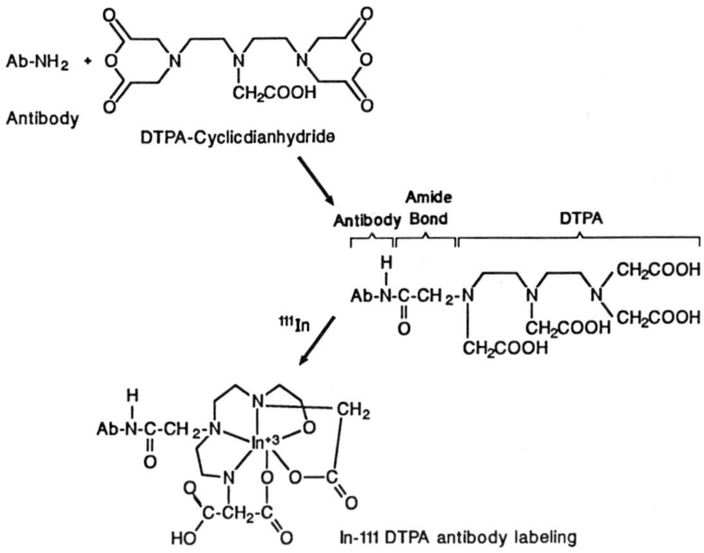

Figure 14. Preparation of ^{111}In-cDTPA-antibody.

Eckelman et al.[5,6] originally prepared EDTA and DTPA dianhydrides by suspending EDTA or DTPA in pyridine and mixing with acetic anhydride. The resulting dianhydrides were then attached to fatty acids to observe the biodistribution effect of a chelating group on a biological compound.

As an alternative to the Sundberg method, a technique commonly used in peptide synthesis was modified by Krejcarek and Tucker[7], to attach DTPA to human serum albumin[7,8], and later to antibodies[9,10,11]. The mixed acid anhydride was prepared by combining DTPA and isobutylchloroformate. The mixed carboxy carbonic anhydride of DTPA was then used without further purification.

Although this method is a convenient method of conjugation, Paik et al.[12] determined the mixed anhydride was unstable, and at low antibody concentrations could not compete effectively with the hydrolysis of the anhydride. He determined the concentration of the anhydride should be determined immediately before use, either by a spectroscopic method utilizing excess benzylamine or a gravimetric method using $Ba(OH)_2$.

Since the mixed acid anhydride was found to be unreliable, Hnatowich et al.[13] further modified the method of Eckelman, using the cyclic anhydride of DTPA (cDTPAA) to radiolabel proteins (figure 14). Cyclic anhydrides have been shown to couple proteins under mild conditions at neutral pH, and are generally more stable to hydrolysis than acyclic anhydrides[14]. Using this method the cyclic anhydride is added as a solid to the protein. Additionally Paik has used cDTPAA to label antibodies[15]. Although cDTPAA is stable to hydrolysis, the method can cause inter and intramolecular cross linking of the Ab. This can be controlled by limiting the cDTPAA/Ab ratio.

The *in vivo* distribution of Ab conjugated by all of these techniques and labeled with In-111 suffer from high liver concentrations (%ID/g). A significant amount of work has ensued to develop methods which would eliminate this problem.

Alternate linking groups have been proposed such as p-isothiocyanato-benzyl-diethylenetriaminepentaacetic acid (SCN-Bz-DTPA), where the protein reactive function is attached at a methylene carbon of the polyamine backbone[16]. The p-isothiocyanatobenzyl moiety has also been attached at the methylene carbon atom of one carboxymethyl arm of DTPA and EDTA, then attached to Ab[17]. Yet, both these new linking groups, when conjugated to Ab and radiolabeled with indium, still show significant liver uptake *in vivo*. The slight reduction in liver uptake noted may be due to the instability of the thiourea bond, where the chelator is attached to the Ab, causing clearance of labeled chelate through the kidneys[18].

In vivo biodistribution of the radiometal complexes of N,N'-bis(2-hydroxybenzyl) ethylenediamine-N, N'-diacetic acid (HBED) and its analogues where rapid liver and blood clearance were observed[19] lead to the selection of Me$_4$HBED as the ligand to be derivatized for protein labeling. This ligand was modified by functionalizing a benzene ring with -NHCOCH$_2$Br, an effective protein linking group. The bifunctional chelate developed was N-(2-hydroxy-3,5-dimethylbenzyl)-N'-(2-hydroxy-5(bromoactamido)benzyl) ethylene-diamine-N,N'-diacetic acid (BrMe$_2$HBED)[20] (figure 15).

This ligand had several deficiencies including chemical instability of BrMe$_2$HBED, as well as 24 hour incubation times needed to acquire high radiolabeling yields. For these reasons N,N'-bis(2-hydroxybenzyl)-1-(4-bromoacetamido-benzyl)-1,2-ethylene-diamine-N,N'-diacetic acid (BrϕHBED) was developed[21].

The increased lipophilicity of the original HBED complexes, leading to rapid clearance of the compound through the hepatobiliary system[18], was not as rapid once the new bifunctional chelates (BrMe$_2$HBED and BrϕHBED) were conjugated to Ab and radiolabeled with In-111. BrϕHBED has shown improved liver clearance over cDTPAA labeling of BB5-antibody[22]. The immunoreactivity of radiolabeled conjugates is higher using either of these ligands than the cDTPAA method of radiolabeling[20,21].

The recent interest in radiolabeling Ab with Cu-67 for therapy has lead to the development of the macrocyclic bifunctional chelating groups. Copper complexes of DTPA and EDTA analogs rapidly decompose in serum. Moi et al. synthesized 6-(p-nitrobenzyl)-1,4,8,11-tetracyclotetradecane-N,N',N",N"'-tetraacetic acid (p-nitrobenzyl

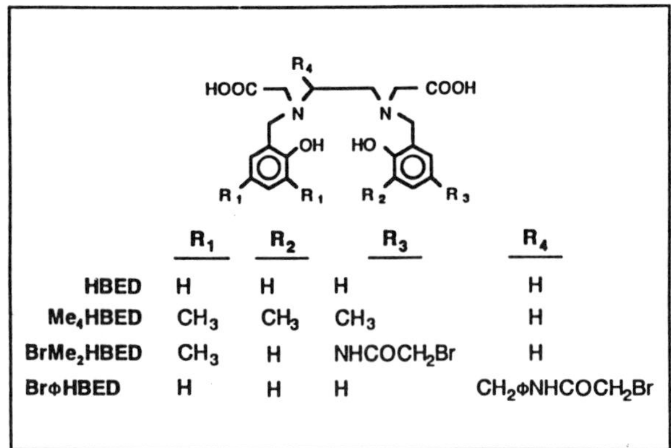

	R$_1$	R$_2$	R$_3$	R$_4$
HBED	H	H	H	H
Me$_4$HBED	CH$_3$	CH$_3$	CH$_3$	H
BrMe$_2$HBED	CH$_3$	H	NHCOCH$_2$Br	H
BrϕHBED	H	H	H	CH$_2\phi$NHCOCH$_2$Br

Figure 15. Structural comparison of the parent multi-dentate ligands, BrMe$_2$HBED, and BrϕHBED.

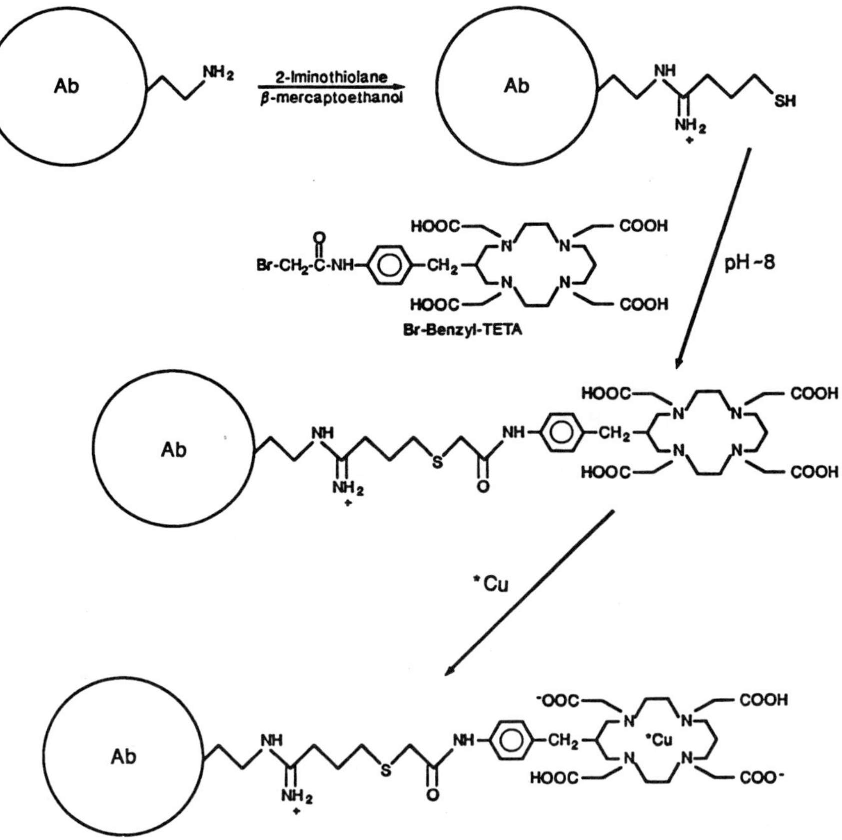

Figure 16. Preparation of Cu-benzyl-TETA-antibody.

TETA) which forms a copper chelate that is quite stable in human serum[23]. A simplified one-step method was then developed by McCall, et.al. for conjugating macrocyclic chelators to Ab[24]. The method employs the reagent 2-iminothiolane to insert a spacer group between the Ab and the macrocycle.

McCall prepared the 6-bromoacetomidobenzyl-1,4,8,11-tetraazacyclotetra-decane-N,N',N'',N''' tetraacetic acid (Br-benzyl-TETA) from the p-nitrobenzyl TETA, using a modification of Muckalas method[25]. The Br-Benzyl TETA has been conjugated to Ab and radiolabeled with copper (figure 16).

All of the techniques described above form stable Ab-metal conjugates. The ideal labeling technique has still not been developed since significant quantities of radiometal still accumulates in non-target organs using any of these methods. The development of a chelating technique to overcome the limitations will extend the clinical utility of both radioimmunoimaging and radioimmunotherapy.

Acknowledgments

This work was supported by Department of Energy grant, number DOE-DE-FG02-87 ER-60512 and the National Institiute of Health grant, number NIH-CA-44728.

REFERENCES

1. Z.D. Grossman, S.F.Rosebough. Clinical Radioimmunoimaging. Grune and Sralton, Inc. 1988.

2. C. Milstein, Monoclonal Antibodies, Scientific Am. 243:66 (1980).

3. M.W. Sundberg, C.F. Meares, D.A. Goodwin, and C.I. Diamanti, Selective binding of metal ions to macromolecules using bifunctional analogs of EDTA, J Med Chem. 17:1304 (1974).

4. S.M. Yeh, D.G. Sherman, and C.F. Meares, A new route to "bifunctional" chelating agents:conversion of amino acids to analogs of ethylenedinitrilotetraacitic acid, J Anal Biochem. 100:152 (1979).

5. W.C. Eckelman, S.M. Karesh, and R.C. Reba, New compounds: fatty acid and long chain hydrocarbon derivatives containing a strong chelating agent, J Pharm Sci.64:704 (1975).

6. S.M. Karesh, W.C. Eckelman, and R.C. Reba, Biological distribution of chemical analogs of fatty acids and long chain hydrocarbons containing a strong chelating agent, J Pharm Sci. 66:225 (1977).

7. G.E. Krejcarek, and K.L. Tucker, Covalent attachment of chelating groups to macromolecules, J Biochem and Biophys Research Comm. 77:581 (1977).

8. S.J. Wagner, and M.J. Welch, Gallium-68 labeling of albumin and albumin microspheres, J Nucl Med. 20:428 (1978).

9. B.A. Khaw, J.T. Fallon, H.W. Strauss, E. Haber, O.A. Gansow. Myocardial infait imaging of Ab to canine cardiac myosin with In-111 DTPA, Science. 209:295 (1980).

10. D.A. Scheinberg, O.A. Gansow, Tumor imaging with radio metal chelate conjugated to Ab, Science. 215:1511 (1982).

11. B.A.Khaw, H.W. Strauss, A. Carvalho, E. Locke, H.K. Gold, E. Haber, Technetium-99m labeling of antibodies to cardiac myosin Fab and to human fibrinogen, J Nucl Med 23:1011 (1982).

12. C.H. Paik, P.R. Murphy, W.C. Eckelman, W.A. Volkert, and R.C. Reba, Optimization of the DTPA mixed-anydride reaction with antibodies at low concentration, J Nucl Med 24:932 (1983).

13. D.J. Hnatowich, W.W. Layne, and R.L. Childs, The preparation and labeling of DTPA-Coupled albumin, J Appl Radiat Isot 33:327 (1982).

14. C.A. Bunton, J.H. Fendler, N.A. Fuller, S. Perry, J. Rocek, The hydrolysis of carboxylic anhydrides. Part VI. Acid hydrolysis of cyclic anhydrides, J Chem Soc London 6174 (1965).

15. C.H. Paik, M.A. Ebbert, P.R. Murphy, C.R. Lassman, R.C. Reba, W.C. Eckelman,

K.Y. Pak, J. Powe, Z. Steplewske, and H. Koprowski, Factors influencing DTPA conjugation with antibodies by cyclic DTPA anhydride, J <u>Nucl</u> <u>Med</u> 24:1158 (1983).

16.   M.W. Brechbiel, O.A. Gansow, R.W. Atcher, J.Schlom, J. Esteban, D.E. Simpson, and D. Colcher, Synthesis of 1-(p-Isothiocyanatobenzyl) derivatives of DTPA and EDTA. Antibody labeling and tumor-imaging studies, J <u>Inorg</u> <u>Chem</u> 25:2772 (1986).

17.   D.A. Westerberg, P.L. Carney, P.E. Rogers, S.J. Kline, and D.K. Johnson, Synthesis of novel bifunctional chelators and their use in preparing monoclonal antibody conjugates for tumor targeting, J <u>Med</u> <u>Chem</u> 32:236 (1989).

18.   G.P. Adams, S.J. DeNardo, S.V. Deshpande, G.L. DeNardo, C.F. Meares, Effect of mass of ^{111}In-benzyl-EDTA monoclonal antibody on hepatic uptake and processing in mice, J <u>Can</u> <u>Res</u> 49:1707 (1989).

19.   C.J. Mathias, Y. Sun, M.J. Welch, M.A. Green, J.A. Thomas, K.R. Wade, and A.E. Martell, Targeting radiopharmaceuticals:comparitive biodistribution studies of gallium and indium complexes of multidentate ligands, J <u>Radiat</u> <u>Appl</u> <u>Instrum</u> 15:69 (1988).

20.   C.J. Mathias, Y. Sun, J.M. Connett, G.W. Philpott, M.J. Welch, and A.E. Martell, A new bifunctional chelate, BrMe$_2$HBED: An effective conjugate for radiometals and antibodies, <u>Inorg</u> <u>Chem</u> 29:1475 (1990).

21.   C.J. Mathias, Y. Sun, M.J. Welch, J.M. Connett, G.W. Philpott, and A.E. Martell, N,N'-Bis(2-hydroxybenzyl)-1-(4-bromoacetamidobenzyl)-1,2-ethylene-diamine-N,N'-diacetic acid: a new bifunctional chelate for radiolabeling antibodies, J <u>Biocon</u> <u>Chem</u> 1:204 (1990).

22.   S.W. Schwarz, C.J. Mathias, J.Y. Sun, W.G. Dilley, S.A. Wells Jr., A.E. Martell, and M.J. Welch, Evaluation of two new bifunctional chelates for radiolabeling a parathyroid-specific monoclonal antibody with In-111, In Press.

23.   M.K. Moi, C.F. Meares, M.J. McCall, W.C. Cole, and S.J. DeNardo, Copper chelates as probes of biological systems: stable copper complexes with a macrocyclic bifunctional chelating agent, J <u>Analytic</u> <u>Biochem</u> 148:249 (1985).

24.   M.J. McCall, H. Diril, and C.F. Meares, Simplified method for conjugating macrocyclic bifunctional chelating agents to antibodies via 2-iminothiolane, J <u>Biocon</u> <u>Chem</u> 1:222 (1990).

25.   V.M. Mukkala, H. Mikola, I. Hemmila, The synthesis and use of activated N-Benzyl derivatives of diethylenetriaminetetraacetic acids: alternative reagents for labeleing of antibodies with metal ions, <u>Anal</u> <u>Biochem</u> 176:319 (1989).

METHODS FOR THE RADIOHALOGENATION OF ANTIBODIES

Michael R. Zalutsky, Pradeep K. Garg,
Ganesan Vaidyanathan and Sudha Garg

Department of Radiology
Duke University Medical Center
Durham, North Carolina, USA

INTRODUCTION

One of the main goals of radiopharmaceutical chemistry is the development of compounds that can be used for the identification or erradication of specific cell populations. Since monoclonal antibodies (MAbs) can be generated, at least in principle, against almost any cellular determinant, there has been a great deal of interest in using MAbs as a mechanism for targeting radionuclides. Although diagnostic and therapeutic investigations with labeled MAbs have focused on their applications in the management of cancer, labeled MAbs also may be useful in the noninvasive diagnosis of infections and heart disease. Numerous problems must be solved before radiolabeled MAbs can make a meaningful impact on the clinical domain, including the development of better MAb labeling methods than those that have been utilized in clinical studies.

For radiolabeling MAbs, two general methods have been used. The first involves reaction of the MAb with a bifunctional chelate or cryptate, followed by complexation of a metallic radionuclide. The existence of radionuclides of multivalent metals with a wide variety of nuclear decay properties makes this approach particularly attractive. As will be described elsewhere in this volume, a considerable effort both in academia and industry has been directed at developing methods for labeling MAbs with nuclides of In, Ga, Tc, Re, Y, Cu and other metals.

The alternate strategy for labeling MAbs is to use halogen nuclides. Nuclear properties of some of the radiohalogens which are being investigated for use as MAb labels are summarized in Table 1. Although many more metallic nuclides are available, the decay characteristics of these halogen nuclides are nearly ideal for most potential diagnostic and therapeutic applications of labeled MAbs. Of the nuclides being utilized for radioimmunoscintigraphy, I-123 probably offers the best combination of gamma ray energy appropriate for nuclear medicine imaging and physical half life compatible with MAb pharmacokinetics. Iodine-123 is ideal for use with single photon emission computed tomography, an imaging technique that has been shown to greatly enhance the sensitivity and specificity of imaging tumors with MAbs (Delaloye et al., 1986). MAbs labeled with F-18, Br-75, Br-76 or I-124 could be used with positron emission tomography (PET), an imaging method offering enhanced spatial resolution and

Applications of Enzyme Biotechnology, Edited by J.W. Kelly and
T.O. Baldwin, Plenum Press, New York, 1991

Table 1. Halogen nuclides for use in antibody labeling

Nuclide	Half-life	Application	Type of Emission
F-18	1.83 hr	PET imaging	positron
Br-75	1.63 hr	PET imaging	positron
Br-76	16.1 hr	PET imaging	positron
I-123	13.0 hr	SPECT imaging	gamma
I-124	4.2 d	PET imaging	positron
I-131	8.0 d	imaging therapy	gamma beta
At-211	7.2 hr	therapy	alpha

quantitative capabilities. For therapeutic applications, factors such as tumor location, size and heterogeneities of antigen expression and hemodynamics will dictate the type of nuclide which would be most appropriate. Beta emitters such as I-131 might be useful for treating relatively large, heterogeneous tumors since the range of their radiation is multicellular. Because of their cellular range of action and high relative biological effectiveness, alpha emitters such as At-211 might be ideal for treating other types of tumors.

RADIOIODINATION OF ANTIBODIES

All clinical studies to date and most investigations in animal models with radiohalogenated MAbs have utilized nuclides of iodine. Radioiodine nuclides offer several advantages for MAb labeling (Larson and Carasquillo, 1988). In contrast to labeling MAbs with radiometals, nuclides of the same element can be used for single photon emission tomographic and conventional imaging (I-123), PET imaging (I-124) and therapy (I-131). This eliminates the difficulties inherent in using a nuclide of a different element with a MAb in an imaging study to predict the suitability of a potential therapeutic application. For example, when some DTPA chelation methods are used, the normal bone uptake observed with In-111 would grossly underestimate the skeletal radiation dose which would be received if Y-90 was used as the label (Hnatowich et al., 1985). Another advantage is the availability of multiple gamma-emitting iodine nuclides (in addition to those listed in Table 1, 60-day I-125), permitting paired-label experiments to be performed (Pressman et al., 1957). This type of study is valuable in elucidating the effect of various parameters on MAb localization since animals or patients can be injected with two different radioiodinated MAbs, allowing direct comparison of various labeling methods, MAbs or routes of injection.

Electrophilic Radioiodination Methods

Electrophilic substitution methods have been used for the radioiodination of MAbs used in patient studies. A variety of oxidants have been used to generate the electrophilic iodinating species, including Chloramine-T (Hunter and Greenwood, 1962), lactoperoxidase (Marchalonis, 1969) and Iodogen (Fraker and Speck, 1978). The conditions used for MAb labeling generally result in the substitution of the iodine _ortho_ to the hydroxyl group on tyrosine residues.

A potential problem with these iodination methods is that the MAb is exposed directly to oxidizing agents which can cause chemical damage to some proteins. For example, Chloramine-T and _N_-chlorosuccinimide have been shown to oxidize methionine, tryptophan and cysteine residues on a variety of proteins (Shechter et al., 1975). It seems likely that the effects of alterations such as these on MAb reactivity will depend on the conformational nature of the MAb of interest. For example, although many MAbs have been labeled using the Iodogen method with satisfactory retention of immunoreactivity, we have also shown that exposing an anti-breast carcinoma MAb to as little as 1 μg of Iodogen decreased immunoreactivity to less than 25% (Hayes et al., 1988).

The primary difficulty in using electrophilic methods for MAb radioiodination is that significant loss of label occurs when the MAb is administered _in_ _vivo_. In animals and patients receiving radioiodinated MAbs and MAb fragments, considerable uptake of activity in stomach and thyroid has been observed (Zalutsky et al., 1985; Hayes et al., 1986). Since these are the tissues in which iodide is primarily sequestered, it is generally assumed that this behavior reflects deiodination of the MAb. This conclusion is also supported by the recovery of 15 to >50% of the injected dose from the urine as nonprotein associated activity during the first 24 hr after radioiodinated MAb administration (Sullivan et al., 1982; Hayes et al., 1986).

It seems likely that the rapid loss of label from radioiodinated MAbs _in_ _vivo_ is mediated by enzymatic processes. Electrophilic methods create iodinated tyrosine residues on the MAb. Since multiple deiodinases exist which are known to deiodinate iodotyrosines and iodothyronines (Smallridge et al., 1981; Gershengorn et al., 1980; Koehrle et al., 1986), MAb dehalogenation is presumably related to the recognition of iodotyrosine residues on the MAb by these enzymes. The liver generally is thought to be the site of MAb deiodination; however, additional anatomic sites may be involved since deiodinases with varying degrees of specificity are also found in the kidney, thyroid and other tissues (Stanbury and Morris, 1958; Leonard and Rosenbert, 1977).

N-Succinimidyl Ester Acylation Agents

If the dehalogenation of MAbs _in_ _vivo_ could be minimized, the advantages inherent in the use of iodine nuclides for labeling MAbs could be more fully exploited. Hypothesizing that decreasing the structural similarity of the MAb iodination site to thyroid hormones would increase retention of label on the MAb, our group has developed a series of agents for protein radioiodination that do not involve iodination _ortho_ to a hydroxyl group on an aromatic ring. In evaluating these new MAb radiolabeling methods, the following criteria have been used: a) the _in_ _vivo_ stability of the bond between the iodine and the MAb must be high; b) the labeling method should not decrease the immunoreactivity or the affinity of the MAb; and c) if possible, the labeled species created in the catabolism of the MAb should be cleared rapidly from the body in order to minimize normal tissue background.

The structure of the Bolton-Hunter reagent (Bolton and Hunter, 1973), _N_-succinimidyl 3-(4-hydroxy-3-[I-125]iodophenyl)propionate, was used as the point of departure for the design of the first radioiodination agent, <u>N</u>-succinimidyl-3-iodobenzoate (SIB) (Zalutsky and Narula, 1987). This approach was undertaken because in general, proteins and peptides labeled using the Bolton-Hunter method exhibit a greater retention of immunological activity than those labeled using other methods (Bolton and Hunter, 1973; Bolton et al., 1979).

Although SIB is conceptually similar to the Bolton-Hunter reagent, two structural changes were incorporated in an attempt to make SIB more useful for <u>in vivo</u> applications. As shown in Figure 1, SIB lacks the phenolic hydroxyl group <u>ortho</u> to the iodine atom which is present in the

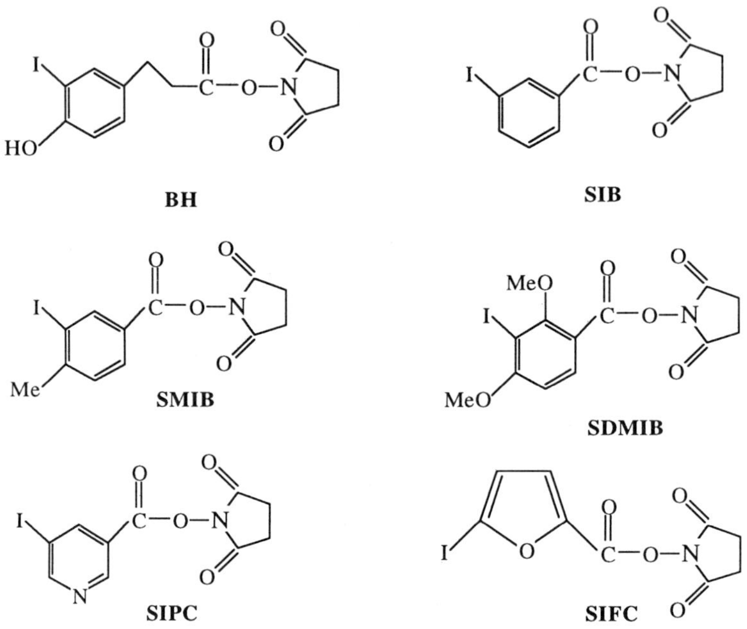

Figure 1. Structures of Some <u>N</u>-succinimidyl Esters Used in the Radioiodination of MAbs.

Bolton-Hunter reagent. This was done to decrease recognition of the iodination site on the MAb by the deiodinases involved in thyroid hormone metabolism. In addition, the two-carbon spacer found in the Bolton-Hunter reagent between the activated ester and the aromatic ring was omitted in order to increase MAb coupling yields by minimizing competitive hydrolysis.

Since SIB does not contain a hydroxyl group to activate the ring for electrophilic substitution, a synthesis utilizing iododestannylation was devised (Zalutsky and Narula, 1987). The alkyl tin ester (ATE), <u>N</u>-succinimidyl 3-(tri-<u>n</u>-butylstannyl)benzoate, was synthesized in three steps from <u>m</u>-bromobenzoic acid (Figure 2). With the exception of the <u>N</u>-succinimidyl 2,4-dimethoxy-3-(trialkylstannyl)benzoates (Narula and Zalutsky,

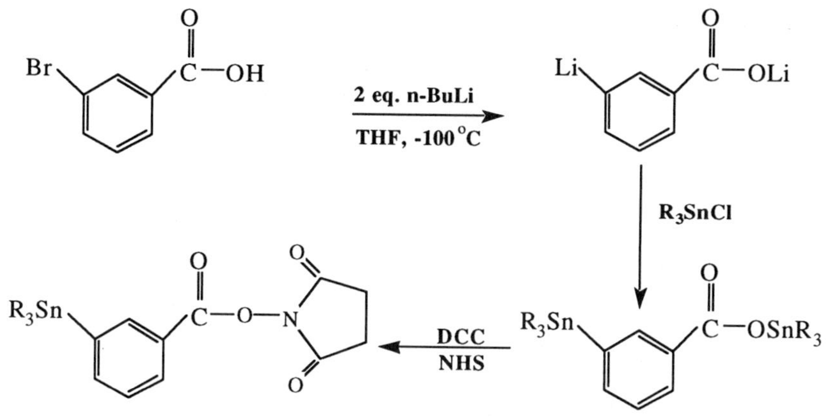

Figure 2. Synthesis of N-succinimidyl 3-(tri-n-butylstannyl)benzoate.

1988; Vaidyanathan and Zalutsky, 1990a), the same general route was used to synthesize the ATE precusors used for radioiodinating the other compounds illustrated in Figure 1.

Multiple MAbs and MAb fragments have been labeled with I-131, I-125 and I-123 using the method outlined in Figure 3. Tert-butylhydroperoxide was used because of its ability to oxidize the radioiodine efficiently in heterogeneous media. SIB was labeled in 85–95% yield after a 30 min reaction at room temperature. Initial studies were performed with SIB purified by passage over a silica gel Sep-Pak column and isolated in 30% ethyl acetate in hexane. Coupling efficiencies for labeling proteins with SIB were dependent on both pH and protein concentration with yields of greater than 60% obtainable at pH 8.5 and a protein concentration of 5 mg/mL.

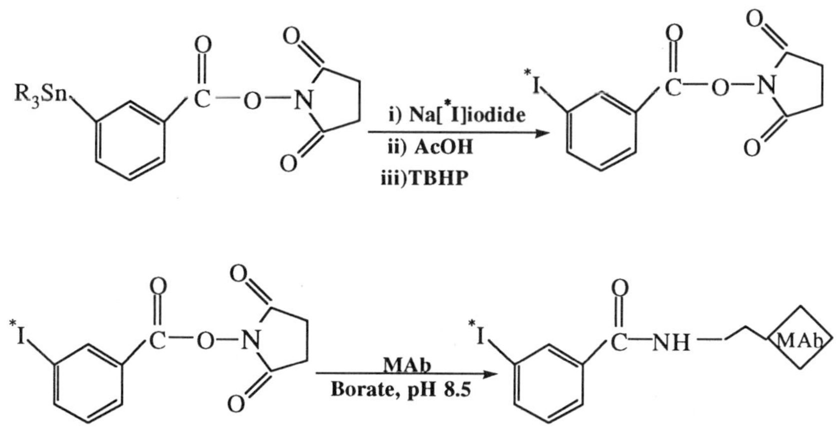

Figure 3. Labeling MAbs using the ATE reagent.

The utility of the ATE method for MAb radioiodination was evaluated initially using the F(ab')$_2$ fragment of OC 125, a MAb directed against ovarian carcinomas (Zalutsky and Narula, 1988). In vitro and in vivo comparisons between OC 125 F(ab')$_2$ labeled using ATE and a conventional method (Iodogen) were performed. Scatchard analyses revealed that the binding affinity to purified CA 125 antigen of the ATE preparation was more than twice that of OC 125 F(ab')$_2$ labeled using Iodogen. With the ATE method, binding affinity of the MAb was dependent on the amount of ATE remaining in the purified SIB preparation. For example, when ATE was present at 240 nmol, the affinity constant was more than 10-fold lower than at 35 nmol.

Since one of our goals was to develop a method for labeling MAbs in which potential labeled catabolites are cleared rapidly, the tissue distribution of m-[I-125]iodobenzoic acid was evaluated (Zalutsky and Narula, 1988). After 1 hr, the whole body activity in normal mice was only 10% of the injected dose, a level more than 6 times lower than that of co-injected [I-131]iodide. Since m-iodobenzoic acid is a likely catabolite of MAbs labeled with SIB, its rapid clearance from normal tissues is a potential advantage of this MAb radioiodination method.

Groups of athymic mice with ovarian carcinoma xenografts were given OC 125 F(ab')$_2$ labeled with I-125 using S[I-125]IB and I-131 using Iodogen. Because of the proclivity of iodide for the thyroid, uptake of radioactivity in this tissue was used as an indicator of deiodination. Use of the ATE method reduced thyroid uptake to <0.1% of the injected dose, a level more than 100 times less than that seen using Iodogen. This observation has been confirmed in subsequent experiments with other MAbs and F(ab')$_2$ fragments and indicates that labeling MAbs with SIB considerably reduces the deiodination of MAbs in vivo.

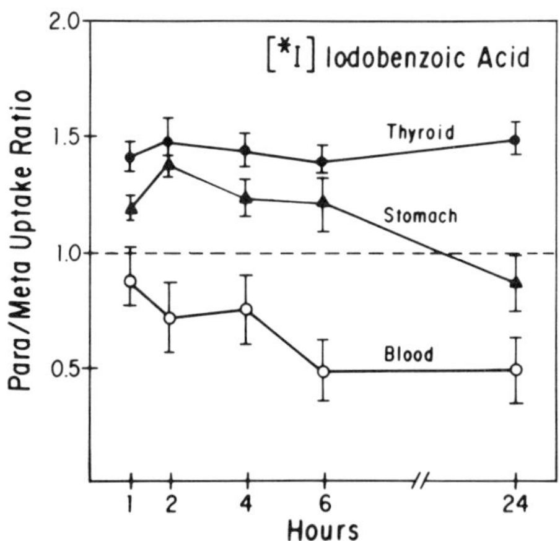

Figure 4.   The p:m Isomer Uptake Ratio for Thyroid, Stomach and Blood.

A subsequent study by Wilbur et al. (1989) reported that use of N-succinimidyl 4-iodobenzoate for labeling MAbs also appeared to decrease their deiodination in vivo. We selected a meta iodophenyl template, beleiving that with an electron withdrawing substituent at position 1, the m isomer would be more inert to substitution of the halide by nuclophiles than o or p isomers. In order to determine if use of the m isomer actually imparted increased stability in vivo, paired-label experiments were performed comparing the _m_ and _p_ isomers as iodobenzoic acids and as intact MAb and F(ab′)$_2$ iodophenyl conjugates (Garg et al., 1989a). As shown in Figure 4, thyroid uptake for the _m_ isomer was 35-48% lower (P <0.005) than the _p_ isomer. Similarly, the thyroid uptake of the _m_ isomer was up to 55% lower in both the MAb and F(ab′)$_2$ experiments, suggesting that use of this isomer offered a small but significant advantage.

In order to better understand the role of several structural factors in the design of MAb radioiodination agents, a direct comparison of SIB and the Bolton-Hunter reagent was performed (Vaidyanathan and Zalutsky, 1990b). In general, protein labeling yields with SIB were about twice those seen with the Bolton-Hunter reagent. A likely explanation is that omitting the two-carbon spacer between the _N_-succinimidyl group and the aromatic ring decreased the rate of ester hydrolysis, thereby increasing the availability of active ester for amide bond formation.

Paired-label studies in normal mice indicated that the thyroid uptake for a MAb labeled using the Bolton-Hunter method was only about twice that of MAb labeled with SIB but only 7% of that observed for MAb labeled with Iodogen. The relatively low rate of dehalogenation using the Bolton-Hunter reagent was surprising since with this method, the radioiodide atom is substituted _ortho_ to a hydroxyl group on an aromatic ring. However, _in vitro_ studies have demonstrated a high degree of structural specificity for the deiodination of aryl iodides by liver, with presence or absence of a hydroxyl group on the ring not being the sole factor determining enzymatic recognition (Dumas et al., 1973).

Optimization of the synthesis of SIB via electrophilic iododestannylation has also been studied (Garg et al., 1989b). Factors which were investigated were nature of the oxidant (_tert_-butylhydroperoxide and _N_-chlorosuccimimide), method of SIB purification (Sep-Pak and HPLC) and bulk of the alkyl substituent on tin (tri-_n_-butyl and trimethyl). As shown in Figure 5, the trimethylstannyl compound gave higher SIB yields than the tri-_n_-butyl derivative with differences most apparent for shorter reaction times. These results are in agreement with those of Wursthorn et al. (1978) who reported that decreasing the bulk of the

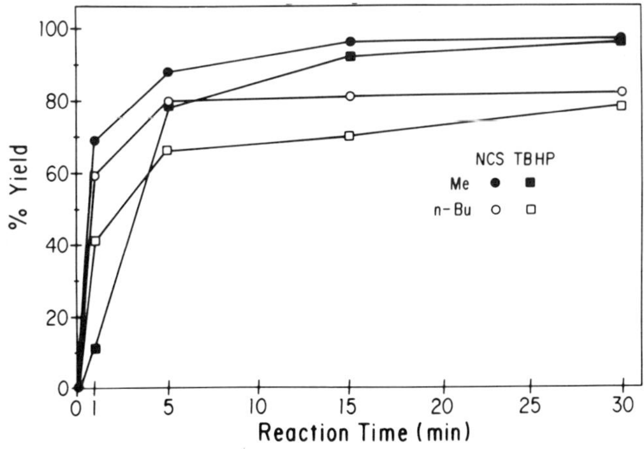

Figure 5. Electrophilic Iododestannylation of _N_-succinimidyl 3-trialkyl-stannylbenzoates Using NCS and TBHP.

alkyl tin substituent increased the rate of iododestannylation. The immunoreactivity and protein labeling efficiency for various combinations of oxidant and SIB purification method were compared using anti-glioma MAb 81C6 as a model system. Optimal results were obtained using TBHP as the oxidant and HPLC for purification of SIB.

The potential utility of the ATE method for the radioiodination of intact MAbs also has been investigated (Zalutsky et al., 1989a). MAb 81C6 was labeled using HPLC-purified SIB and using Iodogen. The immunoreactivity and tissue distribution of the two preparations were compared. In summary, the results of these experiments demonstrated that use of the ATE method for labeling 81C6 a) yielded a MAb with slightly better immunoreactivity and affinity; b) reduced thyroid uptake by 40- to 100-fold; c) increased uptake and retention of radioiodine in tumor by as much as 4- to 12-fold; and d) resulted in superior tumor:normal tissue radiation absorbed dose ratios.

The results which have been obtained labeling MAbs and MAb fragments by reaction with SIB have been most encouraging. Recent studies have demonstrated that use of SIB for labeling a MAb with I-131 significantly increased its therapeutic efficacy in an athymic mouse xenograft model. Nonetheless, other _N_-succinimidyl alkylstannyl esters have been under active investigation not only for use in MAb radioiodination but also for labeling proteins with other halogen nuclides. The structures of some of the other MAb iodination agents which have been synthesized are shown in Figure 1.

The synthesis of _N_-succinimidyl 2,4-dimethoxy-3-(trialkylstannyl)-benzoates have been accomplished (Vaidyanathan and Zalutsky, 1990a). These compounds were used as precursors for the radiosynthesis of _N_-succinimidyl 2,4-dimethoxy-3-iodobenzoate (SDMIB). Yields for SDMIB from the trimethylstannyl and tri-_n_-butylstannyl analogs were nearly identical. Of particular interest was the determination of whether substitution of electron-rich methoxy groups _ortho_ to the iodination site would enhance _in vivo_ stability by decreasing the probability of nucleophilic displacement of the iodine. Paired-label biodistribution studies were performed using a MAb labeled with I-125 using SDMIB and I-131 using SIB. The results indicated that thyroid uptake from SDMIB labeled MAb was 1.4-2.8 times higher than that using SIB but still significantly lower than levels reported in the literature for MAbs labeled using conventional methods. A possible explanation for the greater loss of label from SDMIB compared to SIB is the enzymatic dealkylation of the methoxy groups, resulting in the formation of a hydroxyl group _ortho_ to the iodine atom.

More encouraging results were obtained using _N_-succinimidyl 5-iodo-3-pyridinecarboxylate (SIPC) for MAb labeling (Garg et al., 1991a). Unlike SIB, radioiodination of SIPC did not proceed in high yield at room temperature and required heating at 60-65°C to achieve adequate yields. This presumably reflects the lower nucleophilicity of the pyridine ring and the protonation of the heteroatom under the reaction conditions employed. When _tert_-butylhydroperoxide was used as the oxidant, higher yields generally were obtained with the trimethylstannyl versus the tri-_n_-butylstannyl precursor. Use of _N_-chlorosuccinimide with both stannyl compounds increased the yield of SIPC considerably.

The tissue distribution of 5-[I-131]iodonicotinic acid was determined in normal mice because this compound is a likely catabolite of MAbs labeled using SIPC. Thyroid uptake was between 0.05 ± 0.02% at 1 hr to 0.16 ± 0.04% at 6 hr, suggesting that little deiodination had occurred. Whole body clearance of 5-[I-131]iodonicotinic acid was more rapid than both iodide and 3-[I-125]iodobenzoic acid, a feature that could lead to lower normal tissue background levels when MAbs labeled using SIPC are catabolized. Thyroid uptake levels for a MAb and an F(ab')$_2$ fragment labeled using SPIC were comparable to those observed in the same animals when these proteins were labeled using SIB.

LABELING ANTIBODIES WITH ASTATINE-211

Astatine-211 has a half life of 7.2 hr and can be produced conveniently on a cyclotron by bombarding natural bismuth metal targets with 28-MeV alpha particles. This radiohalogen is of particular interest for radioimmunotherapy because alpha particles are associated with all of its decays. There are several advantages to using alpha particles for certain potential therapeutic applications. The average alpha particle energy for At-211 is 6.8 MeV, nearly a factor of ten higher than those of most beta emitters. The range of these alpha particles in tissue is only about 55-80 μm, limiting their cytotoxic effects to only a few cell diameters. Because of their high decay energy and short range, the alpha particles of At-211 are radiation of high linear energy transfer, a feature which results in a relative biological effectiveness about 8-fold greater than low linear energy transfer beta particles (Barendsen et al., 1966). Examples of tumors which might be treated effectively using alpha emitters such as At-211 are ovarian carcinomas, tumors in the vascular compartment and micrometastases.

Because astatine is a halogen, initial attempts to label proteins with At-211 utilized direct iodination methods. However, proteins labeled using these approaches were deastatinated rapidly both _in_ _vitro_ and _in_ _vivo_ (Aaij et al., 1975; Vaughan and Fremlin, 1978). Presumably, in analogy with protein radioiodination, these electrophilic substitution methods generate astatinated tyrosine residues. Since astatotyrosine is unstable in the presence of oxidants (Visser et al., 1979), it is clear that alternative approaches for labeling proteins with At-211 are required.

We developed a method for labeling proteins with At-211 which involves the synthesis of an astatinated precursor that subsequently is coupled to the protein (Friedman et al., 1977). First, _p_-[At-211]astatobenzoic acid was synthesized from the diazonium salt of _p_-aminobenzoic acid which was purified by ether extraction. The labeled product was reacted with tributylamine and isobutylchloroformate to form an At-211 labeled mixed anhydride which then was added to the protein of interest:

Tissue distribution studies in mice demonstrated that use of this two-step method significantly decreased loss of At-211 from the protein _in_ vivo. Since labeling yields were only 10-15%, the maximum specific activity that could be obtained was about 0.2 mCi/mg. Harrison and Royle (1984) were able to increase overall yields of At-211 labeled protein by a factor of three by using HPLC to purify the p-[At-211]astatobenzoic acid intermediate. Increased protein coupling efficiency as a result of improved purification probably resulted from decreased competition of p-aminobenzoic acid, p-hydroxybenzoic acid and other impurities with p-astatobenzoic acid for amine sites on the protein.

More recently, we have been investigating the feasibility of labeling MAbs with At-211 using N-succinimidyl 3-trialkylstannylbenzoates. Using the 3-tri-n-butylstannyl derivative and tert-butylhydroperoxide, N-succinimidyl-3-[At-211]astatobenzoic acid was produced in about 67% yield from astatide trapped in 0.05 N NaOH (Zalutsky and Narula, 1988b). In contrast with our results for radioiodination, a second labeled peak was seen when the reaction mixture was purified using a silica gel Sep Pak column. We have speculated that the At-211 activity in the additional peak is present as a π-complex between AtX (where X = halogen impurity) and ATE. This possibility is supported by the fact that treatment of this product with tert-butylhydroperoxide in acetic acid results in nearly quantitative formation of N-succinimidyl 3-[At-211]astatobenzoic acid (Narula and Zalutsky, 1989):

$$\text{Sn}(C_4H_9)_3 \xrightarrow{\text{AtX (?)}} {}^{211}\text{At}$$

Our current procedure (Zalutsky et al., 1989b) for labeling proteins with At-211 has incorporated several significant modifications of our initial method. The At-211 now is trapped in chloroform instead of NaOH so that the chemical form of the astatine activity before the addition of oxidant is AtX instead of $At(OH)_2^-$. With this change, yields for the desired At-211 labeled N-succinimidyl ester have been increased and the formation of the labeled π-complex has been eliminated. Because of the greater size of the astatine atom compared to iodine, we speculated that use of the trimethylstannyl precursor would be advantageous. Indeed, yields with this compound consistently were 10-15% higher than observed with the tri-n-butyl analog. Finally, we HPLC purification of N-succinimidyl 3-[At-211]astatobenzoic acid has increased MAb coupling efficiencies.

Several MAbs and MAb F(ab')$_2$ fragments have been labeled with At-211 and their immunoreactivity and tumor localizing capacity have been investigated (Zalutsky et al., 1989b). Using antigen-positive human glioma and antigen-negative rat liver homogenates, specific binding of At-211 labeled 81C6 whole MAb and Mel-14 F(ab')$_2$ were demonstrated. Experiments performed in athymic mice bearing subcutaneous human glioma xenografts indicated that selective and specific tumor uptake of At-211 had been achieved. With At-211 labeled Mel-14 F(ab')$_2$, significantly higher retention of activity in normal tissues was seen than when I-131 was used for labeling. A later study showed that retention of astatine and iodine labels was nearly identical on intact MAbs but not on F(ab')$_2$ fragments (Garg et al., 1990). It appears that differences in the distribution and/or catabolism of intact MAbs and F(ab')$_2$ fragments may necessitate the use of different radioastatination methods.

LABELING ANTIBODIES WITH FLUORINE-18

Positron emission tomography (PET) is an imaging methodology that can provide three-dimensional quantitation of radiotracer uptake _in vivo_. Fluorine-18 is the most commonly used positron-emitting nuclide for clinical studies, primarily as a label for [F-18]fluorodeoxyglucose. If MAbs could be labeled with F-18, then PET might be useful as a diagnostic tool and as a method for more accurately estimating tumor and normal tissue dosimetry prior to labeled MAb radiotherapy.

Kilbourn and coworkers (1987) have reported two methods for labeling proteins with F-18. The first involves the synthesis of methyl 3-[F-18]-fluoro-5-nitrobenzimidate via nucleophilic substitution of [F-18]fluoride for nitro in 3,5-dinitrobenzonitrile, followed by reaction with sodium methoxide in anhydrous methanol. In the second, [F-18]fluoride for nitro exchange was used to prepare 4-[F-18]fluorobenzonitrile which was then converted to 4-[F-18]fluoroacetophenone by reaction with CH_3Li and methanol. Bromination of the methyl group then was performed using $CuBr_2$ to yield 4-[F-18]fluorophenacyl bromide. In both cases, protein labeling was accomplished by incubation of the F-18 labeled acylation agent for 1-2 hr. Although the feasibility of labeling proteins with F-18 was demonstrated, long overall synthesis times resulted in less than ideal radiochemical yields with both approaches.

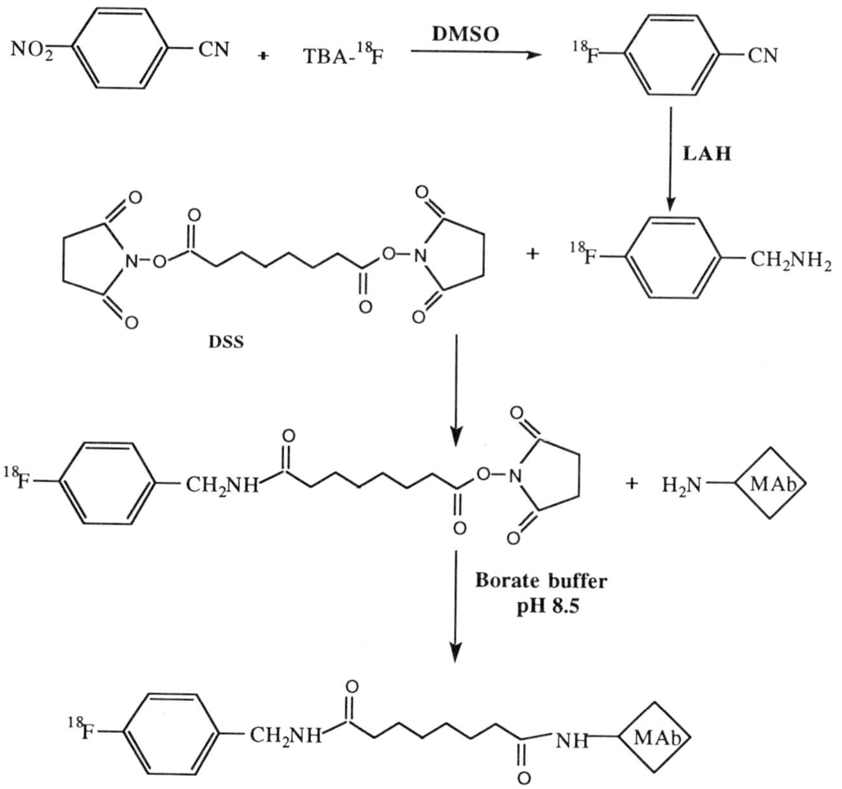

Figure 6. Labeling MAbs with F-18 Using [F-18]SFBS.

We have developed a method for labeling MAbs with F-18 that using
N-succinimidyl 8-{(4'-[F-18]fluorobenzyl)amino}suberate (SFBS) (Garg et
al., 1991b). The reaction scheme is illustrated in Figure 6. 4-[F-18]-
fluorobenzylamine was prepared in two steps from [F-18]fluoride. SFBS
was produced by reaction of this product with disuccinimidyl suberate.
Several MAb fragments have been labeled in 40-45% yield after only a
15 min reaction with SFBS at room temperature. Although HPLC purificaton
of SFBS adds about 15 min to the total preparation time, including this
step results in higher protein coupling efficiency and MAb immunoreactivity. In addition, biodistribution measurements in normal mice showed
that HPLC purification of SFBS decreased liver uptake of an F-18 labeled
Fab fragment by more than twofold.

The immunoreactivities and antigen-mediated localization capabilities of MAb fragments labeled using SFBS currently are being evaluated.
For example, we have reported that the immunoreactivities of F-18 labeled
antimyosin Fab and F(ab')$_2$ fragments, determined using a cardiac myosin
affinity column, were 75% and 89%, respectively (Garg et al., 1991b).
Preliminary experiments in a canine myocardial infarct model using both
F-18 labeled antimyosin fragments have shown selective F-18 uptake in
myosin-rich infarcted myocardium. These results suggest that SFBS may be
a valuable reagent for labeling MAbs with F-18.

ACKNOWLEDGEMENTS

This work was supported by Grant DEFG05-89ER60789 from the Department of Energy and Grants CA 42324, NS 20023 and CA 14236 from the
National Institutes of Health.

REFERENCES:

Aaij, C., Tschrotts, W.R.J.M., Lendner, L., and Feltkamp, T.E.W., 1975,
The preparation of astatine labeled proteins, Int. J. Appl. Radiat.
Isot. 26:25.

Barendsen, G.W., Koot, C.J., van Kersen, G.R., Bewley, D.K., Field, S.B.,
and Parnell, C.J., 1966, The effect of oxygen on the impairment of
the proliferative capacity of human cells in culture by ionizing
radiations of different LET, Int. J.Radiat. Biol. 10:317.

Bolton, A.E. and Hunter, W.M., 1973, The labelling of proteins to high
specific activities by conjugation to a I-125 containing acylating
agent, Biochem. J. 133:529.

Bolton, A.E. Lee-Own, V., McLean, R.K., and Challand, G.S., 1979, Three
different radioiodination methods for human spleen ferritin
compared, Clin. Chem. 25:1826.

Delaloye, B., Bischof-Delaloye, A., Buchegger, F., von Fliedner, V.,
Grob, J.-P., Volant, J.-C., Pettavel, J., and Mach, J.-P., 1986,
Detection of colorectal carcinoma by emission-computerized
tomography after injection of ^{123}I-labeled Fab or F(ab')$_2$
fragments from monoclonal anti-carcinoembryonic antigen antibodies,
J. Clin. Invest. 77:301.

Dumas, P., Maziere, B., Autissier, N., and Michel, R., 1973, Specificite
de l'iodotyrosine desiodase des microsomes thyroidiens et
hepatiques, Biochim. Biophys. Acta 293:36.

Fraker, P.J., and Speck, J.C., 1978, Protein and cell membrane
iodinations with a sparingly soluble chloramide
1,3,4,6-tetrachloro-3α-6α-diphenylglycouril, Biochem. Biophys.
Res. Comm., 80:849.

Friedman, A.M., Zalutsky, M.R., Wung, W., Buckingham, F., Harper, P.V., Sherr, G.H. Wainer, B., Hunter, R.L., Appelman, E.H., Rothberg, R.M., Fitch, F.W., Stuart, F.P., and Simonian, S.J., 1977, Preparation of a biologically stable and immunogenically competent astatinated protein, Int. J. Nucl. Med. Biol. 4:219.

Garg, P.K, Slade, S.K., Harrison, C.L., and Zalutsky, M.R., 1989a, Labeling proteins using aryl iodide acylation agents: Influence of meta vs para substitution on in vivo stability, Nucl. Med. Biol., 16:669.

Garg, P.K., Archer, Jr., G.E., Bigner, D.D., and Zalutsky, M.R., 1989b, Synthesis of radioiodinated N-succinimidyl iodobenzoate: Optimization for use in antibody labelling, Appl. Radiat. Isot., 40:485.

Garg, P.K., Harrison, C.L., and Zalutsky, M.R., 1990, Comparative tissue distribution in mice of the α-emitter ^{211}At and ^{131}I as labels of a monoclonal antibody and F(ab')$_2$ fragment, Cancer Res, 50:3514.

Garg, S., Garg, P.K., and Zalutsky, M.R., 1991a, N-succinimidyl 5-(trialkylstannyl)-3-pyridinecarboxylates: A new class of reagents for protein radioiodination, Bioconjugate Chem., 2:50.

Garg, P.K., Garg, S., and Zalutsky, M.R., 1991b, Fluorine-18 labeling of monoclonal antibodies and fragments with preservation of immunoreactivity, Bioconjugate Chem., 2:44.

Gershengorn, M.C., Glinoer, D., and Robbins, J., 1980, Transport and metabolism of thyroid hormones, In: "The Thyroid Gland", 81, M. DeVisscher, ed., Raven Press, New York.

Harrison, A. and Royle, L., 1984, Preparation of a At-211-IgG conjugate which is stable in vivo, Int. J. Appl. Radiat. Isot. 35:1005.

Hayes, D.F., Zalutsky, M.R., Kaplan, W., Noska, M., Thor, A., Colcher, D. and Kufe, D.W., 1986, Pharmacokinetics of radiolabeled monoclonal antibody B6.2 in patients with metastatic breast cancer, Cancer Res. 46:3157.

Hayes, D.F., Noska, M.A., Kufe, D.W., and Zalutsky, M.R., 1988, Effect of radioiodination on the binding of monoclonal antibody DF3 to breast carcinoma cells, Nucl. Med. Biol., 15:235.

Hnatowich, D.J., Virzi, F., and Doherty, P.W., 1985, DTPA-coupled antibodies labelled with yitrium-90, J. Nucl. Med. 26:503.

Hunter, W.M., and Greenwood, F.C., 1962, Preparation of iodine-131 labelled human growth hormone of high specific activity, Nature 194:495.

Kilbourn, M.R., Dence, C.S., Welch, M.J., and Mathias, C.J., 1987, Fluorine-18 labeling of proteins, J. Nucl. Med. 28:462.

Koehrle, J., Auf'mkolk, M., Rokos, H., Hesch, R.D., and Cody, V., 1986, Rat liver iodothyronine ligand-binding site, J. Biol. Chem. 261:11613.

Larson, S.M., and Carrasquillo, J.A., 1988, Advantages of radioiodine over radioindium labeled monoclonal antibodies for imaging solid tumors, Nucl. Med. Biol. 15:231.

Leonard, J.L., and Rosenbert, I.N., 1977, Subcellular distribution of thyroxine 5'-deiodinase in the rat kidney: A plasma membrane location, Endocrinology 103:274.

Marchalonis, J.J., 1969, An enzymatic method for the trace iodination of immunoglobulins and other proteins, Biochem. J. 113:299.

Narula, A.S. and Zalutsky, M.R., 1988, Synthesis of N-succinimidyl-2,4-dimethoxy-3-(tri-n-butylstannyl)benzoate via regio-specifically generated lithium 2,4-dimethoxy-3-lithiobenzoate, Tetrahedron Lett., 29:4385.

Narula, A.S., and Zalutsky, M.R., 1989, No-carrier-added astatination of N-succinimidyl-3-(tri-n-butylstannyl)benzoate (ATE) via electrophilic destannylation, Radiochimica Acta, 47:131.

Pressman, D., Day, E.D., and Blau, M., 1957, The use of paired labeling in the determination of tumor-localizing antibodies, <u>Cancer Res.</u>, 17:845.

Shechter, Y., Burstein, Y., and Patchornik, A., 1975, Selective oxidation of methionine residues in proteins, <u>Biochem.</u>, 14:4497.

Smallridge, R.C., Burman, K.D., Ward, K.E., Wartofsky, L., Dimond, R.C., Wright, F.D., and Latham, K.R., 1981, 3'5-Diiodothyronine to 3'-monoiodothyronine conversion in the fed and fasted rat: Enzyme characteristics and evidence for two distinct 5'-deiodinases, <u>Endocrinology</u> 108:2336.

Stanbury, J.B., and Morris, M.L., 1958, Deiodination of diiodotyrosine by cell-free systems, <u>J. Biol. Chem.</u>, 233:106.

Sullivan, D.C., Silva, J.S., Cox, C.E., Haagensen, Jr, D.E., Harris, C.C., Briner, W.H., and Wells, Jr., S.A., 1982, Localization of I-131 labeled goat and primate anticarcinoembryonic antigen (CEA) antibodies in patients with cancer, <u>Invest. Radiol.</u>, 17:350.

Vaidyanathan, G., and Zalutsky, M.R., 1990a, Radioiodination of antibodies via <u>N</u>-succinimidyl 2,4-dimethoxy-3-(tri-alkylstannyl)benzoates, <u>Bioconjugate Chem.</u>, 1:387.

Vaidyanathan, G., and Zalutsky, M.R., 1990b, Protein radiohalogenation: Observations on the design of N-succinimidyl ester acylation agents, <u>Bioconjugate Chem.</u>, 1:269.

Vaughan, A.T.M. and Fremlin, J.H., 1978, The preparation of astatine labeled proteins using an electrophilic reaction, <u>Int. J. Nucl., Med. Biol.</u> 5:229.

Visser, G.W.M., Diemer, E.L., and Kaspersen, F.M., 1979, The preparation and stability of astatotyrosine and astatoiodotyrosine, <u>Int. J. Appl. Radiat. Isot.</u> 30:749.

Wilbur, D.S., Hadley, S.W., Hylarides, M.D., Abrams, P.G., Beaumier, P.A., Morgan, A.C., Reno, J.M., and Fritzberg, A.R., 1989, Development of a stable radioiodinating reagent to label monoclonal antibodies for radiotherapy of cancer, <u>J. Nucl. Med.</u> 30:216.

Wursthorn, K.R., Kuivila, H.G., and Smith, G.F., 1978, Nucleophilic aromatic substitution by organostannylsodiums. A second-order reaction displaying a solvent cage effect, <u>J. Am. Chem. Soc.</u> 100:2779.

Zalutsky, M.R., Colcher, D., Kaplan, W., and Kufe, D.F., 1985, Radioiodinated B6.2 monoclonal antibody: Further characterization of a potential radiopharmaceutical for the identification of breast tumors, <u>Int. J. Nucl. Med. Biol.</u>, 12:227.

Zalutsky, M.R. and Narula, A.S., 1987, A method for the radiohalogenation of proteins resulting in decreased thyroid uptake of radioiodine, <u>Appl. Radiat. Isot.</u> 38:1051.

Zalutsky, M.R. and Narula, A.S., 1988a, Radiohalogenation of a monoclonal antibody using an N-succinimidyl 3-(tri-<u>n</u>-butylstannyl)benzoate intermediate, <u>Cancer Res.</u>, 48:1446.

Zalutsky, M.R., and Narula, A.S., 1988b, Astatination of proteins using an N-succinimidyl tri-<u>n</u>-butylstannyl benzoate intermediate, <u>Appl. Radiat. Isot.</u>, 3:227.

Zalutsky, M.R., Noska, M.A., Colapinto, E.V., Garg, P.K., and Bigner, D.D., 1989a, Enhanced tumor localization and <u>in vivo</u> stability of a monoclonal antibody radioiodinated using N-succinimidyl 3-(tri-n-butylstannyl)benzoate, <u>Cancer Res.</u>, 49:5543.

Zalutsky, M.R., Garg, P.K., Friedman, H.S., and Bigner, D.D., 1989b, Labeling monoclonal antibodies and F(ab')$_2$ fragments with the α-particle-emitting nuclide astatine-211: Preservation of immunoreactivity and <u>in vivo</u> localizing capacity, <u>Proc. Natl. Acad. Sci. (USA)</u>, 86:7149.

DIAGNOSIS AND THERAPY OF BRAIN TUMORS UTILIZING

RADIOLABELED MONOCLONAL ANTIBODIES

Herbert E. Fuchs,[1] Michael R. Zalutsky,[2,3] Gary E. Archer,[3] and
Darell D. Bigner[3]

Departments of Surgery (Division of Neurosurgery),[1] Radiology,[2]
and Pathology[3]
Duke University Medical Center
Durham, NC

INTRODUCTION

Monoclonal antibodies (MAbs) have been developed against a wide variety of
tumor-associated antigens and normal tissue antigens, including receptor
molecules, extracellular matrix proteins, enzymes, and hormones. The *in vitro* use
of these MAbs has greatly improved our ability to diagnose a number of diseases,
including many cancers. The extension of MAbs to *in vivo* diagnosis and therapy,
as the "magic bullets" envisioned by Paul Ehrlich[1] at the turn of the century, has,
however, met with more limited success. There are a number of reasons for this,
and currently there is a great effort underway, using animal models to solve these
problems, to pave the way for clinical use of MAbs in patients. To illustrate the
problems inherent in the development of radiolabeled MAbs for clinical use, we
will present our work utilizing MAbs in a variety of brain tumor models.

THE CLINICAL PROBLEM

Primary tumors of the central nervous system (CNS) are among the most
devastating of cancers. The annual incidence of primary brain tumors is 9.2 per
100,000, with over 12,000 new cases each year and over 10,000 deaths per year.
The bimodal peaks of brain tumor incidence in childhood and in middle adult life
effectively increase the impact of these tumors. Malignant gliomas are the most
common primary CNS tumors in adults (representing 15-40% of all intracranial
neoplasms), ranking fourth among males and eighth among females in neoplastic
causes of lost work years.[2] Pediatric intracranial tumors are the second most
common form of cancer in childhood, after leukemia. Medulloblastomas account
for 25% of pediatric brain tumors. Treatment of these tumors with surgical
resection and external radiation therapy, with the possible addition of
chemotherapy, has reached a plateau; more aggressive surgical resection or
radiation therapy may produce unacceptable neurologic deficits or radiation
necrosis of the brain. Clinical trials with adjuvant chemotherapy have to date met
with limited success. Several factors that may be responsible for this poor
therapeutic outcome include cellular phenotypic and genotypic heterogeneity,
diffuse pattern of tumor growth, and lack of specific therapeutic modalities.[3,4]
The diversity of cellular morphology seen in malignant gliomas was first described
by Rudolph Virchow[5] in 1865. Similarly, cultured human glioma cell lines exhibit
phenotypic variability in morphology, growth kinetics, antigen expression, response
to chemotherapeutic agents, expression of biochemical markers and tumorigenicity
in athymic nude mice. Recently, a genotypic basis for this variability has been

recognized; even within an individual tumor, gliomas may have variances in modal chromosome number, marker chromosomes, and DNA content. The recognition of this variability in gliomas and limitations in current therapy has brought about renewed interest in attempts to individualize glioma therapy.

MONOCLONAL ANTIBODIES AND BRAIN TUMORS

The coupling of radionuclides to MAbs directed against tumor-associated antigens has brought about the potential for increased specificity in the diagnosis and treatment of brain tumors. MAbs labeled with γ-emitting nuclides may be utilized to detect tumors in a noninvasive manner. Such radioimmunoscintigraphic techniques may serve as a valuable adjunct to conventional anatomic imaging modalities such as computed tomography or magnetic resonance imaging. The greatest potential application for radioimmunoscintigraphy is, however, in preparation for radioimmunotherapy using MAbs labeled with α- or β-emitting nuclides. This approach offers the potential to deliver curative radiation doses to tumor while minimizing radiation dose to normal tissues. Prior to undertaking clinical trials with radiolabeled MAbs, it is vital to test these reagents in animal systems. The use of human tumor xenografts in immunoincompetent, athymic mice and rats allows MAbs directed against human tumor-associated antigens to be studied in preparation for human clinical trials. In this way, radiolabeled MAbs may be developed in a step-wise progression, identifying and potentially solving problems that may be encountered with their use prior to human studies. The following discussion will describe several aspects of the development of several radiolabeled MAbs for use in the diagnosis and therapy of brain tumors. Preclinical evaluation in animal models, initial clinical trials, and prospects for future progress will be discussed.

The technique of delivering antibodies to tissue *in vivo* is termed immunolocalization.[6,7] Paired-label experiments using ^{125}I-labeled tumor-specific MAb and ^{131}I-labeled nonspecific control immunoglobulin of the same isotype have been used to determine localization indices (LI) as described by Moshakis.[8] This index is used to determine whether localization of MAb is related to specific processes; the higher the LI, the more specific the tumor uptake. Tumor-to-normal tissue ratios (b) are used to demonstrate the selectivity of MAb localization. Once selective and specific localization have been demonstrated, ^{131}I-MAb can be evaluated more extensively for radioimmunoscintigraphic imaging and for radioimmunotherapy of human tumor xenografts in athymic rodents.

$$LI = \frac{\dfrac{^{125}I\,MAb\,(Specific)}{^{131}I\,MAb\,(Nonspecific)}\,Organ}{\dfrac{^{125}I\,MAB\,(Specific)}{^{131}I\,MAb\,(Nonspecific)}\,Blood} \tag{a}$$

$$tumor/tissue\,ratio = \frac{^{125}I\,MAb\,cpm/mg\,tumor}{^{125}I\,MAb\,cpm/mg\,tissue} \tag{b}$$

The specific localization of MAbs within tumors is dependent on a number of factors, including the affinity of antibody for antigen, antigen density in the tumor, and the kinetics of transport of MAb within the various biologic compartments. The degree of localization may also be adversely affected by antigen shedding and modulation. Cross reactivity of MAbs with shared antigens on normal tissues such as bone marrow may not only interfere with localization but also increase toxicity. Delivery of MAb to tumor is also influenced by vascular permeability, blood flow, and extracellular fluid dynamics in the tumor. The blood-brain barrier is another problem not shared by other potential clinical applications of radiolabeled MAbs. This barrier exists at the level of the tight junctions between cerebral capillary endothelial cells. It serves to regulate the passage of low molecular weight ionic compounds as well as higher molecular weight molecules such as proteins and thus could limit delivery of MAb to tumor. The blood-brain barrier has been shown to be heterogeneous in both human gliomas and in experimental animal tumors due to the presence of abnormal blood vessels in gliomas, intratumoral variation in vascular permeability, and altered intratumoral blood flow.[9] Other factors which may influence the delivery of MAbs to tumor include route of delivery (intravenous, intracarotid, intrathecal, and intratumoral), use of smaller, more freely diffusable Fab and F(ab')$_2$ fragments, and transient blood-brain barrier disruption. The following discussion will illustrate these points with specific examples.

Several MAbs have been shown to localize specifically in subcutaneous and intracranial D-54 MG human glioma xenografts in athymic mice and rats. The MAb 81C6 is the best characterized. 81C6 recognizes a 220,000-MW mesenchymal extracellular matrix protein present in gliomas, melanomas, and breast carcinomas[10] (Figure 1). In paired-label studies, higher levels of ^{125}I 81C6, compared with a nonspecific control antibody, were localized in subcutaneous and intracranial human glioma xenografts in athymic mice by 24 to 48 h, with tumor uptake persisting for 5 to 7 days.[11] The specific localization of 81C6 allowed imaging of subcutaneous human glioma xenografts (Figure 2). Similar studies in athymic rats with intracranial human glioma xenografts, an anatomically more realistic model, demonstrated imaging of tumors as small as 20 mg with 81C6.

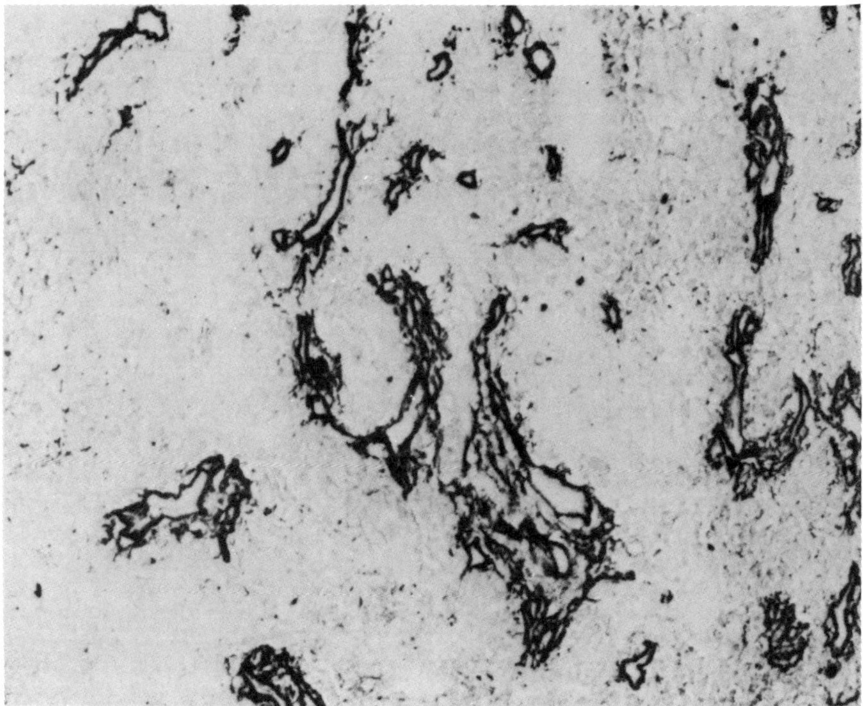

Figure 1. Immunohistological staining of human glioblastoma tissue section using 81C6. Localization is confined to basement membranes associated with abnormal proliferative endothelium and hyperplastic blood vessels, with no detectable localization to tumor cell surfaces, or to endothelial cell luminal surfaces (from Bourdon et al.).[10]

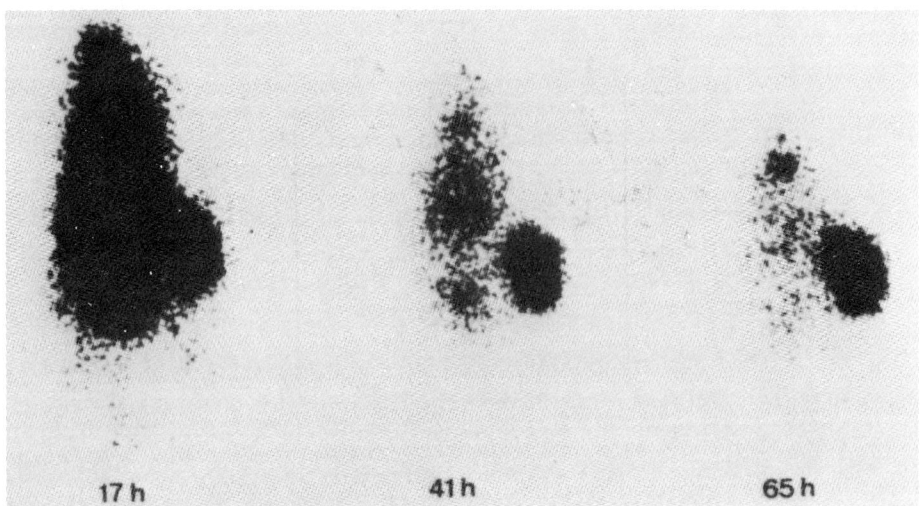

Figure 2. Radioimaging of a subcutaneous U-251 MG human glioma xenograft in an athymic mouse on Days 1 to 3 following intravenous injection of 200 μCi/20 μg [131]I 81C6 using a pinhole collimator and computer image storage. Times shown are in hours (from Bourdon et al.).[11]

Only tumors greater than 300 mg could be imaged with the nonspecific control MAb[12] (Figure 3). Therapy trials with [131]I 81C6 in subcutaneous human glioma xenografts in athymic mice demonstrated significant tumor growth delays with 500 to 1000 μCi doses[13] (Figure 4). Significant survival benefits were seen with [131]I 81C6 in intracranial human glioma xenografts in athymic rats,, using doses up to 2.5 mCi, with several apparent cures.[14] In both studies, the specificity of the response was demonstrated by the lesser or absent response seen with ^{131}I control MAb.

Immunoglobulin fragments such as Fab and F(ab')$_2$, offer many potential advantages over intact MAb as carriers of radionuclides. These smaller fragments are more rapidly cleared from tissues and plasma, resulting in lower background levels for imaging studies, as well as lower normal tissue radiation exposure in therapy studies. Also, the fragments may more easily penetrate the blood-brain barrier to reach the tumor. The absence of the Fc portion of the molecule may reduce the immunogenicity of the murine MAb administered to humans, allowing a greater number of doses to be given. Use of fragments should also result in decreased nonspecific Fc binding to bone marrow and other cells of the reticuloendothelial system.

Fab fragments have a single antigen binding site. The Fab fragment of 81C6 has an affinity that is considerably lower than that of the intact MAb. The plasma clearance of 81C6 Fab fragment is rapid, with a half-life in mice of 7.1 h versus 2.1 days for intact 81C6. The fragment localizes in both subcutaneous and intracranial human glioma xenografts in athymic mice, but at lower levels than intact 81C6.[15] Although dosimetry calculations suggest that the rapid plasma clearance could yield improved tumor/tissue radiation dose ratios with Fab, use of these fragments for therapeutic applications may be difficult because of the much lower absolute magnitude tumor uptake of Fab compared with F(ab')$_2$ or intact IgG. However, as other investigators have demonstrated, Fab fragments are well suited to imaging with short lived nuclides such as ^{123}I and [99m]Tc.

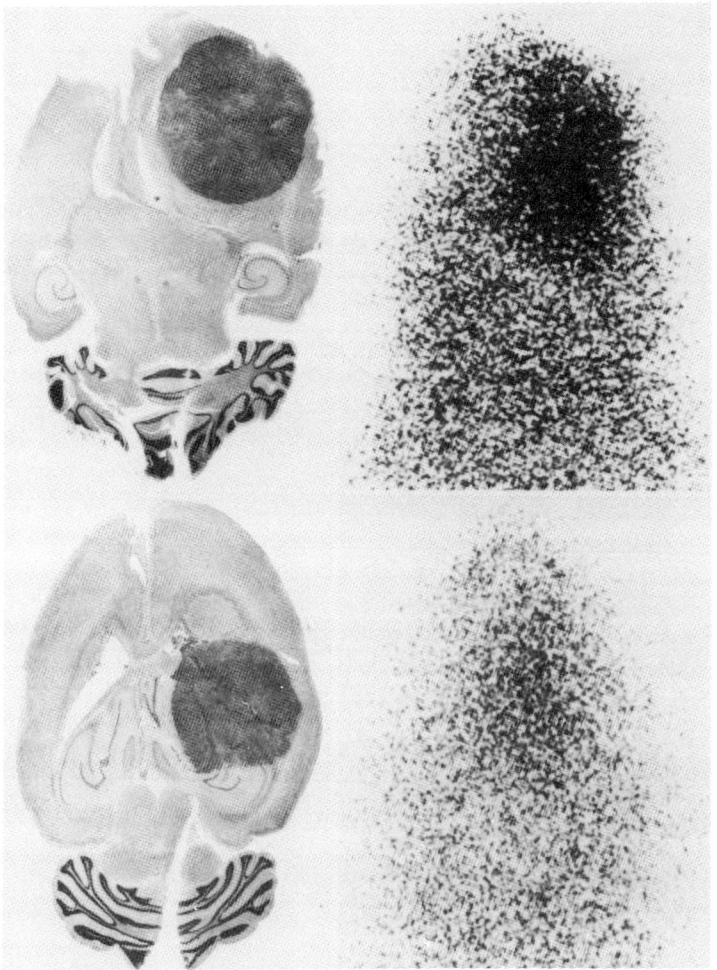

Figure 3.   Comparison of images obtained using ^{131}I-labeled 81C6 (above) and nonspecific control MAb 45.6 (below) on Day 4 following intravenous ^{131}I MAb injection using the intracranial human glioma xenograft model in athymic rats. Corresponding histologic sections documenting tumor location in the right cerebral hemisphere are shown at left (from Bullard et al.).[12]

A higher affinity, divalent F(ab')$_2$ fragment cannot be generated from 81C6 because it is of the IgG$_{2b}$ subclass.   Mel-14, a second MAb which we have investigated extensively, is an IgG$_{2a}$ that can yield F(ab')$_2$ fragments.   Mel-14 recognizes a membrane chondroitin sulfate proteoglycan found in melanomas, gliomas, and some medulloblastomas.[16,17]   Using the athymic mouse-human glioma xenograft model described above, Mel-14 F(ab')$_2$ fragment was shown to localize rapidly in subcutaneous or intracranial tumor as early as 6 to 8 h with more favorable tumor/tissue radiation dose ratios than intact Mel-14.[18]   The therapeutic potential of [131]I Mel-14 F(ab')$_2$ has also been demonstrated in the same intracranial model.[19]

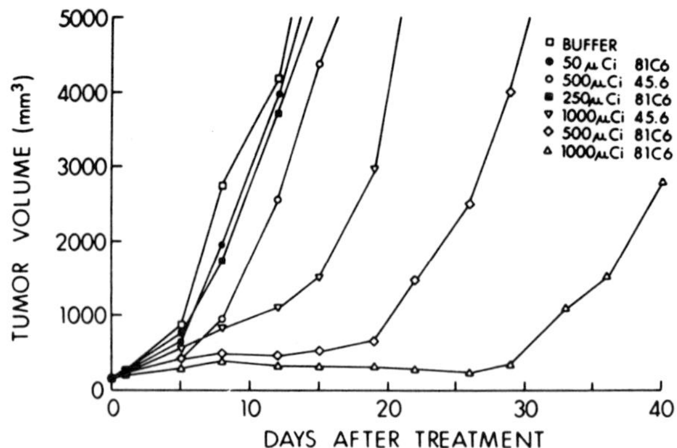

Figure 4. Mean growth curves of subcutaneous xenografts of D-54 MG human glioma in athymic mice following treatment with either [131]I 81C6 or [131]I 45.6 (nonspecific control MAb) demonstrating the growth delay obtained with increasing doses of [131]I 81C6 (from Lee et al.).[13]

In addition to form of MAb, another variable that we have investigated in an attempt to optimize MAb pharmacokinetics is route of administration. For brain tumors, direct intracarotid injection offers two potential advantages over the intravenous route: higher levels in tumor and lower systemic toxicity due to the potential for lower doses to achieve equivalent tumor levels. The potential advantages of nonintravenous delivery have been shown by other investigators using intraperitoneal[20] and intralymphatic[21] administration. In athymic rats with intracranial human glioma xenografts, a 20% increase in delivery of radiolabeled MAb 81C6 to tumor was seen with intracarotid versus injection.[22] However, in paired label studies in patients with anaplastic gliomas,[23] there has been no evidence of intracarotid delivery advantage of MAbs.

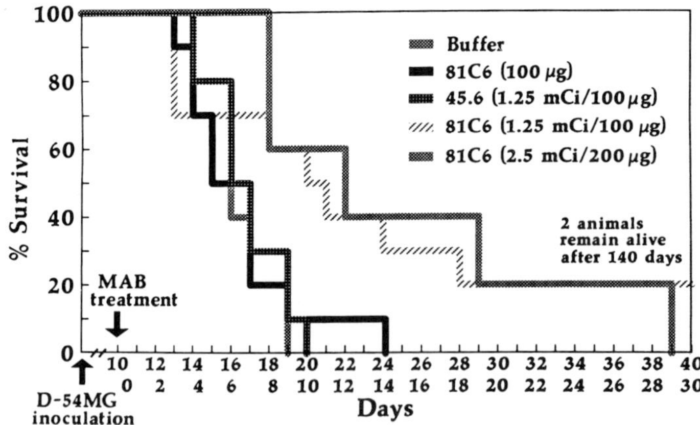

Figure 5. [131]I 81C6 therapy of intracranial human glioma xenografts of D-54 MG in athymic rats, demonstrating significant increases in survival with increasing radiation doses using [131]I 81C6, compared with 45.6, nonspecific control MAb (from Lee et al.).[14]

Based on these results in human glioma xenograft systems, Phase I imaging studies of patients with gliomas have been performed with several MAbs, including 81C6 and Mel-14 F(ab')$_2$ (Figure 5). Paired-label investigations have compared the pharmacokinetics of [131]I 81C6 and isotype-matched control antibody labeled with [125]I. Examination of tissue obtained at craniotomy revealed localization indices for 81C6 as high as 5 for tumor, while values of about 1 were obtained for normal brain.[24] These studies have demonstrated that the uptake and retention of 81C6 in tumor are related to its specificity and are not simply the result of nonspecific accumulation due to alteration in blood-brain barrier by tumor. To date, no toxicity has been observed in patients receiving as much as 100 mg of 81C6 MAb. Further dose escalation toxicity studies are in progress prior to initiating intravenous radioimmunotherapy trials in patients with gliomas.

One promising application of MAb-based radioimmunotherapy is in patients with carcinomatous meningitis. In this type of disease, tumor is spread throughout the spinal fluid around the brain and spinal cord. Leptomeningeal dissemination of tumor may occur with several pediatric brain tumors, including medulloblastoma and ependymoma. Tumor in the CSF also can occur with adult gliomas as well as metastatic tumors, especially lung and breast carcinomas and melanoma, and portends a dismal prognosis.[25,26] Possible factors in the poor outcome of patients with carcinomatous meningitis include inadequate delivery of therapeutic agents due to the presence of the blood-brain barrier as well as a degree of immunologic privilege of the CNS. Conventional external radiotherapy is not effective because CNS and bone marrow toxicity limit doses to less than curative levels of radiation. Intrathecal administration (i.e., direct injection into the CSF space) of radiolabeled MAbs may result in improved delivery to tumor. By coupling a short range α-emitter such as [211]At to MAb, much higher radiation doses may be administered to tumor, while sparing vital normal tissues. To investigate these possibilities, we have developed an animal model of carcinomatous meningitis in the athymic rat. Utilizing a chronic indwelling subarachnoid catheter, a variety of human tumor cell lines have been grown in the CSF.[27] Therapy studies with the rhabdomyosarcoma cell line TE-671 and [211]At 81C6 at doses up to 18 µCi have resulted in significant increases in median survival and 6 out of 10 animals surviving with no evidence of disease after 6 months.[28] Lashford et al.[29] reported the treatment of a group of five patients with leptomeningeal dissemination of tumors such as lymphoma, melanoma, and ependymoma with 11 to 40 mCi ^{131}I MAbs. Imaging studies were performed to document the distribution and retention of radioactivity with the CNS (Figure 6). Four of five demonstrated an objective improvement in response to treatment lasting from 7 months to 2 years. These results are encouraging since

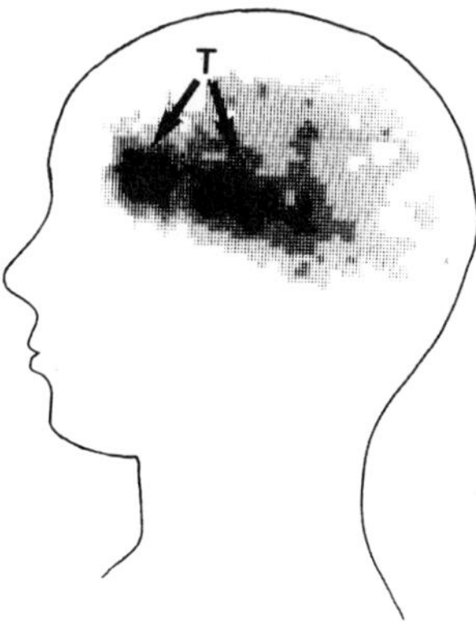

Figure 6. Lateral scintigram obtained at 14 days after intrathecal injection of ^{131}I UJ181.4 showing basal uptake (T) in a patient with neoplastic meningitis from a pineal tumor (from Lashford et al.).[29]

a typical median survival of 2 months is observed in this type of disease. No clinical signs of chronic toxicity have been observed in follow-up for up to 2 years after therapy. A recent report of Moseley et al.[30] described treatment of 15 patients with carcinomatous meningitis with a single intrathecal injection of between 11 and 60 mCi of an ^{131}I radiolabeled MAb chosen for its immunoreactivity with tumor in each case. Toxicity was manifested as nausea, vomiting, and headache in 7 of 15 patients, reversible bone marrow suppression at higher radiation doses in 3 of 8 patients, and seizures in 2 of 15 patients. Of 9 patients evaluable for either a tumor or clinical response, 6 patients demonstrated a sustained clinical response with normalization of spinal fluid cytologies and chemistries for periods between 7 and 26 months.

SUMMARY

The future of radiolabeled MAbs in the diagnosis and therapy of brain tumors appears promising. These preliminary results are paving the way for clinical trials, while great strides are being made in overcoming some of the problems identified in trials to date. Human-mouse chimeric antibodies offer the potential to diminish the immune response to the administered antibody, as only a small portion of the molecule remains of mouse origin. The search continues for more truly tumor-specific antibodies in order to reduce systemic toxicity. A number of tumor markers have been identified that to date have been less than totally specific for tumor, being also expressed on a variety of normal tissues. A recent report by Humphrey et al.[31] describes the development of antibodies to a truly tumor-specific antigen. Many gliomas express amplified and rearranged epidermal growth factor receptors (EGFR) on their cell surfaces. This report describes a totally unique 14 amino acid peptide sequence resulting from the deletion of a portion of the EGFR gene (Figure 7). Antibodies recognizing this unique peptide have been made and MAbs are now in production.

Concurrent with these efforts to develop more optimal MAbs for tumor diagnosis and therapy, other investigators are pursuing better methods for MAb labeling. New developments in labeling MAbs with both halogen and metallic nuclides are described elsewhere in this volume. Combining more tumor-specific MAbs, nuclides with more optimal properties for imaging and therapy, and more stable MAb-radionuclide linkages will hopefully lead to improved reagents for the diagnosis and therapy of brain tumors, and other cancers.

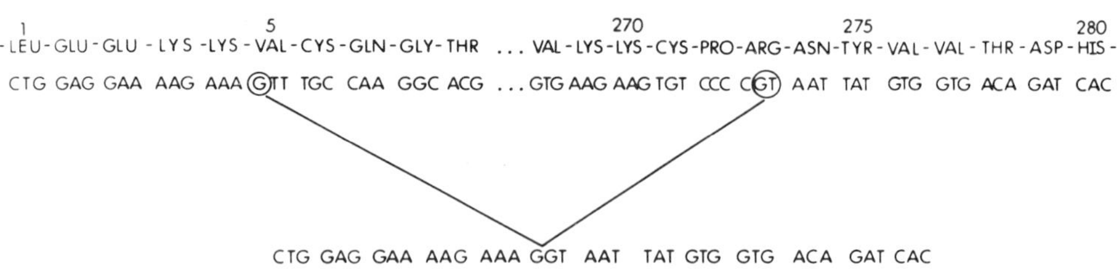

Figure 7. Amino acid sequence of glioma fusion junction peptide of deletion-mutant EGFR. An 802-base pair in-frame EGFR gene deletion (above) results in the fusion of normally distant EGFR gene and protein sequences (below). A glycine residue is created at the fusion point (from Humphrey et al.).[31]

ACKNOWLEDGEMENTS

This work was supported in part by NIH grants CA 11898, CA 42324, CA 14236, and NS 20023.

REFERENCES

1. P. Ehrlich, Collected Studies on Immunology, Wiley, New York (1906).
2. J. L. Murray and L. M. Axtell, Impact of cancer: years of life lost due to cancer mortality, J. Natl. Cancer Inst. 52:3 (1974).
3. D. D. Bigner, Biology of gliomas: potential clinical implications of glioma cellular heterogeneity, Neurosurgery 9:320 (1981).
4. D. Stavrou, Monoclonal antibodies in neuro-oncology, Neurosurg. Rev. 13:7 (1991).
5. R. Virchow, Die krankhalten Geschwuelste, A. Hirshwald, Berlin, (??).
6. W. F. Bale and I. L. Spar, Studies directed toward the use of antibodies as carriers of radioactivity for therapy, Adv. Biol. Med. Phys. 5:285 (1957).
7. D. Pressman, E. D. Day, and M. Blau, The use of paired labeling in the determination of tumor-localizing antibodies, Cancer Res. 17:845 (1957).
8. V. Moshakis, R. A. J. McIlhinney, D. Raghaven, and A. M. Neville, Localization of human tumor xenografts after i.v. administration of radiolabeled monoclonal antibodies, Br. J. Cancer 44:91 (1981).
9. D. R. Groothuis, P. Molnar, and R. G. Blasberg, Regional blood flow and blood-to-tissue transport in five brain tumor models, Prog. Exp. Tumor Res. 27:132 (1984).
10. M. A. Bourdon, C. J. Wikstrand, H. Furthmayer, T. J. Matthews, and D. D. Bigner, Human glioma-mesenchymal extracellular matrix antigen defined by monoclonal antibody, Cancer Res. 43:2796 (1983).
11. M. A. Bourdon, R. E. Coleman, R. G. Blasberg, D. R. Groothuis, and Bigner, D.D., Monoclonal antibody localization in subcutaneous and intracranial human glioma xenografts: paired label and imaging analysis, Anticancer Res. 4:133 (1984).
12. D. E. Bullard, C. J. Adams, R. E. Coleman, and D. D. Bigner, In vivo imaging of intracranial human glioma xenografts comparing specific with nonspecific radiolabeled monoclonal antibodies, J. Neurosurg. 64:257 (1986).
13. Y. S. Lee, D. E. Bullard, M. R. Zalutsky, R. E. Coleman, H. S. Friedman, E. V. Colapinto, and D. D. Bigner, Therapeutic efficacy of murine monoclonal antibody 81C6 in a human glioma xenograft model, Cancer Res. 48:559 (1988).
14. Y. S. Lee, D. E. Bullard, P. A. Humphrey, E. V. Colapinto, H. S. Friedman, M. R. Zalutsky, R. E. Coleman, and D. D. Bigner, Treatment of intracranial human glioma xenografts with ^{131}I-labeled anti-tenascin monoclonal antibody 81C6, Cancer Res. 48:2904 (1988).
15. E. V. Colapinto, Y. S. Lee, P. A. Humphrey, M. R. Zalutsky, H. S. Friedman, D. E. Bullard, and D. D. Bigner, The localisation of radiolabelled murine monoclonal antibody 81C6 and its Fab fragment in human glioma xenografts in athymic mice, Br. J. Neurosurg. 2:173-186 (1988).
16. S. Carrel, R. S. Acolla, A. L. Carmagnola, and J. P. Mach, Common human melanoma associated antigen(s) detected by monoclonal antibodies, Cancer Res. 40:2523 (1980).
17. S. Carrel, M. Schreyer, A. Schmidt-Kessen, and J. P. Mach, Reactivity spectrum of 30 monoclonal antimelanoma antibodies to a panel of 28 melanoma and control cell lines, Hybridoma 1:387 (1982).
18. E. V. Colapinto, P. A. Humphrey, M. R. Zalutsky, D. R. Groothuis, H. S. Friedman, N. de Tribolet, S. Carrel, and D. D. Bigner, Comparative localization of murine monoclonal antibody Mel-14 F(ab')$_2$ fragment and whole IgG$_{2a}$ in human glioma xenografts, Cancer Res. 48:5701 (1988).
19. E. V. Colapinto, M. R. Zalutsky, G. E. Archer, M. A. Noska, H. S. Friedman, S. Carrel, and D. D. Bigner, Radioimmunotherapy of intracerebral human

glioma xenografts with ^{131}I-labeled F(ab')$_2$ fragments of monoclonal antibody Mel-14, <u>Cancer Res.</u> 50:1822 (1990).

20. D. Colcher, J. Esteban, J. A. Carrasquillo, P. Sugarbaker, J. C. Reynolds, G. Bryant, S. M. Larson, and J. Schlom, Comparison of route of administration of radiolabeled monoclonal antibodies in patients with colorectal cancer, <u>J. Nucl. Med.</u> 27:(Abst) 902 (1986).

21. W. B. Nelp, J. F. Eary, R. F. Jones, K. E. Hellstrom, I. Hellstrom, P. L. Beaumier, and K. A. Krohn, Preliminary studies of monoclonal antibody lymphoscintigraphy in malignant melanoma, <u>J. Nucl. Med.</u> 28:34 (1987).

22. Y. S. Lee, D. E. Bullard, C. J. Wikstrand, M. R. Zalutsky, L. H. Muhlbaier, and D. D. Bigner, Comparison of monoclonal antibody delivery to intracranial glioma xenografts by intravenous and intracarotid administration, <u>Cancer Res.</u> 47:1941 (1987).

23. M. R. Zalutsky, R. P. Moseley, J. C. Benjamin, E. V. Colapinto, G. N. Fuller, H. P. Coakham, and D. D. Bigner, Monoclonal antibody and F(ab')$_2$ fragment delivery to tumor in patients with glioma: comparison of intracarotid and intravenous administration, <u>Cancer Res.</u> 50:4105 (1990).

24. R. Moseley, M. R. Zalutsky, H. B. Coakham, R. E. Coleman, and D. D. Bigner, Distribution of [131]I-81C6 monoclonal antibody (MAb) administered via carotid artery in patients with glioma, <u>J. Nucl. Med.</u> 28:603-604 (1987).

25. H. J. G. Bloom, Medulloblastoma in children: increasing survival rates and further prospects, <u>Int. J. Radiat. Oncol. Biol. Phys.</u> 8:2023 (1982).

26. W. R. Wasserstrom, J. P. Glass, and J. B. Posner, Diagnosis and treatment of leptomeningeal metastases from solid tumors, <u>Cancer</u> 49:759 (1982).

27. H. E. Fuchs, G. E. Archer, O. M. Colvin, S. H. Bigner, J. M. Schuster, G. N. Fuller, L. H. Muhlbaier, S. C. Schold, H. S. Friedman, and D. D. Bigner, Activity of intrathecal 4-hydroperoxycyclophosphamide in a nude rat model of human neoplastic meningitis, <u>Cancer Res.</u> 50:1954 (1990).

28. M. R. Zalutsky, P. K. Garg, J. M. Schuster, H. E. Fuchs, G. E. Archer, S. Garg, and D. D. Bigner, Radioimmunotherapy of leptomeningeal tumor using AT-211 monoclonal antibody. 9th International Congress of Radiation Research. <u>Radiat. Res.</u>, in press (1991).

29. L. S. Lashford, A. G. Davies, R. B. Richardson, S. P. Bourne, J. A. Bullimore, H. Eckert, J. T. Kemshead, and H. B. Coakham, A pilot study of ^{131}I monoclonal antibodies in the therapy of leptomeningeal tumors, <u>Cancer</u> 61:857 (1988).

30. R. P. Moseley, A. B. Davies, R. B. Richardson, M. Zalutsky, S. Carrel, J. Fabre, N. Slack, J. Bullimore, B. Pizer, V. Papanastassiou, J. T. Kemshead, H. B. Coakham, and L. S. Lashford, Intrathecal administration of ^{131}I radiolabelled monoclonal antibody as a treatment for neoplastic meningitis, <u>Br. J. Cancer</u> 62:637 (1990).

31. P. A. Humphrey, A. J. Wong, B. Vogelstein, M. R. Zalutsky, G. N. Fuller, G. E. Archer, H. S. Friedman, M. M. Kwatra, S. H. Bigner, and D. D. Bigner, Anti-synthetic peptide antibody reacting at the fusion junction of deletion-mutant epidermal growth factor receptors in human glioblastoma, <u>Proc. Natl. Acad. Sci. USA</u>, 87:4207 (1990).

OXYGENATION BY METHANE MONOOXYGENASE:

OXYGEN ACTIVATION AND COMPONENT INTERACTIONS

Wayne A. Froland, Kristoffer K. Andersson, Sang-Kyu Lee, Yi Liu
and John D. Lipscomb

Department of Biochemistry, Medical School
University of Minnesota
Minneapolis, MN 55455

INTRODUCTION

Methanotrophic bacteria possess the unique ability to utilize methane as the sole source of carbon and energy. Indeed, methane is the only carbon source capable of sustaining vigorous and long term growth of these organisms[1]. The methanolytic activity of methanotrophs can be ascribed to the elaboration of a unique enzyme, methane monooxygenase[2] (MMO), which catalyzes the following reaction:

$$\text{Methane} + O_2 + \text{NADH} + H^+ \rightarrow \text{NAD}^+ + H_2O + \text{Methanol}$$

In subsequent, successive reactions, carbon originally derived from methane is oxidized enzymatically[3,4] through formaldehyde, formic acid, and finally to CO_2. Although the MMO catalyzed reaction requires the utilization of energy in the form of NADH, the energy gleaned from the three subsequent oxidation reactions is sufficient to sustain growth. The NADH utilized by the MMO catalyzed reaction is regenerated during the oxidations of formaldehyde and formic acid by specific dehydrogenases. Carbon for growth is derived by diversion of a fraction of the formaldehyde through assimilation pathways specific to the individual strains of methanotrophs[4]. Methanotrophs have been subclassified as Type I and Type II based on the different carbon assimilation pathways and on significant differences in cellular morphology[1].

MMO has been shown to occur in both soluble and particulate forms in many methanotrophs[5,6]. These forms have different substrate specificities and appear to be different gene products. They are regulated such that only one form is present in the bacterium at a time. The regulation is poorly understood, but it has been proposed to be controlled by the copper concentration of the growth media[5,7]. Although some progress has been made on the characterization of the particulate form, it has proven to be exceptionally labile after cell lysis and has not been obtained in pure form. In contrast, the soluble form of MMO has been purified from several sources[8-14] and will be the focus of this discussion.

The reaction catalyzed by MMO is energetically the most demanding of any oxygenase. In the course of the reaction, O_2 is cleaved and one oxygen atom inserted into a very stable C-H bond of methane. Similar, albeit less demanding, hydrocarbon oxidation reactions are catalyzed by cytochrome P450 monooxygenase[15], a heme containing enzyme. Significantly, our studies[14,16] and those of other laboratories[9,17,18] have shown that the reactive site of MMO does not contain heme or any other cofactor previously associated with monooxygenases. Thus, MMO uses a heretofore unrecognized strategy for producing the activated oxygen species capable of reacting with hydrocarbons.

Another unique aspect of the soluble MMO is the enormous range of accessible substrates[19,20]. All unsubstituted linear and most unsubstituted branched hydrocarbons, both saturated and unsaturated, up to approximately C_8 in length are readily oxidized to alcohols or epoxides. In addition, many saturated, unsaturated, and aromatic cyclic compounds serve as good substrates. Heterocyclic compounds, ethers, and halogenated hydrocarbons extend the substrate list still further. It is clear from the initial studies conducted in many laboratories that the true catalytic potential of the enzyme has only begun to be tapped. The substrate range of MMO may be the broadest of any known enzyme. Moreover, the substrates are abundant, and the products are of great economic significance.

In their natural ecological niche at the interface between anaerobic and aerobic environments, methanotrophs scavenge the roughly 1 billion tons of methane produced annually by anaerobic bacteria before it reaches the atmosphere[21,22]. This function is of considerable environmental import because of the potent "greenhouse" effect of methane exacerbated by its 5-10 year life time in the atmosphere[23]. Our studies have shown that the large substrate range of soluble MMO can be exploited in other ways relevant to the environment. For example, halogenated solvents such as trichloroethylene, a pollutant present in soils and ground water throughout the world, are excellent substrates for MMO[24]. The products are largely short lived epoxides that spontaneously breakdown in aqueous media to volatile or nontoxic end products.

These diverse aspects of MMO chemistry and reactivity have raised the interest of many segments of the scientific community in recent years. The studies described here are directed toward achieving a more complete understanding of the complex structure of the enzyme and its cofactors, as well as the roles these cofactors play in catalysis at the molecular level.

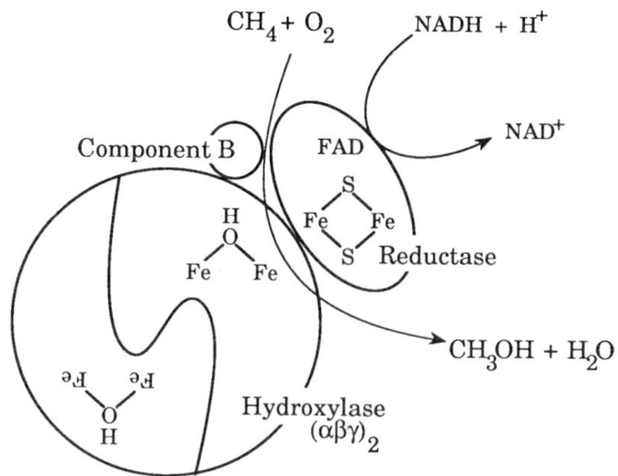

Fig. 1 Protein components and cofactors of MMO.

ENZYME STRUCTURE AND SPECTROSCOPY

As illustrated in Fig. 1, the well characterized MMO enzymes from *Methylococcus capsulatus* Bath[8,9,10] (a Type I methanotroph) and *Methylosinus trichosporium* OB3b[13,14] (a Type II methanotroph) both consist of 3 protein components. Our studies have lead to the purification of the components from the latter bacterium. This preparation exhibits a specific activity many times that of the other preparations that have been reported, and we will discuss only this system here in detail. Recently, the protein sequences of the components from both the *M. capsulatus*[25] and *M. trichosporium*[26] systems have been deduced from the gene sequences. A very high degree of homology was noted suggesting that results determined from either system are relevant to the other. In previous reviews of the *M. trichosporium* system[27,28], we have summarized the methods that resulted in the high specific activity preparation as well as the physical studies which lead to the current view of the protein structure and cofactor content.

The three protein components of the *M. trichosporium* MMO are a 40 kDa reductase containing one FAD and a [2Fe-2S] iron sulfur cluster, a 16 kDa cofactorless component B, and a 245 kDa hydroxylase containing 2 μ-oxo-bridged dinuclear iron clusters. The hydroxylase has a quaternary structure consisting of 3 types subunits apparently assembled as $(\alpha\beta\gamma)_2$.

The cofactor content of the components was determined by a combination of chemical analysis and spectroscopy. The spectroscopic studies that lead to the discovery of the μ-oxo-bridged dinuclear iron cluster of the hydroxylase were particularly extensive and will be briefly reviewed here, in part to establish the nomenclature associated with this critical component. In the hydroxylase as isolated, both irons of the μ-oxo-bridged cluster are in the Fe^{3+} state ([Fe(III)-O-Fe(III)]). Mössbauer spectroscopy showed the irons to be fundamentally high spin (S = 5/2) ferric[16]. Nevertheless, the spectrum at low temperature appeared as a quadrupole doublet rather than the expected six line "magnetic" spectrum characteristic of mononuclear high spin Fe^{3+}. This suggested that the two irons were antiferromagnetically coupled to form a diamagnetic (S = 0) state. A diamagnetic ground state was confirmed through the use of high field Mössbauer spectroscopy at 2-4 K. As expected for such a state, the oxidized enzyme was EPR silent, but on the introduction of one electron per cluster to form the mixed valent state ([Fe(II)-O-Fe(III)]), an intense EPR spectrum was observed[9,16-18]. The *g*-values for this spectrum (1.94, 1.86, 1.75) have only been observed for μ-oxo-bridged dinuclear iron clusters with antiferromagnetically coupled (resultant spin, S = 1/2) irons[29]. Mössbauer spectra showed a rich magnetic spectrum consistent with this assignment. Following the introduction of a second electron per cluster to form the fully reduced state ([Fe(II)-O-Fe(II)]), the Mössbauer spectrum again appeared as a quadrupole doublet with parameters of high spin Fe^{2+}. The EPR spectrum in the *g* < 2 region was no longer observed, but a new signal was observed near *g* = 16. The signal was enhanced when the microwave field of the EPR instrument was aligned parallel with the fixed magnetic field[16,30]. This demonstrated that the signal arose from an integer spin state of the cluster. We subsequently developed a description of the theory supporting the spectroscopic origin of this type of center and used it to simulate and quantitate the spectrum[30]. The simulation showed that the signal can only be accounted for under the assumption that the irons are coupled ferromagnetically (S = 4). Quantitation of the spectrum showed that it accounted for most of the iron in the sample. Thus, it could be used as a sensitive probe of this important state of the hydroxylase.

EXAFS spectra measured of the mixed valent state of the hydroxylase isolated from *M. capsulatus*[31] and *Methylobacterium species*[18] showed Fe-Fe distances similar to inorganic model complexes for μ-hydroxo-bridged iron clusters. Moreover, no short Fe-O bond was observed suggesting that the bridging oxygen was protonated or substituted in some way. This is supported by the high field Mössbauer spectrum of the oxidized *M. trichosporium* hydroxylase at higher temperatures (50 K) which revealed the presence of a low lying excited state consistent with weak antiferromagnetic coupling and, by implication, a substituted oxo-bridge[16]. Likewise, the coupling constant measured for the mixed valent state from the temperature dependence (2-25 K) of the half saturation microwave power[32] is much lower than those determined for other proteins known to have an unsubstituted oxo-bridge[29].

Beyond characterization of the μ-oxo-bridged cluster, these spectroscopic techniques have proven to be very useful probes of the state of the cluster during catalysis. For example, it is relevant to determine which state of the cluster reacts with O_2 during the catalytic cycle. A simple approach to this question was initiated by titrating the hydroxylase with a reductant to maximize the mixed valent state. Under the conditions used, the redox potentials of the oxidized to mixed valent and mixed valent to fully reduced states dictated that about equal amounts of the cluster were in the oxidized, mixed valent, and reduced states. The mixed valent and fully reduced states each gave their characteristic EPR spectra. On admission of O_2, the $g = 16$ signal of the fully reduced state immediately disappeared, while the $g < 2$ signal of the mixed valent was essentially unaffected[14]. Thus, the fully reduced state appeared to be the O_2 reactive form, a conclusion strongly supported by the turnover experiments described below.

CATALYTIC ROLES OF THE COMPONENTS

It was readily demonstrated that each of the three components of MMO had a role in catalysis. Combination of each component individually with NADH, organic substrate, and O_2 failed to yield any oxygenated product[14]. The reductase was the only component to become reduced, but it reoxidized without affecting the substrate. When combined in pairs, only the combination of the reductase and hydroxylase yielded any product. However, the rate of this reaction was maximally increased 150-fold by the addition of the component B. This simple experiment indicates that each component has a role, but it fails to reveal which component is the site of the monooxygenase chemistry. Such a site might be found on either of the components we have termed "hydroxylase" and "reductase", both of which contain metallocofactors. The reductase also contains a flavin, a type of cofactor commonly found in oxygenases. Moreover, the possibility must be considered that an active site could be formed at the interface of two components involved in a protein-protein complex. To better localize the catalytic site, a different experimental strategy was designed[14]. Several hundred milligrams of each isolated component was stoichiometrically reduced with a chemical reductant, mixed with organic substrate, and then exposed to O_2. It was postulated that if the component were capable of catalyzing the oxygenation reaction by itself, a single turnover would occur to yield the same product as the NADH coupled, complete MMO system. The single turnover experiment was successful only in the case of the fully reduced hydroxylase component showing that this component, and presumably the μ-oxo-bridged cluster within this component, is the site of monooxygenase chemistry. The mixed valent state of the hydroxylase showed a much lower yield of product. In fact, the amount of product observed could be accounted for by the amount of fully reduced hydroxylase unavoidably present when the mixed valent state is made due to the proximity of the redox potentials of the states as described above. In considering the overall mechanism of MMO, this was a very enlightening observation because it showed that once 2 reducing equivalents had been transferred to the hydroxylase, the other components were no longer required to achieve catalysis. Thus, the fully reduced μ-oxo bridged cluster of the hydroxylase should be the focus of efforts to comprehend and model MMO chemistry. The technique also greatly simplified the potential direct application of the enzyme for commercial synthesis because it reduced the number of proteins required from three to one and showed that chemical, photochemical, or electrochemical reductants could replace NADH. The involvement of the μ-oxo-bridged dinuclear iron cluster in monooxygenase chemistry is unprecedented in biology. Other protein and enzymes which contain μ-oxo-bridged dinuclear iron cofactors such as hemerythrin[33], purple acid phosphatases[34], uteroferrin[35], and ribonucleotide reductase[36,37] apparently do not have oxygenase activity. Thus, MMO presents a new role for μ-oxo-bridged dinuclear iron as well as a new class within the oxygenase family.

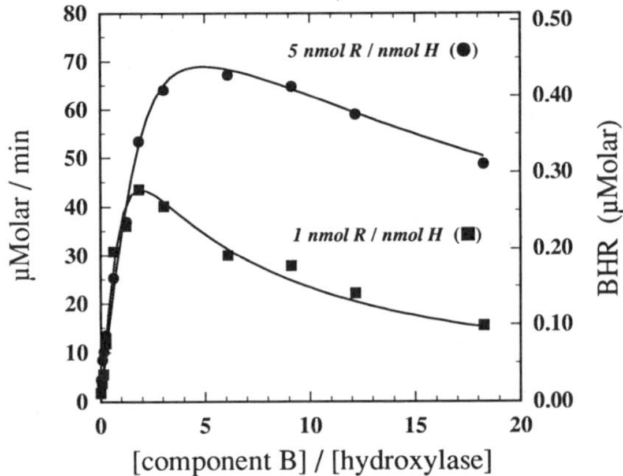

Fig. 2. The effect of the ratio of component B to hydroxylase on the steady state turnover of MMO. Initial velocities of furan oxidation were measured as O_2 uptake rate for 0.25 μM hydroxylase (H) (0.5 μM active sites) and concentrations (per hydroxylase site) of component B (B) and reductase (R) shown. Solid lines are simulations based on the scheme and K_d values shown in Fig. 3. Adapted from reference 32.

The single turnover experiment indicated that the roles of the reductase and component B must be to facilitate electron transfer to the hydroxylase in some way. This is a reasonable role for the reductase given its complement of cofactors that are commonly found in the reductase components of the other systems. However, the nature of the effect component B, devoid of cofactors, could have on electron transfer is more enigmatic. An extensive investigation of the physical interaction of the reductase and component B with each other and the hydroxylase has been conducted[32]. As illustrated in Fig. 2, the effect of component B on the rate and yield of product formation is very complex. When the hydroxylase was present at typical assay concentrations, less than stoichiometric concentrations of component B strongly activated the reaction, while concentrations a few fold higher than stoichiometric resulted in gradual decrease in turnover rate. The point at which the rate was maximized depended upon the absolute concentrations of all three components. As the absolute concentration of the reductase increased, more component B was required to maximize the rate.

The kinetic model shown in Fig. 3 was proposed to account for these observations[32]. In this model, the catalytically relevant species is a complex involving all three components when the system is coupled to NADH oxidation. The inhibitory effects of component B were proposed to arise from the formation of a second complex with the hydroxylase which would in some way block catalysis. An inactive complex between component B and the reductase was also proposed to account for the effect of reductase on the amount of component B required to maximize the rate. Dissociation constants for each of the equilibria required by this kinetic model were estimated by fitting the experimental data. Subsequently, the formation of several of these complexes was observed and quantitated by physical and chemical techniques. Complexes between the component B and reductase and between the hydroxylase and reductase were detected

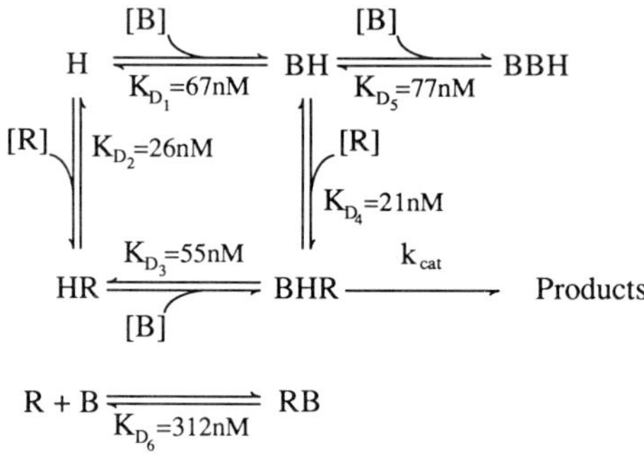

Fig. 3. Kinetic model for MMO turnover coupled to NADH oxidation. Dissociation constants shown were derived from fits to experimental data. Adapted from reference 32.

through observation of the quenching of tryptophan fluorescence of the component B or hydroxylase by the reductase. The K_d values measured from fluorescence titrations were identical within experimental error to those predicted by the kinetic model. A complex between the hydroxylase and component B was detected through component B induced perturbations of the EPR spectra of the mixed valent and fully reduced states of the hydroxylase. At the relatively high concentrations required for EPR, the titration with component B was linear and exhibited a sharp endpoint at approximately one component B per μ-oxo bridged dinuclear iron cluster or two B components per hydroxylase. When component B was provided in large excess, no further changes in EPR line shape occurred, but reduction of the hydroxylase was strongly inhibited.

A final indication of the formation of the predicted complexes came through chemical cross-linking of the components with 1-ethyl-3-(3-dimethylaminopropyl)-carbodiimide (EDC)[32]. This reagent cross-links amino groups with carboxylate groups by facilitating the elimination of the elements of water. As such, no part of the EDC is incorporated and the cross-link is termed "zero length". As summarized in Fig. 4A, cross-linking occurred between the α and β subunits of the hydroxylase as well as between the two β subunits suggesting that these subunits are in contact in the quaternary structure of the hydroxylase. When the hydroxylase was mixed with EDC and either the reductase or component B, cross-linking occurred. The reductase was cross-linked to the β subunit and the component B was cross-linked to the α subunit.

The fact that component B both perturbs the EPR spectrum of the μ-oxo bridged cluster and apparently binds to the α subunit suggests that the cluster and, presumably, the active site are present in this subunit. Cross-linking of a mixture of all three components resulted in all the same cross-linking patterns except that as the concentration of component B was increased above 2:1 relative to the hydroxylase, less and less of the reductase-hydroxylase cross-link formed. A 6-fold excess of component B was sufficient to completely inhibit formation of the reductase-hydroxylase cross-link. Thus, it appears that the inhibitory effects of excess component B on catalysis may stem, in part, from its ability to block the association of the reductase with the hydroxylase. This could be effected by simple physical occlusion of the reductase binding site by a component B dimer bound in the component B binding site (Fig. 4B). Indeed, a cross-linked component B dimer was also identified in the cross-linking investigation.

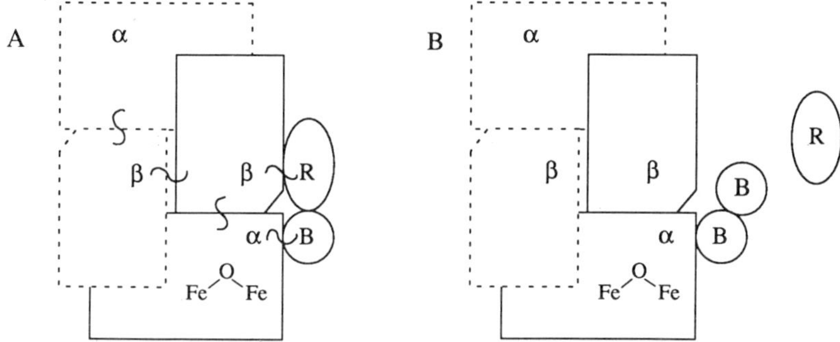

Fig. 4. Cross-linking of MMO by EDC. *A*: The indicated cross-links were detected between MMO components and between the subunits of the hydroxylase. Cross-linked proteins were identified by apparent molecular weight determined from denaturing SDS gel electrophoresis and by N-terminal amino acid sequencing of electro-eluted bands. *B*: Proposal to account for the inhibitory effects of component B most apparent at B:H ratios greater than 2:1. Adapted from reference 32.

While the experiments just described fail to reveal the details of the role of the component B, they strongly suggest that it acts through formation of strong and specific complexes with the hydroxylase. Although a complex is not required for monooxygenase chemistry to occur (when electrons are supplied by reductants other then the MMO reductase), complex formation does perturb the structure or environment of the μ-oxo-bridged cluster responsible for catalysis. The effect of this perturbation could be to facilitate transfer of electrons to the cluster from the reductase and/or to ensure that the two electrons required to activate the hydroxylase are transferred essentially simultaneously. The second role of component B, to prevent electron transfer from the reductase, may also involve complex formation. This effect is most readily detected at concentrations of component B high enough to saturate the complex responsible for the activation effects of component B. A different type of complex is probably involved in which component B may prevent the approach of the reductase to the appropriate location on the hydroxylase. Another rather different role for component B in catalysis has recently come to light through the catalytic studies described in the next sections.

CATALYTIC CYCLE OF METHANE MONOOXYGENASE

Despite the obvious structural differences between MMO and cytochrome P450, the reaction chemistries they catalyze are strikingly similar. In cases where the same substrate is turned over by both enzymes, the same products generally result. However, in cases where more than one product is observed from these substrates, the two enzymes usually give different product distributions. The current proposal for the mechanistic cycle of cytochrome P450 is illustrated in Fig. 5[15,38]. The critical intermediate of this cycle is an oxo-Fe(IV) porphyrin π cation radical species generated by the heterolytic cleavage of O_2 bound to the heme system reduced by two electrons relative to its resting state[38,39]. Such cleavage produces water and a highly reactive oxygen atom with six valence electrons. This "oxene" oxygen could abstract a hydrogen atom from a C-H bond to yield a heme bound hydroxyl and a substrate radical. Reabstraction of the hydroxyl by the radical would yield the product and recycle the enzyme to the resting state. The classic experiments that lead to this proposal have been repeated for MMO by many laboratories including our own. For example, we have shown that halogenated ethylenes such as trichloroethylene (TCE) are turned over rapidly to form epoxides as the primary products[24]. However, a minor fraction of the products were observed to be aldehydes resulting from halogen or hydrogen migration. In the case of TCE, 5% of the product was chloral (2,2,2-trichloroacetaldehyde) formed by chloride migration. Such atom migration reactions generally imply the intermediate formation of a carbocation. Generation of a carbocation is consistent with the abstraction of an electron from the π system of TCE by an electron deficient species such as an oxene of the type proposed in the catalytic cycle. Cytochrome P450 also oxidizes TCE with the formation of TCE epoxide and chloral[40]. In this case, however, a much larger fraction of the product was found to be chloral suggesting that the enzyme active site can influence the product distribution. In a different type of experiment, Rataj et al. have recently shown that *exo, exo, exo, exo* d_4 norbornane is converted to d_3 and d_4 *endo* and *exo* alcohols during MMO turnover[41]. This is most consistent with epimerization of a substrate radical intermediate of the type postulated in the mechanism. The same experiment was used by Groves et al. to provide the first evidence for the radical rebound mechanism of cytochrome P450[42,43]. Thus, MMO and cytochrome P450 appear to use the same type of activated oxygen species and general mechanistic strategy, but different approaches have evolved to generate the reactive oxygen species.

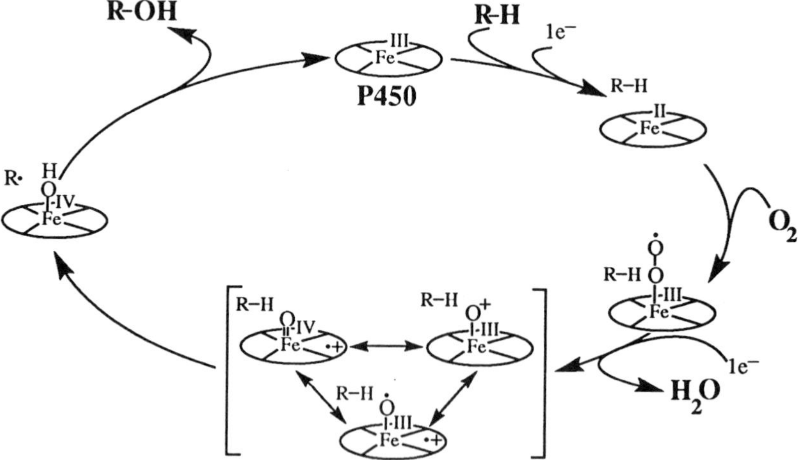

Fig. 5. The proposed mechanistic cycle of cytochrome P450.

Our current proposal[14,24] for the mechanistic cycle of MMO is illustrated in Fig. 6. In contrast to the cytochrome P450 cycle, the MMO cycle appears to progress through both single electron transfer reactions before a form that can react with O_2 is generated. It is postulated that the resulting fully reduced hydroxylase component can catalyze heterolytic cleavage of the O_2 to generate an electron deficient "oxene" species analogous to that of the cytochrome P450 reaction cycle. This species would be stabilized by internal one electron transfer from each of the iron atoms of the μ-oxo bridged dinuclear iron cluster to the bound oxygen to yield an [Fe(IV)-O-Fe(IV)]-oxene species. In this proposal, the second iron can be thought of as fulfilling the role of the heme of P450 by supplying the second electron required to stabilize the oxene. To the extent that the μ-oxo bridged dinuclear iron cluster provides a less effective stabilizing environment for the oxene than the heme, it will be a better hydroxylating reagent. This may account for the unique ability of MMO to attack methane.

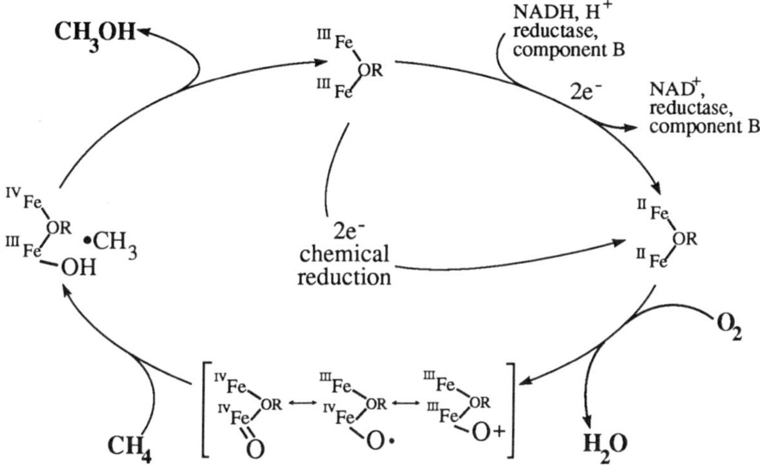

Fig. 6. Proposal for the mechanistic cycle of MMO. The hydroxylase component is represented by its μ-oxo bridged dinuclear iron center. The R-group shown on the bridging oxygen has not been identified and may be a proton, alkyl or other moiety. Adapted from reference 24.

HYDROGEN PEROXIDE COUPLED CATALYSIS

Substantial support for the mechanism of cytochrome P450 derived from the observation that both the single electron transfer reactions and the addition of molecular oxygen could be omitted if oxygen and reducing equivalents were supplied together in the form of a peroxide[38,44,45]. This reaction has been termed the "peroxide shunt" of cytochrome P450. Many organic peroxides in addition to hydrogen peroxide could serve in the shunt[44]. Peracids and single atom oxygen transfer reagents were also found to be functional. The enzyme itself was inactivated by such turnover, but the overall relevance of the proposed catalytic cycle was established.

Our proposal for the mechanism of MMO would suggest that a peroxide shunt analogous to that of cytochrome P450 should apply. Accordingly, we have recently demonstrated that addition of a substrate and H_2O_2 to the oxidized hydroxylase component in the absence of the reductase and component B leads to O_2 independent product formation[46]. Table 1 shows representatives from several of the classes of MMO substrates that are turned over in the H_2O_2 coupled hydroxylase system to give the same products as observed for the reconstituted three component system coupled to NADH.

Table 1.   Products from H_2O_2 Coupled MMO Hydroxylase
Catalyzed Reactions[a]

Substrate	Products (% distribution)
Methane	Methanol (100)
Ethane	Ethanol (100)
Propane	1-Propanol (15) 2-Propanol (85)
Isopentane	2-Methyl-1-butanol (25) 3-Methyl-1-butanol (11) 3-Methyl-2-butanol (24) 2-Methyl-2-butanol (40)
Cyclohexane	Cyclohexanol (100)
Propene	Propene oxide[b] (100)
Nitrobenzene	p-Nitrophenol (42) m-Nitrophenol (53) o-Nitrophenol (5)

[a]Reaction conditions: 16 mg hydroxylase, 100 mM H_2O_2 in 0.5 ml 25 mM
Na-K phosphate buffer pH 7.5, 30 °C.   Adapted from reference 46.
[b] Conditions: 5.4 mg hydroxylase, 10 mM H_2O_2

Control experiments showed that the reaction was dependent on the presence of functional hydroxylase, and that the hydroxylase could not be replaced by added iron salts or iron plus reducing reagents. Also, scavengers for hydroxyl radicals (mannitol) and superoxide (superoxide dismutase) did not affect the reaction indicating that it occurred in the active site of the hydroxylase. Accordingly, specific inhibitors of the enzyme, such as ethanol, also inhibit the H_2O_2 coupled reaction.

The H_2O_2 coupled reaction has been characterized in more detail for a few substrates. It was found that the H_2O_2 coupled hydroxylase turns over propene to propene oxide at a rate dependent on both H_2O_2 and enzyme concentration as shown in Fig. 7, *top*. Approximately 50% yield was observed after the H_2O_2 was exhausted. This implies that the hydroxylase has an inherent peroxidase activity; such activity could be observed in the absence of substrate. The products of this peroxidase reaction have not yet been identified. At 10 mM H_2O_2 concentration and 120 μM hydroxylase, the reaction proceeds to completion in about 30 minutes (about 20 turnovers considering the enzyme has two active sites and the overall yield is ~50%). The hydroxylase recovered from this reaction retained at least 70% of its original activity and the EPR spectra of the mixed valent and fully reduced states were unaltered. Thus, in contrast to cytochrome P450, MMO hydroxylase is only slowly inactivated by H_2O_2 coupled turnover.

The H_2O_2 coupled reaction exhibited saturation kinetics for both the organic substrate and H_2O_2 (Fig. 7, *bottom*) consistent with the formation of an enzyme complex. The observed K_m value for the organic substrate was different for each substrate. However, the K_m value for hydrogen peroxide was approximately the same (~265 mM) for the two substrates shown in Fig. 7. Extrapolation to infinite H_2O_2 concentration predicated a maximum velocity similar to that observed for NADH coupled turnover of the complete system. These observations are very supportive of the relevance of an intermediate at the oxidation level of a peroxide in the catalytic cycle. However, the high K_m value of H_2O_2 relative to that of O_2 in the natural cycle suggests that access to the active site by H_2O_2 is restricted. Larger organic peroxides and single atom transfer reagents such as sodium periodate have not been observed to function in this system.

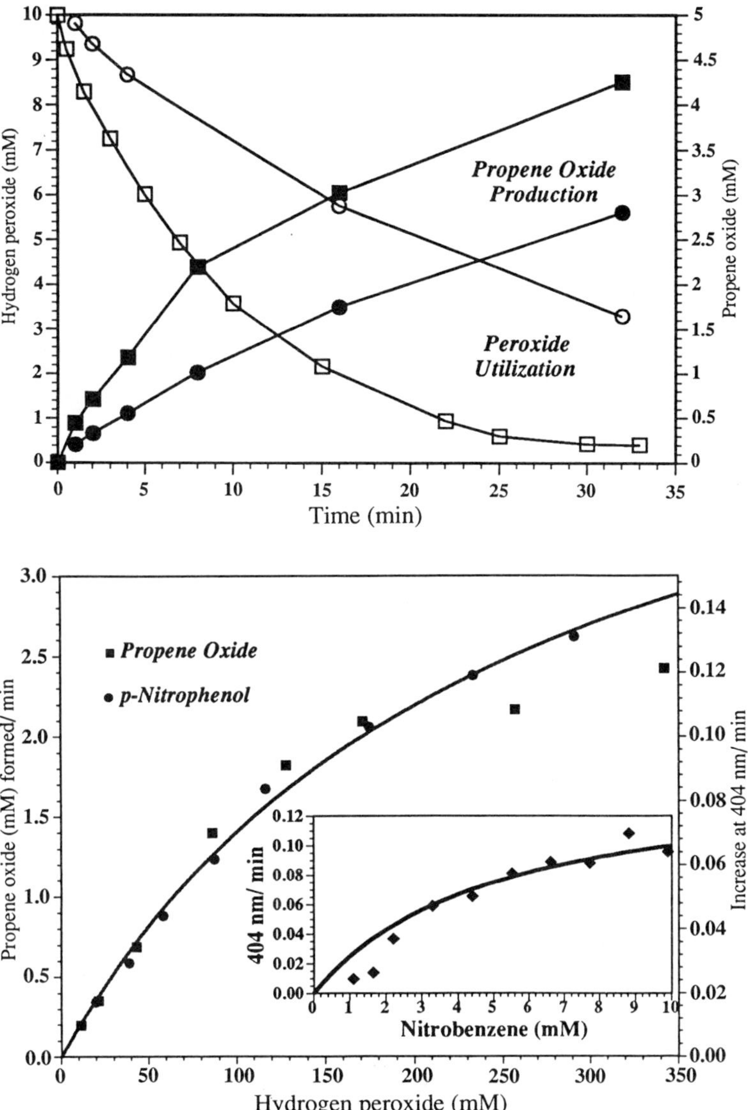

Fig. 7. Hydrogen peroxide coupled turnover by MMO hydroxylase component. *Top:* Time dependence of the oxygenation of propene (1 atm) by hydroxylase at 32 °C. Filled symbols represent propene oxide production. Open symbols represent H_2O_2 utilization. The following reactions are represented: ($\square$, $\blacksquare$) 10 mM H_2O_2 with 123 µM hydroxylase; ($\bullet$, $\bigcirc$) 10 mM H_2O_2 with 44 µM hydroxylase.

Bottom: Dependence of the initial rate of product formation on H_2O_2 and substrate. *Main figure:* Initial velocity of propene oxide formation ($\blacksquare$, 44 µM hydroxylase, 1 atm propene), and *p*-nitrophenol formation ($\bullet$, 22 µM hydroxylase, 8.6 mM nitrobenzene, monitored at 404 nm). Solid curve is a hyperbolic fit with K_m for H_2O_2 = 265 mM. *Inset:* Initial velocity of *p*-nitrophenol formation with 250 mM H_2O_2 and 22 µM hydroxylase. Adapted from reference 46.

These observations allow our proposed MMO mechanism to be augmented by a peroxide shunt as illustrated in Fig. 8. As in the case of the fully reduced hydroxylase single turnover experiments described above, the ability to catalyze hydroxylation reactions in the absence of the reductase and component B demonstrates that the active site of MMO is located on the hydroxylase. The fact that the H_2O_2 coupled reaction is catalytic and proceeds through many turnovers may facilitate the use of this enzyme for synthetic applications.

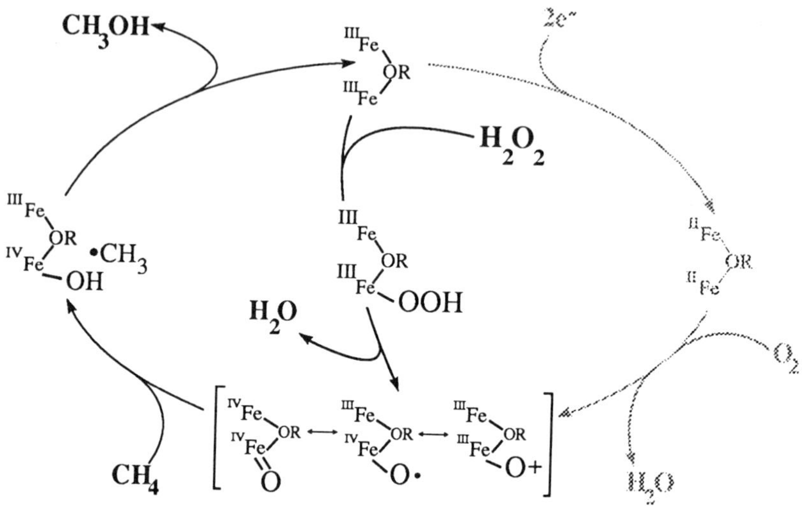

Fig. 8. Proposal for peroxide shunt of the mechanistic cycle of MMO.

PRODUCT DISTRIBUTION

The substrate range of MMO includes many molecules that can potentially be hydroxylated in more than one position. In general, the enzyme will catalyze such multiposition hydroxylation reactions (or hydroxylation in combination with epoxidation or allylic migration reactions). Any given molecule will only undergo only a single reaction unless the product concentrations become very high. The product distribution from these reactions of branched alkanes shows a strong preference for primary and secondary carbons over tertiary carbons suggesting that accessibility of these carbons rather than ease of hydrogen atom abstraction is the major selection criterion. Moreover, the preference for primary and secondary carbons suggests that the consummation of the reaction occurs very rapidly after hydrogen atom abstraction so that intramolecular radical rearrangements are limited prior to "rebound" of the hydroxyl group.

Table 2 shows the product distribution observed for oxidation of isopentane by the reconstituted MMO system coupled to NADH and by the hydroxylase component alone coupled to hydrogen peroxide.* Although the products of the reactions are the same, the product distributions are quite different. In the case of the H_2O_2 coupled system, the product distribution is shifted significantly toward the tertiary product and is similar to that observed for small molecule metal catalysts. Interestingly, when component B is added to the H_2O_2 coupled hydroxylase system, the product distribution is shifted significantly toward the primary carbon hydroxylation characteristic of the reconstituted

---

*W. A. Froland, K. K. Andersson, and J. D. Lipscomb, manuscript submitted.

Table 2.   Regioselectivity of Isopentane Oxidation
by MMO and Small Molecule Catalysts

Reaction Conditions	Distribution of Isopentanols (%)		
	primary	secondary	tertiary
Complete MMO System plus NADH / O_2	80	9	11
MMO Hydroxylase Comp. plus H_2O_2	36	24	40
MMO Hydroxylase and Comp. B (1:1) plus H_2O_2	61	19	20
Fe(II)(DPAH)$_2$[a] plus PhNHNHPh / O_2	21	29	50
Fe(II)(PA)$_2$[b] plus PhSeSePh / H_2O_2	25[c]	35	40

[a] DPAH = (6-carboxy-2-carboxylato)pyridine; C. Sheu and D. T. Sawyer, *J. Am. Chem. Soc.* 112:8212 (1990).
[b] PA = picolinato; C. Sheu et al., *J. Am. Chem. Soc.* 111:8030 (1989).
[c] Products determined as phenylselenium derivatives.
Data from: W. A. Froland, K. K. Andersson, and J. D. Lipscomb, manuscript submitted.

MMO system. Conversely, when the reconstituted system is utilized, but component B is omitted, the product distribution is shifted toward the tertiary product. Thus, it appears that component B plays a heretofore unrecognized role in catalysis. While it is not the site of monooxygenase chemistry, it can direct that chemistry for some substrates so as to determine the nature of the product distribution. Such a role may simply involve repositioning of the substrate as a consequence of the formation of the strong complex between the hydroxylase and B components. Alternative, more complex mechanisms by which this role could be effected are possible and are currently being investigated.

CONCLUDING DISCUSSION

The studies of MMO conducted thus far have shown that it is a three protein component system. Nevertheless, the monooxygenase chemistry occurs on the hydroxylase component. The only cofactor present in the hydroxylase is a μ-oxo bridged dinuclear iron center suggesting that this cofactor plays a role in oxygen activation, cleavage, and insertion. This type of cofactor has not been reported for any other oxygenase, and, thus, MMO apparently uses a new strategy to effect monooxygenase chemistry. We have speculated that the cofactor is a good choice for this enzyme because the presence of two metal ions gives it the potential to both promote O_2 cleavage and stabilize the resulting electron deficient oxygen atom. The first function is facilitated by the ability to accept two electrons from natural or artificial donors, while the second function may be facilitated by the ability to donate two electrons to stabilize the oxene proposed to be generated by heterolytic cleavage of O_2. Two types of experiments have recently been conducted by other research groups which demonstrate the potential of μ-oxo-bridged dinuclear iron clusters to participate in these types of reactions. In the first type of experiment, inorganic model compounds containing μ-oxo-bridged dinuclear iron clusters have been shown to catalyze monooxygenase-like chemistry for a variety of hydrocarbons[47-51]. Unfortunately, no model compound has been convincingly demonstrated to catalyze the oxidation of methane to methanol. However, rapid oxidation of substrates such as cyclohexane can now be achieved with small molecule catalysts of this type[51]. In the second type of experiment, hydrogen peroxide has been shown to react with an inorganic model compound containing μ-oxo-bridged dinuclear iron cluster to generate a transient species containing Fe(IV)[52]. This species is analogous to the Fe(IV) species proposed in our catalytic cycle except that the μ-oxo-bridge is broken in the model complex and the oxene is apparently stabilized by abstraction of an electron from the chelate ligands rather than from the second iron of the cluster.

The studies presented here have demonstrated that MMO can be used to catalyze hydrocarbon oxidation reactions in several ways. The reconstituted, NADH coupled system catalyzes the oxidation of many hundreds of hydrocarbons at rates considerably greater than other biological systems such as cytochrome P450. Moreover, MMO is a single enzyme system with a broad substrate range whereas cytochrome P450 is actually a family of enzymes, each member of which having a relatively narrow substrate range. As an alternative to the reconstituted system, we have shown that the hydroxylase component can act by itself to catalyze the full range of MMO oxidations if it is first fully reduced to the *diferrous state* by a nonenzymatic electron donor and the exposed to O_2. Finally, we have shown that the hydroxylase component in the *diferric state* can catalyze hydrocarbon oxidations in the absence of the other components and O_2 if hydrogen peroxide is supplied as the oxygen and electron donor. Moreover, the product distribution can be altered by controlling the concentration of component B. These reactions of MMO and its individual components have been useful in developing a hypothesis for the molecular mechanism of the enzyme. Furthermore, they offer the potential to utilize either the enzyme or a small molecule catalyst based on the enzyme structure for biosynthetic applications.

ACKNOWLEDGEMENT

This work was supported by grant GM-40466 from the National Institute of General Medical Sciences and by a contract from Amoco Biotechnology Corporation.

REFERENCES

1. R. Whittenbury, K. C. Phillips, and J. F. Wilkinson, *J. Gen. Microbiol.* 61:205 (1970).
2. J. Colby and H. Dalton, *Biochem. J.* 171:461 (1978).
3. J. Colby, H. Dalton, and R. Whittenbury, *Ann. Rev. Microbiol.* 33:481 (1979).
4. C. Anthony, "The Biochemistry of the Methylotrophs" Academic Press, London (1982).
5. S. H. Stanley, S. D. Prior, D. J. Leak, and H. Dalton, H. *Biotech. Lett.* 5:487 (1983).
6. D. Scott, D. J. Best, and I. J. Higgins, *Biotech. Lett.* 3:641 (1981).
7. K. J. Davis, A. Cornish, and I. J. Higgins, *J. Gen. Micro.* 133:291 (1987).
8. J. Green, and H. Dalton, *J. Biol. Chem.* 260:15795 (1985).
9. M. P. Woodland and H. Dalton, *J. Biol. Chem.* 259:53 (1984).
10. J. Colby and H. Dalton *Biochem. J.* 177:903 (1979).
11. R. N. Patel, *Arch. Biochem. Biophys.* 252:229 (1986).
12. R. N. Patel, and J. C. Savas, *J. Bacteriol.* 169:2313 (1987).
13. B. G. Fox and J. D. Lipscomb, *Biochem. Biophys. Res. Comm.* 154:165 (1988).
14. B. G. Fox., W. A. Froland, J. Dege, and J. D. Lipscomb, *J. Biol. Chem.* 264:10023 (1989).
15. R. E. White and M. J. Coon, *Ann. Rev. Biochem.* 49:315 (1980).
16. B. G. Fox, K. K. Surerus, E. Münck, E., and J. D. Lipscomb, *J. Biol. Chem.* 263:10553 (1988).
17. M. P. Woodland, D. S. Patil, R. Cammack, and H. Dalton, *Biochim. Biophys. Acta,* 873:237 (1986).

18. R. C. Prince, G. N. George, J. C. Savas, S. P. Cramer, and R. N. Patel, *Biochim. Biophys. Acta* 952:220 (1988).

19. I. J. Higgins, D. J. Best, and R. C. Hammond, *Nature* 286:561 (1980).

20. J. Green and H. Dalton, *J. Biol. Chem.* 264:17698 (1989).

21. D. H. Enhalt and U. Schmidt, *Pure Appl. Geophys.* 116:452 (1978).

22. D. H. Enhalt, *in:* "Microbial Production and Utilization of Gases", H. G. Schlegel, G. Gottschalk, N. Pfennig, eds., pp. 13-22, Goltze Publishers, Göttingen (1976).

23. B. Hileman, *Chem. & Eng. News* 67:25 (1989).

24. B. G. Fox, J. G. Borneman, L. P. Wackett, and J. D. Lipscomb, *Biochemistry* 29:6419 (1990).

25. A. C. Stainthorpe, V. Lees, G. P. C. Salmond, H. Dalton, and J. C. Murrell, *Gene*, 91:27 (1990).

26. D. L. N. Cardy, V. Laidler, G. P. C. Salmond, and J. C. Murrell, *Molecular Microbiology* 5:335 (1991).

27. B. G. Fox, W. A. Froland, and J. D. Lipscomb, *in:* "Gas Oil and Coal Biotechnology I" C. Akin and J. Smith, eds., pp. 197-214, Institute of Gas Technology Press, Chicago, (1990).

28. B. G. Fox and J. D. Lipscomb, *in:* "Biological Oxidation Systems", C. C. Reddy, G.A. Hamilton, and M.K. Madyastha, eds., Vol. 1, pp. 367-388, Academic Press, San Diego, (1990).

29. P. Bertrand, B. Guigliarelli, and J. P. Gayda, *Arch. Biochem. Biophys.* 245:305 (1986).

30. M. P. Hendrich, E. Münck, B. G. Fox, and J. D. Lipscomb, *J. Amer. Chem. Soc.* 112:5861 (1990).

31. A. Ericson, B. Hedman, K. O. Hodgson, J. Green, H. Dalton, J. G. Bentsen, R. H. Beer, and S. J. Lippard, *J. Amer. Chem. Soc.* 110:2330 (1988).

32. B. G. Fox, Y. Liu, Y., J. Dege, and J. D. Lipscomb, *J. Biol. Chem.* 265:540 (1990).

33. R. E. Stenkamp, L. C. Sieker, L. H. Jensen, *J. Am. Chem. Soc.* 106:618 (1984).

34. B. C. Antanaitis and P. Aisen, *Adv. Inorg. Biochem.*, 5:111 (1983).

35. B. C. Antanaitis, P. Aisen, and H. R. Lilienthal, *J. Biol. Chem.* 258:3166 (1983).

36. P. Reichard and A. Ehrenberg, *Science* 221:514 (1983).

37. P. Nordlund, B-M. Sjöberg and H. Eklund, *Nature* 345:593 (1990).

38. T. J. McMurry and J. T. Groves, *in:* "Cytochrome P-450 Structure, Mechanism and Biochemistry", P. R. Ortiz de Montellano, ed., pp 1-28. Plenum Press, New York (1986).

39. G. A. Hamilton, *in:* "Molecular Mechanisms of Oxygen Activation" O. Hayaishi, ed., pp 405-451, Academic Press, New York (1974).

40. R. E. Miller and F. P. Guengerich, *Biochemistry* 21:1090 (1982).

41. M. J. Rataj, J. E. Kauth, and M. I. Donnelly, *J. Biol. Chem.* 266: (1991), in press.

42. J. T. Groves, G. A. McClusky, R. E. White, and M. J. Coon, *Biochem. Biophys. Res. Commun.* 81:154 (1978).

43. J. T. Groves and G. A. McClusky, *J. Am. Chem. Soc.* 98:859 (1976).

44. E. G. Hrycay, J-Å. Gustafsson, M. Ingelman-Sundberg, and L. Ernster, *FEBS Lett.*, 56:161 (1975).

45. A. D. Rahimtula and P. J. O'Brien, *Biochem. Biophys. Res. Commun.*, 60:440 (1974).

46. K. K. Andersson, W. A. Froland, S-K. Lee, and J. D. Lipscomb, *New J. Chem.* 15: (1991), in press.

47. J. B. Vincent, J. C. Huffman, G. Christou, Q. Li, M. A. Nanny, D. N. Hendrickson, R. H. Fong, and R. H. Fish, *J. Am. Chem. Soc.* 110:6898 (1988).

48. B. P. Murch, F. C. Bradley, and L. Que, Jr., *J. Am. Chem. Soc.* 108:5027 (1986).

49. N. Kitajima, H. Fukui, and Y. Moro-Oka, *J. Chem. Soc., Chem. Comm.* 7:485 (1988).

50. D. H. R. Barton, E. Csuhai, D. Doller, N. Ozbalik, and G. Balavoine, *Proc. Natl. Acad. Sci., U. S. A.* 87:3401 (1990).

51. R. A. Leising, R. E. Norman, and L. Que, Jr., *Inorg. Chem.* 29:2553 (1990).

52. R. A. Leising, B. A. Brennen, L. Que, Jr., B. G. Fox, and E. Münck, E., *J. Am. Chem. Soc.* 113:3988 (1991).

STRUCTURE AND MECHANISM OF ACTION OF THE ENZYME(S) INVOLVED IN METHANE

OXIDATION

Howard Dalton

Department of Biological Sciences
University of Warwick
Coventry, CV4 7AL, England.

INTRODUCTION

It is generally accepted that life on this planet evolved around
3.5×10^9 years ago. At that time it appears that the planet was
sufficiently cool to permit the permanent existence of water, the
atmosphere being comprised principally of methane, ammonia, and hydrogen
with traces of carbon dioxide, hydrogen sulfide, dinitrogen and a few
noble gases. Clearly the primitive life forms were bacterial in nature
and utilized either methane or carbon dioxide or simple heterotrophic
substrates produced by electrical discharge from these 'primitive' gases
under anaerobic conditions for growth. Once the highly toxic oxygen
molecule appeared in reasonable amounts in the atmosphere (around 1.7 x
10^9 years BP) then organisms evolved which could utilize the large
amount of energy released by the aerobic oxidation of organic substrates
to CO_2 and water. In present day terms the existence of aerobic
methane-oxidizers had been known for over 100 years but only aroused
polite interest until Jackson Foster, here in Texas, revitalized
activity in these bugs by reisolating several strains in the late 50's[1].
This work prompted others (notably Whittenbury and colleagues) to devise
facile isolation and cultivation techniques which led to over 100 new
strains being clearly identified[2]. Subsequent work at the biochemical
level in several laboratories have shown that the aerobic methane
oxidation pathway is as follows:-

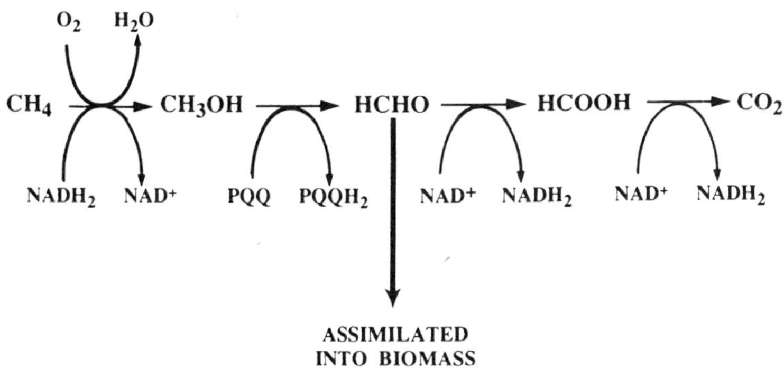

Applications of Enzyme Biotechnology, Edited by J.W. Kelly and
T.O. Baldwin, Plenum Press, New York, 1991

This highly exergonic reaction culiminating in CO_2 ($\Delta G^{o'}$ = -379 KJ/mol), permits healthy growth of bacteria in simple media with air and methane and it's from these organisms that our current understanding of methane oxidation emantes. Methanotrophic bacteria, through the exigency of the enzymic steps outlined above, have been able to execute the oxidation of methane in a controlled manner via a series of individual reactions. The net result is that carbon from methane can be trapped at the level of formaldehyde either for biosynthesis or for energy generation by further oxidation to CO_2.

Of immediate interest, however, is the first step catalysed by the enzyme methane monooxygenase. Methane is converted to methanol in one step, the source of oxygen in the product coming from dioxygen with the second atom ending up in water[3]. The water-soluble enzyme system has now been well characterized from a number of sources and appears to be comprised of three proteins, the properties of which are discussed below. Clearly if one is able to unravel the physical and kinetic complexities of this biological reaction, which functions well at ambient temperature and pressure, then it might be possible to devise robust chemical catalysts which mimic the direct action of the enzyme but which are stable over a long time period (the biological catalyst has a half life measured in terms of minutes at 45°C).

The existing catalytic process to make methanol from methane is an indirect three-stage process involving the production of synthesis gas (reaction 1), a balancing reaction (reaction 2) to increase the ratio of CO to H_2 from 1:3 to 1:2 such that reaction 3 can proceed optimally.

$$CH_4 + H_2O \xrightarrow[\text{700-900°C 1-25 bar}]{\substack{\text{15-20% Ni catalyst} \\ \text{on } Al_2O_3 \text{ or } SiO_2}} CO + 3H_2 \qquad (1)$$

$$CO_2 + H_2 \xrightarrow{\text{Ni catalyst}} CO + H_2O \qquad (2)$$

$$CO + 2H_2 \xrightarrow[\text{250-280°C 70-110 bar}]{\text{Cu/Zn catalyst}} CH_3OH \qquad (3)$$

Overall this indirect process is endergonic whereas the direct route:-

$$CH_4 + \tfrac{1}{2}O_2 \rightarrow CH_3OH \qquad (4)$$

is exergonic and is similar to the biological reaction.

There is no involvement of oxygen in the industrial process and in some respects resembles the sort of biological reaction that may well have occurred on this planet before the advent of oxygenic photosynthesis.

Indeed there is a considerable body of evidence to suggest that biological methane oxidation does occur in anaerobic environments. Most of this work has concentrated on *in situ* studies in which $^{14}CH_4$ is fed to a marine sediment or anoxic water column and the resultant $^{14}CO_2$ evolved is used as a measure of anaerobic methane oxidation. Generally speaking the principal electron acceptor is sulphate which would suggest that *Desulfovibrio* might be the causative agent. Unfortunately no one has successfully demonstrated that any characterized strain of this genus is able to effect this reaction although Panganiban *et al* (1979)[4] did find that sediment samples from Lake Mendota in Wisconsin could oxidize methane anaerobically only when sulphate was used as an electron acceptor and either acetate or lactate were the electron donors. There was no incorporation of methane carbon into the cell. Studies in the author's laboratory by Dr. Mark Hocknull has led to the enrichment of several strains of anaerobic methane oxidizers which also only use sulphate as the electron acceptor and lactate as the electron donor. Interestingly the oxidation is not inhibited by acetylene[5], a potent inhibitor of anaerobic methanotrophy, indicating a fundamental mechanistic difference from the aerobic process. Like the Panganiban *et al*. system methane carbon is not incorporated into cellular constituents and appears quantitatively as carbon dioxide. This inability to grow with methane as the carbon source anaerobically should occasion little surprise. To effectively grow the oxidation should yield sufficient energy to permit the synthesis of ATP. The reaction

$$CH_4 + 2H_2O \rightarrow CO_2 + 4H_2 \qquad\qquad (5)$$

has a $\Delta G^{o'}$ of +130 KJ/mol. However coupling this to the reduction of sulphate, for example, would give a net $\Delta G^{o'}$ of -22 KJ/mol. Also the ΔG value for reaction (5) could be made negative by removal of the hydrogen to values below 10^{-6} M which in practice could be readily accomplished by an indigenous low K_m hydrogenase or another hydrogen-utilizing organism.

These organisms, once cultivated in quantities sufficient to allow biochemistry to be performed on them, will certainly provide new insights into methane activation showing as they do a strong similarity with methane activation through steam reformation although no evidence for CO formation in the biological system has yet been adduced. We are currently studying these slow-growing organisms with a view to establishing their enzymic mechanism of methane oxidation.

<u>Aerobic biochemical methane oxidation</u>

Two different enzyme systems have been shown to be active in aerobic methane oxidation. One is a membrane-bound form of the enzyme which has been shown to be present in all but one of the aerobic species so far studied. Many species, however, have been shown to contain an alternative form of the enzyme which is soluble. In those species where the two forms, referred to as pMMO and sMMO respectively, can be isolated it appears that the level of copper ions in the environment is responsible for dictating which form predominates[6]. At low copper to

biomass ratios sMMO is present; at high copper to biomass ratios pMMO
predominates. It is the pMMO which is probably that form of the enzyme
which is most important *in vivo* since in the environment the biomass
concentrations would rarely, if ever, achieve those found in laboratory
fermenters and therefore the Cu : biomass level would be higher.
Furthermore the carbon conversion efficiency, which is a measure of how
well the cell converts methane into biomass, is almost 40% higher in
Methylococcus capsulatus (Bath) in cells containing pMMO compared with
cells containing sMMO[7]. It is therefore tempting to speculate that the
soluble form may be an unusual form which is manifest only in
laboratory-grown cultures and plays no role in methane oxidation *in
situ*. Nevertheless it is the soluble form which has been the most
intensively studied from which our principal physical and mechanistic
information has been deduced. Such a situation is faintly reminiscent
of the two forms of dinitrogen reductase component of nitrogenase - one
a molybdenum-containing protein exhaustively studied for over 20 years;
the other a recently isolated and hence poorly-studied vanadium-
containing protein which is probably the most important form in
organisms *in situ* bearing in mind the relative abundance of molybdenum
and vanadium ions in the environment[8,9].

Particulate enzymes from *Methylococcus capsulatus* (strain M)[10] and
Methylosinus trichosporium OB3b[11] have been reported to be readily
solubilized and purified yielding systems containing either two (*M.
capsulatus* M) or three (*M. trichosporium* OB3b) separable proteins. The
corresponding system from *M. capsulatus* (Bath) has been solubilized but
could not be resolved into individual components[12]. There is little
mechanistic data available for any of these enzymes. The enzyme system
from *M. trichosporium* is somewhat nebulous in nature and defies
purification in other laboratories; *M. capsulatus* M is an organism
restricted to the Soviet Union and is not available elsewhere and is
also extremely unstable and the *M. capsulatus* Bath system is also
unstable when solubilized. One consistent feature of the three enzyme
systems was the presence of copper ions. In the case of the enzyme from
the Bath strain copper ions appeared also to be required for maximum
activity.

The upshot of this work is that attention has been focussed on the
truly soluble enzyme from two organisms principally; *M. capsulatus*
(Bath) and *M. trichosporium* OB3b. The latter is the subject of the
paper by Professor Lipscomb and will not be considered here. The former
has been the subject of study in my laboratory for over fourteen years
and has given considerable information on the activation of methane.

Cells grown to a low copper : biomass ratio yield soluble enzyme
exclusively. The enzyme can be resolved into three proteins[13,14,15].
The hydroxylase (protein A) has an M_r of 220,000 and is comprised of two
α (M_r = 54 kDa), two β (M_r = 42 kDa) and two γ (M_r = 17,000) subunits.
Analysis of the subunit size from the cloned gene sequence suggests
values of 60,636, 44,726 and 19,844 for the α, β and γ subunits whereas
values by electrospray mass spectrometry (ESMS) indicate values of
44,004 and 19,715 for the β and γ subunits; the α subunit size could not
be determined by this method. Many determinations of a wide variety of
samples have indicated a value of 2 g atoms of non heme iron per protein
although reconstituted protein in which the iron atoms were reintroduced
into the apoprotein did give slightly higher values[16]. The reductase
(protein C) is a single polypeptide of M_r = 38,000 (38,550 from gene

sequence, 38,930 by ESMS). It contains an FAD and Fe_2S_2 cluster which
will interact with the electron donor NAD(P)H and passes electrons onto
a variety of acceptors including the hydroxylase. Protein B is a
regulatory protein which functions by binding tightly with protein A in
the enzyme complex to effect oxygenation of methane to methanol. In the
absence of protein B electron transfer to A from C still occurs but now
methane is not oxidized and all the oxygen appears in water[14]. The
molecular mass measurements on protein B have given curious results. A
value of 17 KDa was found by gel filtration, 16,020 from gene sequence,
and 31,705 by ESMS which indicates that the protein may exist as a
dimer. Furthermore in its monomeric form protein B appears to exist in
two forms, B and B', the latter being a truncated form with a molecular
mass 1 KDa less than B which is inactive in the MMO assay system[17]. It
is not unreasonable to expect that the regulatory protein B would itself
be regulated by proteolytic cleavage but at this stage the identity of
the cleavage enzyme of the sequence of events leading to its expression
are unknown.

In the functioning complex all three proteins are necessary for
activity. FAD center of the reductase is readily reduced by $NADH_2$ which
will accept up to two electrons[18]. One electron can then be transferred
to the Fe_2S_2 center with a second order rate constant of 3.6×10^6 $M^{-1}.s^{-1}$
at $18^{\circ}C$. Electrons are then passed on to protein A with rate
constant of 6.5×10^5 $M^{-1}.s^{-1}$. Removal of the Fe_2S_2 center permits the
reductase to pass electrons on to other acceptors (ferricyanide, DCPIP
etc.) but was inactive in the MMO system[19]. Protein C appears to act as
a transformase by passing electrons one at a time to the hydroxylase
from what is essentially a two electron donor ($NADH_2$). Stopped flow
studies have indicated that the cycle shown in Fig. 1. is the sequence
of events for electron donation from $NADH_2$ to the hydroxylase via the
reductase.

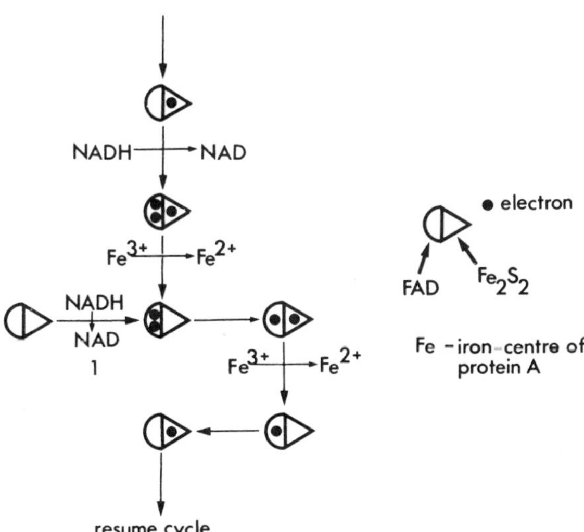

Fig. 1.   Catalytic cycle of protein C from *Methylococcus capsulatus*
(Bath) showing electron distribution between FAD and Fe_2S_2
centers.

As indicated above protein B acts as the regulator of activity. It is important that, in the absence of methane in the environment (not an uncommon occurrence), the enzyme should not consume precious NADH in a wasteful NADH oxidase reaction. Protein B appears to act as a 'sensor' for methane. When methane is present then the complex functions normally, and produces methanol and water. In the absence of methane protein B shuts down electron flow from the reductase to the hydroxylase such that NADH is not wastefully oxidized in a non-productive reaction. If protein B is absent then $NADH_2$ is used to reduce dioxygen exclusively to water[14].

In the active complex the hydroxylase is the site of methane oxidation. The protein contained two non-heme iron atoms which, from the epr evidence, appeared to be in an environment similar to that observed in hemerythrin, purple acid phosphatase or the B2 protein of ribonucleotide reductase. In each case the iron atoms were antiferromagnetically coupled via a μ-oxo bridge. The half reduced (or mixed valence) form of the enzyme was epr active giving rise to a relatively axial epr spectrum with principal g values of $g_z = 1.98$, $g_y = 1.88$ and $g_x = 1.95$, the g_{ave} being 1.87. The shape and amplitude of the signal changed little when reacted with dioxygen which suggested that this form of the protein did not interact directly with this substrate[20]. Subsequently a low field epr signal around g = 15 was noted during reductive titration of the hydroxylase from *Methylosinus trichosporium*[21] which had been previously noted in the azide adduct of deoxyhemerythrin[22]. This signal appeared upon full reduction of the protein as the g = 1.85 signal disappeared. Furthermore the signal at g = 15 disappeared upon addition of oxygen indicating that it was the fully reduced form (Fe^{II}/Fe^{II}) of the hydroxylase[23] that interacted with dioxygen. A similar low field signal was also observed in the *M. capsulatus* Bath hydroxylase but with a g value around 12.3. This ferromagnetically-coupled diferrous system may now be readily observed by epr and can provide a useful means of following the reduced catalytically-competent form of the enzyme.

The nature of the binuclear bridge was confirmed by EXAFs studies[24] in which no short Fe-O distance could be fitted to the data and it was concluded, based on model studies, that the bridge was more likely to be μ-hydroxo or μ-alkoxy rather than μ-oxo.

<u>Kinetic mechanism of the MMO reaction</u>

Steady state kinetics have established that the reaction sequence is of the concerted-substitution type which closely resembles the cytochrome P450 cycle in which hydrocarbon substrate first binds to the enzyme followed by reduction with NADH to give the first ternary complex[25]. This then binds to O_2 to produce a second ternary complex which eventually breaks down to methanol and water (Fig. 2). Pre-steady state kinetics have shown that the electron transfer reactions are all much higher that the turnover number of the enzyme[18] and that protein B shuts down electron transfer from C to A in the absence of methane (5.8 s^{-1} in the presence of methane, 0.35 s^{-1} in the absence of methane). The high kinetic isotope effect observed also confirms that it is the final step (i.e. C-H bond breakage) which is rate limiting in the oxidation process.

Substrate specificity studies

There is little doubt that the binuclear center in the hydroxylase
is intimately involved in the catalytic oxidation of methane - removal
of 90% of the iron from the hydroxylase reduced activity of the complex
by 90% and abolished the epr spectrum[16]. Both of these could be
restored when the iron was reintroduced using chelated iron in the
presence of dithiothreitol. How the binuclear center is involved in
methane activation needs to be addressed. At present there is no direct
evidence for the direct activation of methane by the enzyme through the
formation of an Fe-C bond although studies on the Gif system, which
could be a mimic for MMO, does suggest the involvement of such a species
as a transient intermediate in hydrocarbon activation[26]. The well-
studied mechanisms of hydrocarbon activation have relied upon oxygen

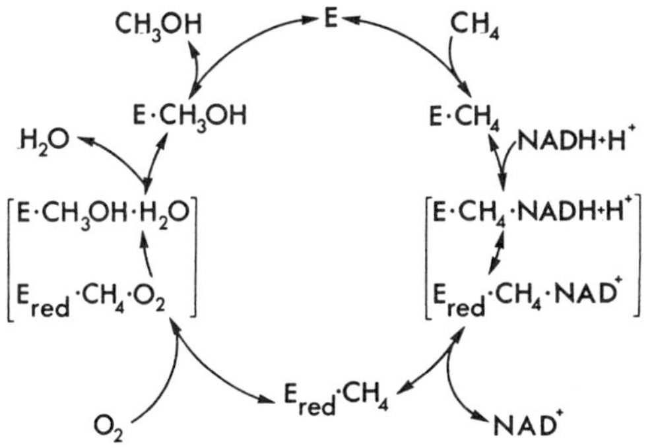

Fig. 2   The catalytic cycle of methane monooxygenase E = enzyme;
Ered = reduced enzyme [ ] = ternary complex.

activation to effect C-H bond activation e.g. the Fenton reagent or high
valent metal-oxy compounds which are powerful oxidants capable of
hydrogen abstraction from hydrocarbons. The cytochrome P450 mimics
clearly indicate that a high valent metal oxo species is involved via
radical chemistry. It is this latter system which has proved to be of
value in helping unravel the complexities of the MMO system.

Early studies using crude or whole cell systems had clearly
indicated that MMO was extremely catholic in its substrate specificity
being able to activate C-H C=C, C≡C and N-H bonds in a wide variety of
organic compounds[27]. Subsequent studies using purified proteins has
confirmed and extended this range[28]. Oxidation of straight chain n-
alkanes (C_5-C_7) showed a preference for attack at secondary positions.
Oxidation of branched chain n-alkanes generally showed a preference for
either tertiary (if available) or secondary positions. In some cases,

e.g. 2,2'-dimethyl butane oxidation steric hindrance at the two position caused oxidation to occur at the C_4 primary position exclusively. The preference for attack at tertiary and secondary over primary positions would indicate that a radical-type non concerted mechanism may be operative in the MMO system. It was particularly interesting to note that the cyclic alkane, adamantane, was attacked with a C_2/C_3 selectivity of 1.0 further strengthening the view of a preference of tertiary over secondary positions. In cytochrome P450 systems the selectivity for this substrate is generally about 0.15 but for the GifIII system this ratio is around 1.15 leading the Texas group[26] to draw a close analogy between the Gif and MMO systems. Indeed this similarity is quite close for a number of substrates (Fig. 3). The Gif system does not oxidize methane, epoxidize alkenes or sulfoxidize sulfides, all of which are characteristic of MMO.

The haloalkenes dichloro- and trichloro-ethylene are substrates which led to the inactivation of MMO presumably through the formation of reactive radical intermediates[28]. Furthermore this inactivation was associated with a loss of iron from the hydroxylase which could not be replaced by the usual reconstitution procedure. Presumably a fairly substantial alteration to the active site had occurred. The oxidation of monohalo benzenes have also been useful as mechanistic probes which _via_ Hammett plots have indicated that a charged intermediate is involved in the rate limiting step[28].

These observations coupled with the scrambling of stereochemistry observed with _cis_ cyclohexanes and cyclohexenes[29], allylic rearrangements in the oxidation of β-pinene and methylene cyclohexane and the opening of cyclopropyl rings during hydroxylation strongly indicate that a non-concerted reaction mechanism is apparent in MMO.

In many respects this is strongly reminiscent of the mechanism of action of cytochrome P450 in which a strong electrophile is generated to abstract hydrogen from the substrate. The oxene species on the iron of the porphyrin ring system in P450 is generated from cleavage of the bound peroxy species. This high valent species is presumed to be stabilized by the porphyrin ring system to generate a porphyrin radical cation on an $Fe^{4+}O^-$ complex. In the case of MMO no such ring system is present but stabilization could be achieved through delocalization of electron density from the binuclear cluster[30].

It is possible therefore to propose a scheme for methane oxidation based on our current knowledge as outlined above. In this scheme (Fig. 4) it is proposed that methane binds to the oxidized form of the hydroxylase at a site close to the binuclear iron center but does not interact directly with it. It is possible that this site is highly hydrophobic and that displacement of water by methane would provide the entropy for binding. Following binding of methane $NADH_2$ effects reduction of the hydroxylase in two one-electron steps to produce the fully reduced form. The half-reduced form (3) shows the g = 1.85 signal whereas the fully reduced form has a low field signal around g = 12. Dioxygen binding to this form of the enzyme is extremely rapid (3 x 10^7 M^{-1} s^{-1}) and implies that there is direct binding to the iron without having to displace strongly bound ligands. The resultant peroxy species (5) may then be stabilized by the hydroxo bridge. Subsequent cleavage of the O-O bond can then occur via homolytic cleavage to produce the hydroxyl radical which then abstracts hydrogen from methane in a Fenton-like reaction to form a methyl radical and water. This subsequently collapses with the resultant $Fe^{4+}O^-$ and a proton to form methanol. Another possibility is that heterolytic cleavage prevails with protons to form water and the high valent oxene species (6). These are of

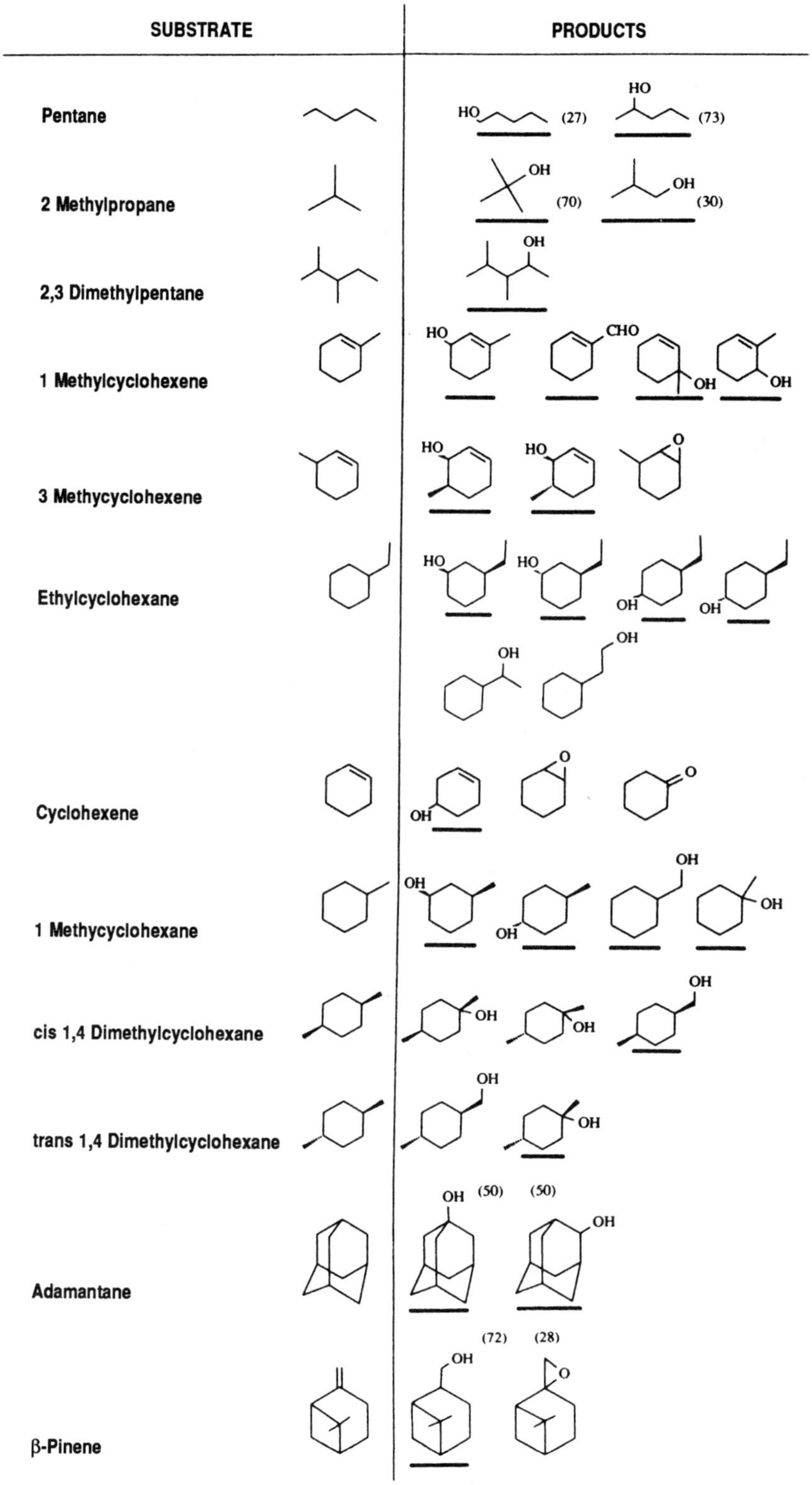

Fig. 3   Substrates and products of the soluble MMO from *Methylococcus capsulatus* (Bath). Underlined products are also identified in Barton's Gif system.

course hypothetical intermediates and have not been physically
demonstrated in this system.

Whatever its nature the highly electrophilic hydrogen-abstracting
species (6) would need to be stabilized possibly by the second iron in
the bridge which could exist as a tautomer with Fe^{IV}-OH-Fe^{IV}=O. Again
the electrophilic species abstracts hydrogen from methane to form the
methyl radical which could then recombine with the captive hydroxyl
radical via an oxygen rebound mechanism. This latter route is

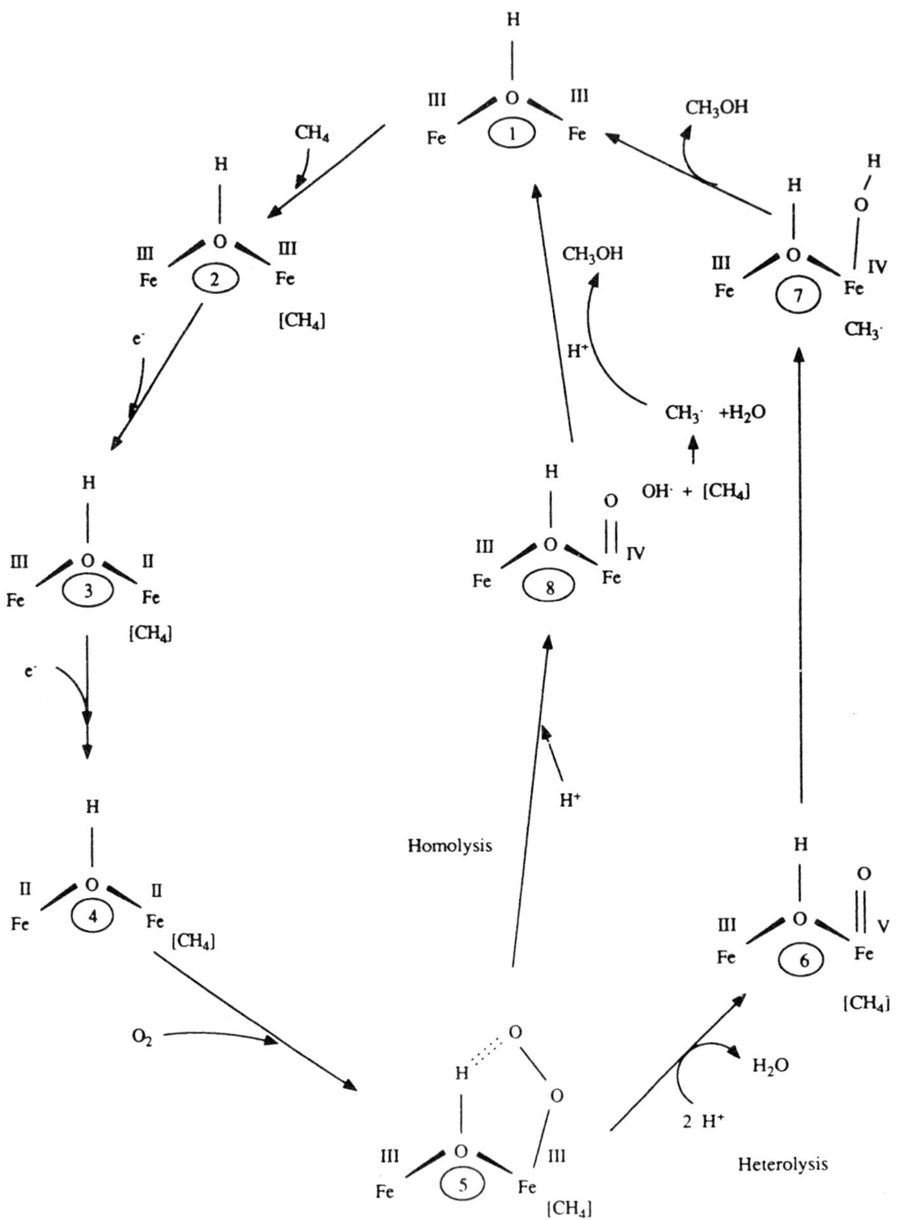

Fig. 4   Postulated intermediates at the active site of methane
          monooxygenase during catalytic turnover.

thermodynamically less favourable than homolytic cleavage but has the
added advantage that free hydroxyl radicals are not produced at the
active site where they may inflict serious damage to the enzyme.

<u>Molecular biology and the nature of the active site ligands</u>

Cloning of the structural genes for MMO has now been achieved for
two species, *M. capsulatus* (Bath) and *M. trichosporium* (OB3b)[17,31,32].
The arrangement of the genes on the PCH4 plasmid is shown in Fig. 5.
Subcloning of a fragment containing mmoB has led to high level
functional expression of this protein in *Escherichia coli* and further
subcloning is underway to express the hydroxylase (X, Y, and Z) and
reductase (C) separately in a similar host. The intervening mmoB region
has hindered a clean expression of the hydroxylase.

Analysis of the cloned fragments of the MMO region has indicated a
sequence identity of over 90% in the α subunit of the hydroxylase. It
is this subunit that is believed to contain the binuclear center.
Furthermore the recently published[33] X-ray crystal structure of the B2
protein of ribonucleotide reductase (RNR) has identified those ligands
which are coordinated to that particular binuclear center. Inspection
of the amino acid sequence of eleven different RNR's has shown that 18
residues out of 375 are conserved between these proteins (Britt-Marie
Sjoberg, personal communication) which include residues associated with
the iron species (Fig. 6). Alignment of the RNR protein sequences with
the deduced MMO protein sequence has revealed particularly interesting

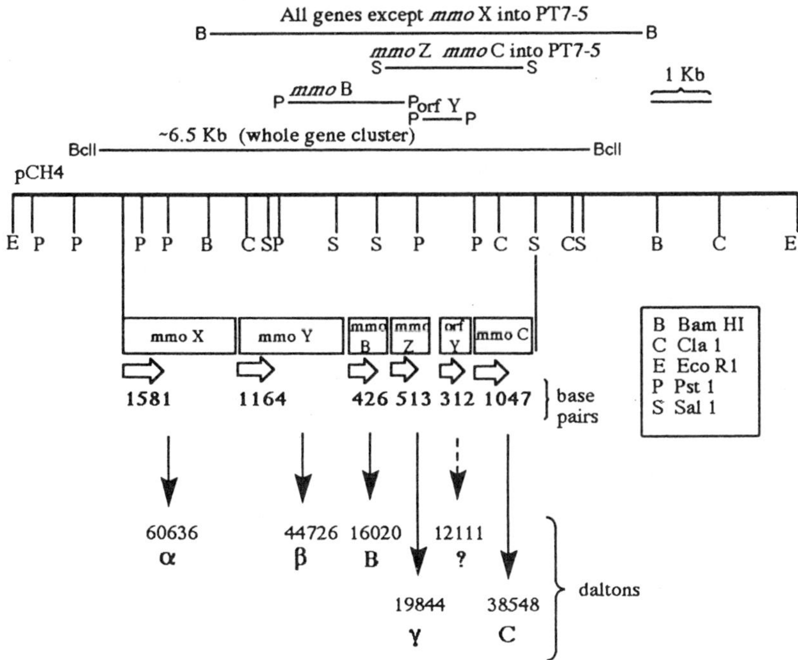

Fig. 5.   Restriction map and arrangement of gene cluster of the 11.9 kb
insert of *M. capsulatus* (Bath) DNA of plasmid pCH4.

analogies. In MMO there are two regions with the sequence asp-glu-X-X-his (143-147 and 242-246) which could be aligned with the glu-X-X-his (115-118 and 238-241) regions in RNR known to be involved as ligands to the Fe-Fe center. The regions identified in MMO are the only regions that contain glu and his separated by two other residues such that it is possible that they might be involved in metal binding. Removal of one iron atom from MMO has revealed the presence of two titratable histidine residues from the hydroxylase protein. Identification of the sequence associated with these residues is underway.

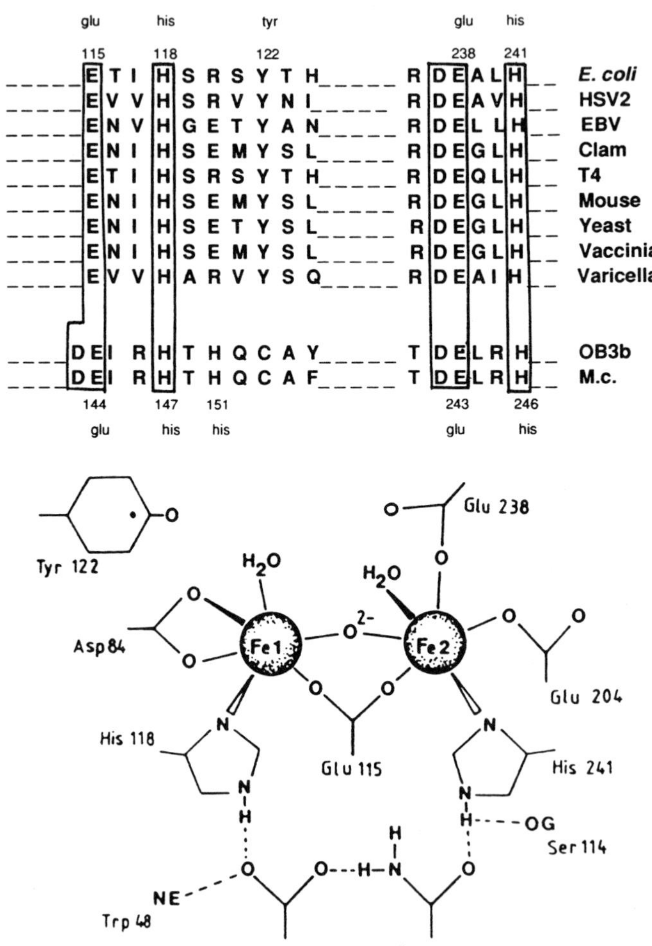

Fig. 6. Sequence homology between the iron-binding sites in the B2 protein of ribonucleotide reductase and the α subunit of protein A of methane monooxygenase.

    Acknowledgements. I would like to thank all the research workers
in my group over the past fifteen years who have each contributed their
part in enabling me to present this summary of our work. Grateful
thanks must also go to our sponsors in that time, the Science and
Engineering Research Council, British Petroleum, British Gas and the Gas
Research Institute (Chicago, Illinois).

References

1.   Leadbetter, E. R. and Foster, J. W., 1958, Arch. Microbiol.,
      30:91-118.
2.   Whittenbury, R., Phillips, K. C. and Wilkinson, J. F., 1970, J.
      Gen. Microbiol., 61:205-218.
3.   Higgins, I. J. and Quayle, J. R., 1970, Biochem. J., 118:210-218.
4.   Panganiban, A. T., Patt, T. E., Hart, W. and Hanson, R. S., 1979,
      Appl. Env. Microbiol., 37:303-309.
5.   Prior, S. D. and Dalton, H., 1985, FEMS Microbiol. Lett., 29:105-
      109.
6.   Stanley, S. H. S., Prior, S. D., Leak, D. J. and Dalton, H., 1983,
      Biotechnol. Lett., 5:487-492.
7.   Leak, D. J. and Dalton, H., 1986, Appl. Microbiol. Biotech.,
      23:470-476.
8.   Bishop, P. E., Jarlenski, D. M. L. and Hetherington, D. R., 1980,
      PNAS USA, 77:7342-7346.
9.   Bishop, P. E. and Joerger, R. D., 1990, Ann. Rev. Pl. Physiol. Pl.
      Mol. Biol., 41:109-125.
10.   Akent'eva, N. F. and Gvozdev, R. I., 1988, Biokhimiya 53:91-96.
11.   Tonge, G. M., Harrison, D. E. F. and Higgins, I. J., 1977,
      Biochem. J., 161:333-344.
12.   Smith, D. D. S. and Dalton, H., 1989, Eur. J. Biochem., 182:667-
      671.
13.   Colby, J. and Dalton, H., 1979, Biochem. J., 177:903-908.
14.   Green, J. and Dalton, H., 1985, J. Biol. Chem., 260:15795-15801.
15.   Woodland, M. P. and Dalton, H., 1984, J. Biol. Chem., 259:53-60.
16.   Green, J. and Dalton, H., 1988, J. Biol. Chem., 263:17561-17565.
17.   Pilkington, S. J., Salmond, G. P. C., Murrell, J. C. and
      Dalton, H., 1990, FEMS Microbiol. Lett., 72:345-348.
18.   Green, J. and Dalton, H., 1989, Biochem. J., 259:167-172.
19.   Lund, J. and Dalton, H., 1985, Eur. J. Biochem., 147:291-296.
20.   Woodland, M. P., Patil, D. S., Cammack, R. and Dalton, H., 1986,
      Biochim. Biophys. Acta., 873:237-242.
21.   Fox, B. G., Surerus, K. K., Munck, E. and Lipscomb, J. D., 1988,
      J. Biol. Chem., 263:10553-10556.
22.   Reem, R. C. and Solomon, E. I., 1987, J. Amer. Chem. Soc.,
      109:1216-1226.
23.   Bentsen, J. G., Lippard, S. J., DeWitt, J., Hedman, B.,
      Ericson, A., Hodgson, K. O., Green, J. and Dalton, H., 1989,
      Poster preparation at 'Metals in Biology' Gordon Research
      Conference Ventura, California, USA.
24.   Ericson, A., Hedman, B., Hodgson, K. O., Green, J., Dalton, H.,
      Bentsen, J. G., Beer, R. H. and Lippard, S. J., 1988, J. Amer.
      Chem. Soc., 110:2330-2332.
25.   Green, J. and Dalton, H., 1986, Biochem. J., 236:155-162.
26.   Barton, D. H. R., Csuhai, E., Doller, D., Ozbalik, N. and
      Balavoine, G., 1990, Proc. Natl. Acad. Sci. USA, 87:3401-3404.

27.   Colby, J., Stirling, D. I. and Dalton, H., 1977, <u>Biochem. J.</u>, 165:395-402.
28.   Green, J. and Dalton, H., 1989, <u>J. Biol. Chem</u>., 264:17698-17703.
29.   Leak, D. J. and Dalton, H., 1987, <u>Biocatalysis</u>, 1:23-36.
30.   Ortiz de Montellano, P. R., 1986, In: 'Cytochrome P450, Structure, Mechanism and Biochemistry' (P. R. Ortiz de Montellano, ed.), Plenum Press, New York, pp. 217-271.
31.   Stainthorpe, A. C., Lees, V., Salmond, G. P. C., Dalton, H. and Murrell, J. C., 1990, <u>Gene</u>, 91:27-34.
32.   Stainthorpe, A. C., Murrell, J. C., Salmond, G. P. C., Dalton, H. and Lees, V., 1989, <u>Arch. Microbiol</u>., 152:154-159.
33.   Nordlund, P., Sjöberg, B. M. and Eklund, H., 1990, <u>Nature</u>, 345:593-598.

STUDIES OF METHANE MONOOXYGENASE AND ALKANE OXIDATION MODEL COMPLEXES

Amy C. Rosenzweig, Xudong Feng, and Stephen J. Lippard

Department of Chemistry
Massachusetts Institute of Technology
Cambridge, Massachusetts 02139

INTRODUCTION

Among the recently delineated class of non-heme iron oxo proteins is the hydroxylase component of methane monooxygenase, an enzyme that catalyzes the conversion of methane to methanol according to eq. 1.[1] Methane monooxygenases (MMOs) are found in methanotrophic bacteria

$$CH_4 + NADH + H^+ + O_2 \rightarrow CH_3OH + NAD^+ + H_2O \qquad (1)$$

that use methane as their sole source of carbon and energy.[2] In this article we discuss mainly the results of studies that have been carried out on MMOs from the organisms *Methylococcus capsulatus* (Bath) and *Methylosinus trichosporium* OB3b. The soluble MMOs from both of these organisms contain two proteins in addition to the hydroxylase, a reductase with associated FAD and Fe_2S_2 prosthetic groups and a smaller polypeptide, designated protein B, that is believed to play a role in regulating electron transfer between the reductase and hydroxylase components.[3,4] The relative roles of these proteins in the overall MMO system are displayed in Figure 1. Most catalysts that effect the hydroxylation of alkanes by dioxygen are also able to catalyze the direct oxidation (autoxidation) of the reductant with dioxygen. The MMO system avoids this potential problem by physically isolating the hydroxylase and reductase functionalities on different proteins.

Recently our laboratory has undertaken a major program to investigate the proteins of methane monooxygenase and to develop models for the diiron center in the hydroxylase component. Our goals for MMO are to determine the structure of the dinuclear iron core in the hydroxylase by EXAFS spectroscopy (in collaboration with the research group of K. O. Hodgson at Stanford University) and by other methods, to understand the EPR, Mössbauer, optical and vibrational spectra of the diiron center in relation to the structure, to elucidate the chemical properties of the diiron core, and eventually to establish the molecular mechanism of the hydroxylation reaction. Initially our focus has been on MMO from *Methylococcus capsulatus* (Bath), the organism and proteins having been originally provided to us by H. Dalton and subsequently grown and isolated at MIT. Parallel studies have also been carried out with model complexes of the hydroxylase dinuclear iron center, the objectives being to reproduce the spectroscopic and functional properties of the core.

One purpose of the present article is to review our current knowledge of the structure and redox properties of the diiron center in the MMO hydroxylase enzyme from *Methylococcus capsulatus* (Bath). The emphasis has been placed on recent contributions from our own laboratory, most of which are unpublished at the time of this writing. Relevant studies on the model systems will then be described and related to work on the protein and from other laboratories. Mechanistic speculations concerning the hydroxylation of alkanes by MMO and model systems will be made to serve as a working hypothesis for interpretation of current and future experimental results.

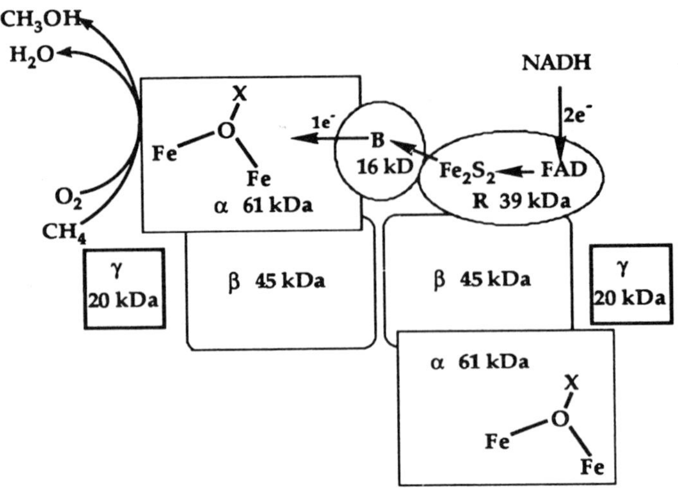

Figure 1. Schematic representation of the MMO proteins showing the relative locations of the α, β, and γ subunits of the hydroxylase enzyme as revealed by chemical crosslinking experiments,[5] the substrate binding sites, the path of electron flow, and product release. The molecular weights depicted are derived from the sequence of the *M. capsulatus* (Bath) genes.[6]

METHANE MONOOXYGENASE

The System. The methane monooxygenase enzyme systems from two organisms, *Methylococcus capsulatus* (Bath) and *Methylosinus trichosporium* OB3b, have been purified, and are currently under investigation in several laboratories, including our own.[3,4,7] The systems from the two organisms are very similar. Three proteins are involved in the oxidation of CH_4 to CH_3OH (Fig. 1). The hydroxylase, which is the site of O_2 activation, is a nonheme iron protein. It exists as a dimer comprised of two copies each of three subunits, with an $\alpha_2\beta_2\gamma_2$ configuration (α MW 61 kDa; β, MW 45 kDa; γ, MW 20kDa).[3,4] EPR, EXAFS, and Mössbauer spectroscopic data suggest that the hydroxylase contains two catalytic dinuclear Fe cores,[4,7,8] similar but not identical to the diiron oxo centers found in the proteins hemerythrin, ribonucleotide reductase, and purple acid phosphatase.[9] Protein B (MW 16 kDa), which is required for activity, is a regulatory protein with no metal or prosthetic group.[10] The third protein, the reductase (MW 39 kDa), transfers electrons from NADH to the hydroxylase. The reductase contains one mole of FAD and one mole of Fe_2S_2 cluster per mole of protein.[11]

The formation of specific complexes among these components has recently been demonstrated for the *M. trichosporium* OB3b MMO system.[5] Protein B binds to the hydroxylase, perturbing the structure of the dinuclear Fe center, as revealed by EPR spectroscopic studies. Evidence for complexation between the reductase and protein B and between the reductase and the hydroxylase has been obtained from fluorescence spectroscopy. In addition, chemical crosslinking experiments have shown that protein B binds to the α subunit of the hydroxylase, that the reductase interacts with the β subunit of the hydroxylase, and that the α and β subunits are close to each other. It is likely that similar complexes form in the *M. capsulatus* MMO. A schematic diagram for the *M. capsulatus* MMO system reflecting this expectation is presented in Fig. 1. In this diagram, boxes or ovals that intersect one another represent components that have been chemically cross-linked in the *M. trichosporium* OB3b system.[5]

The Genes. Since all three MMO proteins are required for activity, it is not surprising that the genes for the three proteins are organized in a common operon. The cloning of a 12 kb EcoR1 restriction fragment from *M. capsulatus* genomic DNA which contains the genes for the reductase, the three subunits of the hydroxylase, and two open reading frames, *orfX* and *orfY*, has been reported.[6] More recently, it has been discovered in our laboratory[12] and elsewhere[13] that *orfX* corresponds to the gene for protein B. Identification of *orfX* as the protein B gene, *mmoB*, demonstrates that all three MMO proteins are clustered in the *M. capsulatus* genome and form a MMO operon (Fig. 2).

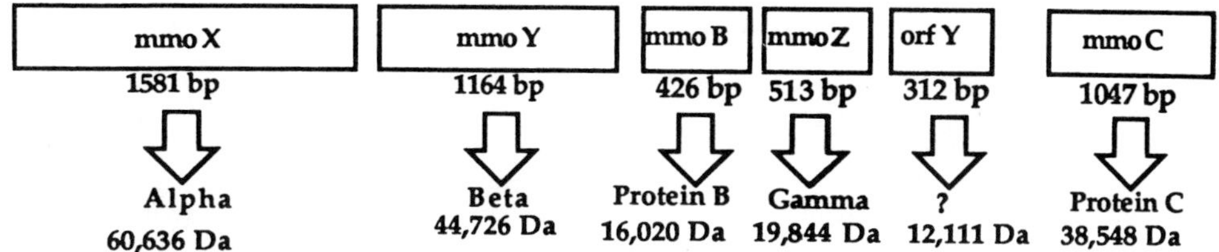

Figure 2. Schematic diagram of the MMO operon based on references 6 and 13.

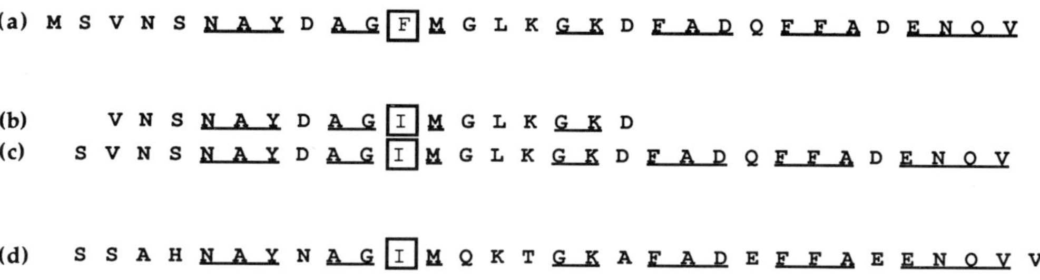

Figure 3. (a) Predicted amino acid sequence for *M. capsulatus* (Bath) protein B from gene sequence.[13] (b) N-terminal amino acid sequence.[13] (c) N-terminal amino acid sequence obtained in our laboratory.[12] (d) N-termnial amino acid sequence of *M. trichosporium* OB3b protein B.[5]

The predicted amino acid sequences from the gene and the amino acid sequences obtained from N terminal sequencing are compared in Fig. 3. At amino acid 11, the protein sequence contains an isoleucine while the gene sequence contains a phenylalanine. This discrepancy is present in two different amino acid sequences for the *M. capsulatus* protein B as well as in the sequence for the *M. trichosporium* OB3b protein B. While the protein B sequences from the two different organisms are fairly homologous, they are not identical, indicating that there may be an evolutionary divergence between the two species.

The location of *mmoB* implies that the production of protein B may be closely coupled with that of the hydroxylase. The *mmoB* gene lies between the genes encoding subunits α and β of the hydroxylase. Its translational start codon is separated by 21 base pairs from the stop codon for the β subunit, while the protein B stop codon lies 10 base pairs upstream from the translational start codon for the γ subunit. In the absence of recognizable and separate promoter regions upstream of the protein B gene that would initiate transcription of the gene independently of the other subunits,[6] it seems likely that protein B and subunits β and γ are transcribed as one mRNA transcript. In this manner, the cell can couple production of both proteins.

The isolation, cloning, and sequencing of the *M. capsulatus* MMO genes has been reported[6], but expression of the genes in *E. coli* has not yet appeared in the literature. We have recently achieved a high level of expression of protein B in *E. coli*,[14] a result which will allow us to pursue site-directed mutagenesis studies on protein B. Through site-directed mutagenesis, it will be possible to identify regions of protein B which are involved in the complexation of protein B and the hydroxylase observed by EPR spectroscopy and chemical crosslinking experiments.[5] The hydroxylase has not yet been expressed in *E. coli*, but possible alignments of the sequence of the α subunit with that of the B2 subunit of ribonucleotide reductase[15, 16] (Figure 4) suggest several amino acids as possible ligands to the diiron center, and thus as potential targets for site-directed mutagenesis. Such studies will help to choose among models such as A and B.

EXAFS Studies of the Hydroxylase. Preliminary EXAFS studies on oxidized hydroxylase samples from *M. capsulatus* and *M. trichosporium* OB3b[17] which were photoreduced by the X-ray beam to the mixed valent Fe(II)Fe(III) form indicated that the diiron center does not contain an oxo bridge. A detailed EXAFS investigation of the *M. capsulatus* hydroxylase in all three oxidation states has now been completed.[7] The EXAFS of the oxidized hydroxylase resembles that observed for the hydroxo bridged model compound $[Fe_2(OH)(OAc)_2(HB(pz)_3)_2]^{+}$[18] and is not at all similar to that observed for the oxo-bridged complex $[Fe_2O(OAc)_2(HB(pz)_3)_2]$,[19] suggesting that the Fe center in the protein resembles the Fe center in the hydroxo bridged model compound. The EXAFS of the mixed valent samples from *M. capsulatus* and *M. trichosporium* OB3b are similar, indicating that the diiron cores in the hydroxylases from the two organisms are structurally similar. For the reduced hydroxylase, the Fourier transformed EXAFS spectra lacked a peak due to Fe-Fe backscattering. First shell fits give an average Fe-O/N distance of 2.04 Å for the oxidized hydroxylase, of 2.06 - 2.09 Å for the semimet form, and of 2.15 Å for the fully reduced form. No evidence for a short (1.80 Å) Fe-O distance was found. The absence of such a feature further supports the conclusion from comparison of the EXAFS spectra with those of model compounds that the hydroxylase does not contain an oxo bridge linking the two iron atoms.

Parameters from the $[Fe_2(OH)(OAc)_2(HB(pz)_3)_2]^{+}$ model complex were used to perform second shell fits on the data. For the oxidized and semimet samples, two Fe⋯Fe minima were found at ~3.0 Å and ~3.4 Å, depending on the initial Fe⋯Fe distance used in the fit. From a detailed analysis of the

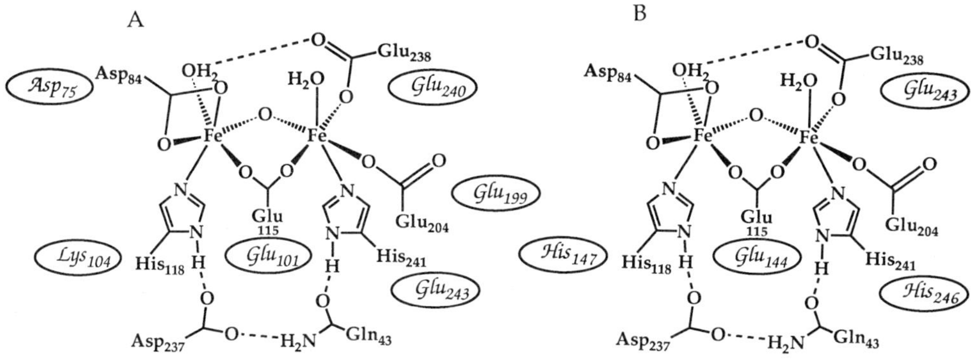

A   B

Figure 4. Possible alignments of the amino acid sequence of the α subunit of MMO and the sequence of the B2 subunit of *E. coli* ribonucleotide reductase. The amino acid residues on the α subunit which can be matched with the Fe binding residues in the B2 diiron core are shown in the ovals. Alignment A was obtained by using programs in the GCG (Genetics Computer Group) Sequence Analysis Software Package for VAX/VMS computers. Alignment B was obtained by visual inspection of the two sequences.[16]

fits, we concluded that the 3.4 Å Fe⋯Fe distance is the appropriate interpretation of the data and that the 3.0 Å minimum is most likely due to correlation between Fe-Fe and Fe-C parameters. In addition, the longer distance corresponds well to the position and magnitude of the second shell peak in the Fourier transformed spectrum for both the protein samples and the model complexes.[20] These conclusions are strongly supported by the the first shell fits which indicated that no oxo bridge is present. An Fe⋯Fe distance for the diferric hydroxylase from *Methylobacterium* CRL-26 of 3.05 Å has been reported in a previous EXAFS study.[21] While the hydroxylases from the two different species may differ in the structure of the dinuclear iron center, it is possible that this discrepancy results from differences in methods of data analysis.

In analyzing the results, we have discovered that the best Fe···Fe distance for the protein data depends on the model compound used to obtain Fe-Fe backscattering parameters. When cross fits were performed on the two model complexes [Fe$_2$O(OAc)$_2$(HB(pz)$_3$)$_2$] and [Fe$_2$(OH)(OAc)$_2$(HB(pz)$_3$)$_2$]$^+$, two Fe···Fe distances were found for each: the correct distance and a distance ~0.4 Å away. If the wrong model is used in the fit, for example the hydroxo model parameters to fit the oxo compound data, the result is then biased toward the wrong Fe···Fe distance. This model dependence of second shell fits is significant and must be taken into account in future EXAFS studies of iron oxo proteins.

Table 1 compares the Fe···Fe and average Fe-O/N distances for the dinuclear center in oxidized *M. capsulatus* hydroxylase with those in various model complexes. According to the EXAFS data, viable models for the protein core include a singly bridged (μ−oxo)diiron(III) center, a μ-hydroxo tribridged center, a μ-phenoxo tribridged center, and a μ-alkoxo dibridged center. Based on Fe···Fe and Fe-O,N distances, μ-oxo di- and tribridged cores are very unlikely candidates for the MMO

Table 1. Structural Features of Diiron Centers in MMO and Model Complexes[a]

	Fe-μ-O (Å)	Fe···Fe (Å)	Fe-O,N (Å)	Ref
M. capsulatus (Bath) Hydroxylase Fe(III)Fe(III)		3.42	2.04	7
μ-oxo monobridged [Fe(tsalen)]$_2$O·py	1.78	3.53	2.06	23
μ-hydroxo tribridged [Fe$_2$(OH)(OAc)$_2$(HB(pz)$_3$)$_2$](ClO$_4$)	1.96	3.44	2.05	18
μ-phenoxo tribridged (Me$_4$N)[Fe$_2$(5-Me-HXTA)(OAc)$_2$]	2.01	3.44	2.05	24
μ-alkoxo dibridged [Fe$_2$(HPTB-Et)(OAc)$_2$(O)$_2$]	2.01	3.49	2.13	25
μ-oxo dibridged [Fe$_2$O(OAc)(TPA)$_2$](ClO$_4$)$_3$·2H$_2$O	1.80	3.24	2.13	26
μ-oxo tribridged [Fe$_2$O(OAc)$_2$(HB(pz)$_3$)$_2$]	1.78	3.15	2.13	19

[a]Ligand abbreviations are as follows: tsalen = 1,2-bis(thiosalicylideneamino)ethane, OAc = acetate, HB(pz)$_3$ = hydrotris(pyrazolyl)borate, 5-Me-HXTA = N,N'-(2-hydroxy-5-methyl-1,3-xylylene)bis(N-carboxymethylglycine), HPTB = 1,3-bis[N,N-bis(2-benzimidazolylmethyl)amino]-2-hydroxypropane, TPA = tris(2-pyridylmethyl)amine.

hydroxylase active site. The spectroscopic and magnetic properties of the hydroxylase provide further insight into possible structures for the Fe center.

Spectroscopic and Magnetic Properties of the Hydroxylase. The three oxidation states of the hydroxylase have been studied by optical, EPR, and Mössbauer spectroscopy. The optical spectrum of the diferric hydroxylase from both *M. capsulatus* and *M. trichosporium* OB3b lacks any distinct features beyond 300 nm.[4,7] The absence of significant visible absorption indicates that the μ-oxo and μ-phenoxo models are unlikely candidates for the MMO Fe center. The hydroxo-bridged model compound [Fe$_2$(OH)(OAc)$_2$(HB(pz)$_3$)$_2$]$^+$ exhibits a broad absorption band with λmax = 375 nm,[18] a feature not present in the hydroxylase. Another hydroxo-bridged model complex, [Fe$_2$(OH)(OAc)$_2${CpCo[OP(OEt)$_2$]$_3$}$_2$]$^+$, has no distinct optical features beyond 300 nm, except for a ligand band at 325 nm.[22] Therefore, the presence of a μ-hydroxo bridge in the hydroxylase iron center cannot be ruled out based on optical properties.

The Mössbauer spectroscopic parameters and spin exchange (J) values for MMO and relevant model compounds are summarized in Table 2. The Mössbauer spectrum for the oxidized hydroxylase differs from those of well defined 6-coordinate oxo- or hydroxo-bridged diiron(III) centers. The quadrupole splitting observed for diferric iron oxo proteins and model complexes is ΔE_Q = 1.5 - 1.8 mm/s,[29] while for $[Fe_2(OH)(OAc)_2(HB(pz)_3)_2]^+$, ΔE_Q = 0.25 mm/s.[18] The values obtained for the *M. capsulatus* hydroxylase, ΔE_Q = 1.05 mm/s,[7] and for the *M. trichosporium* OB3b hydroxylase, ΔE_Q = 1.07 mm/s,[4] suggest that the diiron center may contain an unusual bridge. The observed quadrupole splitting parameter most closely resembles that observed for the monobridged oxo complex $[Fe(tsalen)_2]O \cdot py$.[23] One possible explanation for the quadrupole splitting in the diferric hydroxylase is that the iron atoms are hydroxo-bridged and pentacoordinate.

The mixed valent form of the hydroxylase exhibits an EPR signal at g_{av} = 1.83, similar to values observed in other dinuclear nonheme iron proteins in the Fe(II)Fe(III) state. A microwave power saturation study was carried out on a photoreduced *M. capsulatus* EXAFS sample, and a coupling

Table 2. Mössbauer (δ, ΔE_Q) and Exchange Coupling (J) Parameters of Diiron Centers in MMO and Model Complexes[a]

	$Fe^{III}Fe^{III}$			$Fe^{II}Fe^{III}$	$Fe^{II}Fe^{II}$		Ref
	δ (mm/s)	ΔE_Q (mm/s)	J (cm^{-1})	J (cm^{-1})	δ (mm/s)	ΔE_Q (mm/s)	
M. Capsulatus (Bath) hydroxylase	0.50	1.05		-32	1.30	3.014	7
M. trichosporium OB3b hydroxylase	0.50	1.07		-30	1.30	3.14	4
$[Fe(tsalen)]_2O \cdot py$	0.43	1.10	-100				23
$[Fe_2O(OAc)_2(HB(pz)_3)_2]$	0.52	1.60	-121				19
$[Fe_2OH(OAc)_2(HB(pz)_3)_2]^+$	0.47	0.25	-17				18
$\{Na[Fe^{II,III}(acen)_2]_2O\}_2$				-30calc			27
$[Fe_2(OCH)_4(BIPhMe)_2]$					1.26	2.56	28
					1.25	3.30	

[a]Ligand abbreviations are given in Table 1 except for the following: acen = N,N'-ethylenebis(acetylacetone iminate), BIPhMe = bis(1-methylimidazol-2-yl)phenylmethoxymethane.

constant J = -32 cm^{-1} was determined.[7] This value is consistent with that measured for the *M. trichosporium* hydroxylase, J = -30 cm^{-1}.[5] The weak antiferromagnetic exchange indicated by this J value is consistent with hydroxo, alkoxo, or monodentate carboxylato bridging in the diiron center. Preliminary electron spin echo (ESE) spectroscopic data obtained for the Fe(II)Fe(III) *M. capsulatus* hydroxylase indicates that there is at least one nitrogen ligand coordinated to the Fe center.[30]

The diferrous, Fe(II)Fe(II), form of the *M. capsulatus* hydroxylase exhibits a low field EPR signal at g = 15 similar to the low field feature reported for the *M. trichosporium* OB3b hydroxylase[4] and for the diferrous model complexes $[Fe_2(OCH)_4(BIPhMe)_2]^{31}$ and $[Fe_2(BPMP)(OPr)_2]BPh_4$.[32] The Mössbauer spectrum of the diferrous hydroxylase has a quadrupole splitting ΔE_Q = 3.014 mm/s, similar to the ΔE_Q = 3.30 mm/s measured for the pentacoordinate Fe atom in the asymmetric model compound $[Fe_2(OCH)_4(BIPhMe)_2]$.[28] This result suggests that pentacoordinate Fe atoms could occur in the active site of reduced MMO hydroxylase.

In summary, the structural and spectroscopic data for the diiron center in the MMO hydroxylase can be interpreted in terms of the model presented in Fig. 5. This structure will serve as a working hypothesis for understanding other properties of the protein. Its validity will require additional structural information, preferably an X-ray crystal structure determination of the holoprotein.

Redox Properties of the Hydroxylase. The reduction potentials of the *M. capsulatus* hydroxylase have recently been determined in our laboratory.[33] Reduction of the diferric iron center can be represented by eq. 2. The g_{av} = 1.83 EPR signal of the mixed valent hydroxylase was quantitated

$$Fe(III)Fe(III) \xrightarrow{E_1^\circ} Fe(II)Fe(III) \xrightarrow{E_2^\circ} Fe(II)Fe(II) \qquad (2)$$

at different potentials, and values of 48 and -135 mV vs NHE for E_1° and E_2°, respectively, were obtained. In the presence of the substrate propylene, the reduction potentials were slightly lower, 30 mV for E_1° and -156 mV for E_2°. Therefore, substrate may perturb the diiron center in some way, slightly affecting the redox potentials of the iron core.

The reduction potentials of the hydroxylase may provide important clues as to the nature of the bridges in the diiron center.   The potentials indicate that the Fe(II)Fe(III) form in MMO is stable with respect to disproportionation. By contrast, the mixed valent state of hemerythrin is unstable to disproportionation since E_1° (110 mV) is less than E_2° (310 mV vs NHE).[9] Interestingly, hemerythrin contains an oxo bridge in the diferric form while MMO does not. Furthermore, the phenoxo dibridged model complex [Fe$_2$(H$_2$bab)$_2$(DMF)$_2$(N-MeIm)] (H$_4$Hbab = 1,2-bis(2-hydroxybenzamido)benzene) exhibits reduction potentials of E_1° = -9 and E_2° = -259 mV vs NHE.[34] The separation of the two potentials in this compound, 250 mV, is close to that observed in the hydroxylase, 183 mV. In addition, this compound contains all oxygen ligands.

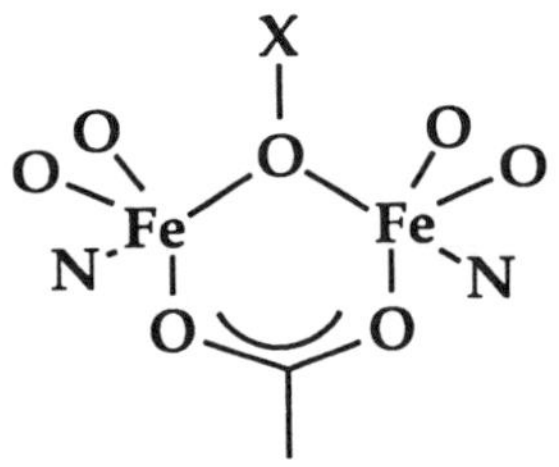

Figure 5. Postulated model for the structure of the diiron core in the hydroxylase component of methane monooxygenase. The group X is H, alkyl, C(O)R, but not aryl.

In the presence of the reductase and protein B, but no substrate, no reduction of the iron center occurred even at potentials as negative as -200 mV. Thus, protein B and reductase completely inhibit electron transfer to the diiron core in the absence of substrate. This inhibition of electron transfer to the iron center is consistent with previous reports that hydroxylase and reductase oxidize NADH without substrate turnover, but that the addition of protein B results in NADH consumption only when substrate is oxidized.[10] A striking change occurred when the substrate propylene was added to all three proteins: the hydroxylase was reduced to the Fe(II)Fe(II) state at potentials as high at 150 mV. Substrate therefore serves as the trigger for the hydroxylation reaction. Electron transfer only takes place in the presence of substrate, and when substrate is present the diferric hydroxylase is reduced at a potential 0.25 mV higher than in the absence of substrate. The dramatic effects of substrate on the three methane monooxygenase proteins can be interpreted in several ways. The lack of reduction of the hydroxylase in the presence of reductase and protein B could either be a kinetic effect, such as blocking access of reductant to the iron center or a thermodynamic effect, such as changing the iron coordination environment. When substrate binds, this effect is reversed, and the hydroxylase becomes more electron deficient and easier to reduce. This change in the iron center could be conformational or could be due to proton transfer or hydrogen bond formation in the iron coordination sphere. Clearly, the MMO system has been tuned by nature so as to preserve reducing equivalents in the absence of substrate and to generate a high energy activated iron core in the presence of substrate.

Model Complexes for the Redox Properties of the MMO Hydroxylase. An examination of the reduction potentials for the Fe(II)/Fe(III) transition in a series of mononuclear model compounds reveals some interesting trends that may help to understand the observed redox properties (Fig. 6).[35,36] As indicated in the figure, increasing the number of nitrogen donor ligands around the iron center from 0 to 6 results in an increasingly more positive reduction potential. In the presence of substrate, protein B, and the reductase, the reduction potential of the diiron core becomes more positive. It is therefore conceivable that the iron center may gain a nitrogen ligand in the presence of substrate. Alternatively, and perhaps more reasonable, the iron core may lose an oxygen ligand, which would also result in an increased reduction potential. Proton transfer or hydrogen bond formation could also account for the increased electrophilicity of the iron core. This change in the diiron center which results in the increased reduction potential might involve a conformational change which would allow electron transfer to occur among the reductase and protein B and the diiron core.

The potentials in Fig. 6 are also in accord with the previous structural analysis of the MMO hydroxylase core (Fig. 5). The postulated structure of the iron core contains just 1 nitrogen donor ligand and 4 oxygen ligands coordinated to each iron atom. The measured potential, $E_2^{\circ} = -0.135$ V, is clearly in the range observed for the model compounds with primarily oxygen donor ligands. This agreement further suggests that the environment of the iron center is relatively hydrophobic, which would have to be the case in order for the absolute potential values to mimic those of the organic solvents in which the reduction potentials of the model complexes in Fig. 6 were measured.

MODEL CHEMISTRY FOR THE HYDROXYLASE OF MMO

The synthesis and study of models for the diiron center in the hydroxylase protein of methane monooxygenase constitutes another major focus of activity in our laboratory. One specific goal is to understand the basic reaction chemistry of bridged diiron(II) cores with dioxygen. The reduced form of the MMO hydroxylase is capable of reacting with dioxygen and alkane to produce alcohol in a single turnover experiment,[4] and the chemistry that mimics this reaction is important to delineate. Of particular interest are to work out the stoichiometry of the reaction of O_2 with diiron(II) complexes and to identify intermediates by using spectroscopic, kinetic, and structural methods. A second objective is to develop catalytic systems to hydroxylate alkanes as functional models of the MMO hydroxylase. Here our guidelines have been to use biologically relevant ligands, to run reactions with dioxygen at ambient pressure and temperature, to find systems that are selective for alcohol rather than ketone formation with a high yield based on reductant, and to achieve chemo-, regio- and stereoselectivities that match those of MMO. Ultimately, the aim of these studies is to elucidate the detailed molecular

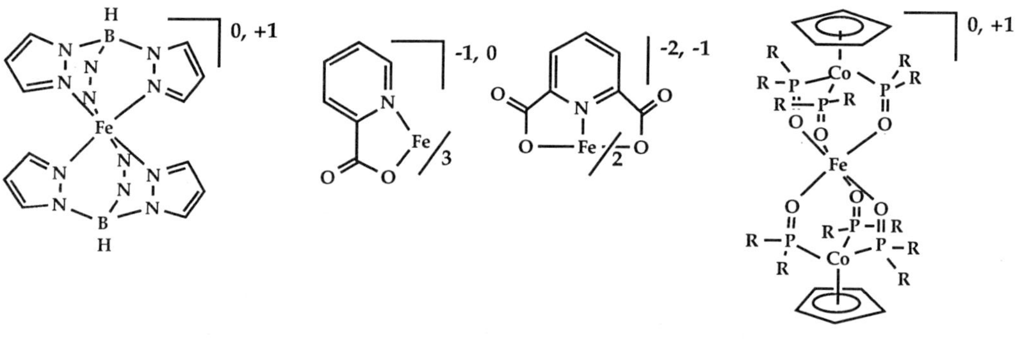

	FeN$_6$	FeN$_3$O$_3$	FeN$_2$O$_4$	FeO$_6$
$E_{1/2}$ (V vs. NHE)	+0.47	+0.21	-0.07	-0.40
Solvent	CH$_3$CN	DMF	DMF	CH$_3$CN
Ref	19	35	35	36

Figure 6. Redox potentials of Fe(II)/Fe(III) couples for mononuclear iron complexes

mechanism of the hydroxylation chemistry and to use the insights gained to unravel the parallel steps in the enzyme catalyzed hydroxylation reactions.

Reaction of a Diiron(II) Complex with Dioxygen. Recent work in our laboratory led to the synthesis and characterization of a dinuclear iron(II) complex that reacts with dioxygen according to the scheme shown in Fig. 7.[31] The source of the bridging atom in the product, $[Fe_2O(O_2CH)_4(BIPhMe)_2]$, was found to be exclusively dioxygen through the use of resonance Raman spectroscopy and ^{18}O-labelled O_2. The symmetric Fe-O-Fe stretching vibration shifted from 520 to 502 cm^{-1} when the labelled dioxygen was employed. Through the use of manometric titrations, it was established that both atoms of the dioxygen molecule are incorporated into the product. The presence of a mixed valent Fe(II)Fe(III) intermediate in the dioxygen reaction was detected by fast freeze EPR spectroscopy. Power saturation studies indicated antiferromagnetic coupling with a J value of -31 (2) cm^{-1}, identical to the value found in the mixed valent form of the MMO hydroxylase. A novel peroxide bridged intermediate was suggested for the reaction mechanism. Peroxide-bridged polyiron(III) complexes have been previously characterized by X-ray crystallography[37] and solution spectroscopy.[38,39]

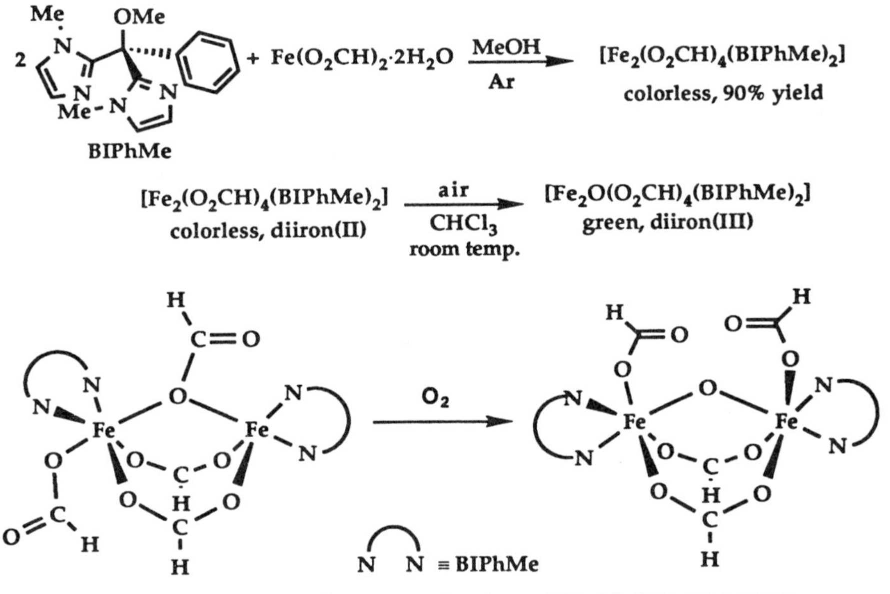

Figure 7. Synthesis and O_2 reaction chemistry of $[Fe_2(O_2CH)_4(BIPhMe)_2]$.

A Catalytic Model System for the Hydroxylation of Alkanes. A functional model system for the MMO hydroxylase has been developed in our laboratory over the last several years.[40] The simple dinuclear $(\mu-oxo)$diiron(III) complex, $[Fe_2OCl_6]^{2-}$ **(1)**, together with a suitable reductant such as tetramethyl reductic acid (TMRA) or ascorbic acid (AA) (Fig. 8), are capable of catalyzing the air oxidation of alkanes with high selectivities. In a typical reaction, **1** is first allowed to react with the reductant in a 1:20 mole ratio in acetone or acetonitrile in the absence of air, generating a purple complex. Dried air is then introduced into the reaction vessel and the formation of the products is monitored periodically by gas chromatography. Figure 9 shows the formation of the products as a function of time at 32 °C using cyclohexane as substrate for both **1**/TMRA and **1**/AA systems.

Figure 8. Reductants and ligands used in the catalytic hydrogenation system.

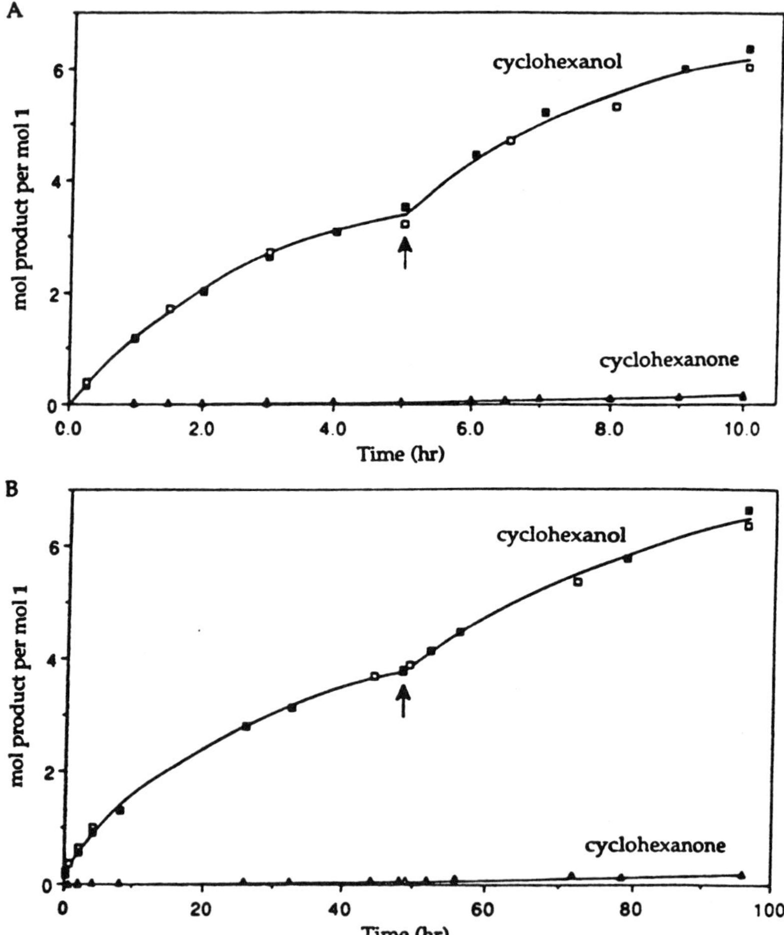

Figure 9. Formation of cyclohexanol and cyclohexanone as a function of time in the presence of air and the catalyst systems 1/TMRA and 1/AA with cyclohexane as substrate (data from ref. 40).

When AA is used as the reductant, the reaction proceeds much more slowly than with the 1/TMRA system because of the limited solubility of both AA and its iron complex in acetone or acetonitrile. After a period of 5 hr, 3.5 mol of cyclohexanol and only 0.08 mol cyclohexanone are produced per mol of 1 in the 1/TMRA system, which affords a 17% yield based on the reductant. The formation of the products slows appreciably after 5 hours and the color of the solution turns brown. Addition of another 20-fold excess of reductant (based on 1) regenerates the purple color of the catalyst and affords 70-80% of the original activity over another 5 hours. The catalytic cycle can be maintained by continuously adding in the reductant.

The relatively good yield of alcohol based on the reductant and the high selectivity for the formation of alcohol over ketone are two of the distinctive features of this system. Among most of known catalytic systems employing non-heme multinuclear iron catalysts with zinc as the reductant, the reported product yield per mole of reductant is usually low ($\sim$1.5%).[41,42] One exception is the *Gif* system ($Fe_3O(OAc)_6Py_{3.5}/Zn/Py/CH_3COOH/O_2$) where the yield based upon the reductant (Zn) is as high as 17.5% when cyclohexane is used as the substrate.[43] The system, however, produces much more cyclohexanone than cyclohexanol. The electron efficiency of the reductant in our system is fairly high, considering the strong competition between the autoxidation of the reductant and the oxidation of the alkane for the activated dioxygen in the system. Figure 10 displays the possible autoxidation products of TMRA.[44] The triketone is the major product formed under our reaction conditions, as determined by GC/MS analysis. It arises both from consumption of electrons in the hydoxylation as well as through the autoxidation reaction.

The high selectivity for the formation of alcohol in this system ($\sim$50:1) closely mimics the function of MMO. This selectivity is an important feature that distinguishes the present chemistry from that of the well-known Fenton system, where the same amounts of alcohol and ketone are formed during the oxidation of alkanes with hydroxyl radical as the active species.[45] Control experiments have shown that the alcohol and ketone are formed via two independent pathways. These products do not interconvert under our reaction conditions.

Two mononuclear iron complexes, $FeCl_2$ and $[FeCl_4]^-$, were also studied under the same reaction conditions to compare their activities with that of 1 as catalysts for alkane oxidation. The results demonstrated not only a significant decrease in the reaction rate and total yield in the case of these monomers but also a different distribution pattern of oxidation products. During the same period of time, only 1/4 as much cyclohexanol was produced in the oxidation of cyclohexane by using $FeCl_2$ and $[FeCl_4]^-$. The specificity for the formation of alcohol over ketone was also much lower.

Other substrates were also employed to examine the chemo- and regioselectivity of the system (Table 3). Benzylic C-H bonds in ethylbenzene were very reactive and more ketone formation occurred (1:2.5 ketone:alcohol) at this position compared to that of the normal secondary aliphatic C-H bands. Interesting regioselectivity was observed when methylcyclohexane was used as substrate. The ratio of tertiary vs. secondary position oxidation (after statistical correction for the numbers of bonds on each position) was 2.2:1, much lower than that observed for a typical hydroxyl radical attack on these positions[46] and very similar to that reported for MMO.[47]

The product distribution pattern in the oxidation of alkenes provides valuable mechanistic information about the reaction mechanism. The participation of oxygen-derived free radicals usually results predominantly in the formation of allylic alcohols while epoxide formation is best accounted for by oxo transfer chemistry.[48,49] When cyclohexene was used as substrate, a ratio of cyclohexenol/cyclohexenone /cyclohexene oxide of about 10/1/2 was observed and this ratio remained the same throughout the reaction as long as excess amounts of reductant were present. In addition, a radical

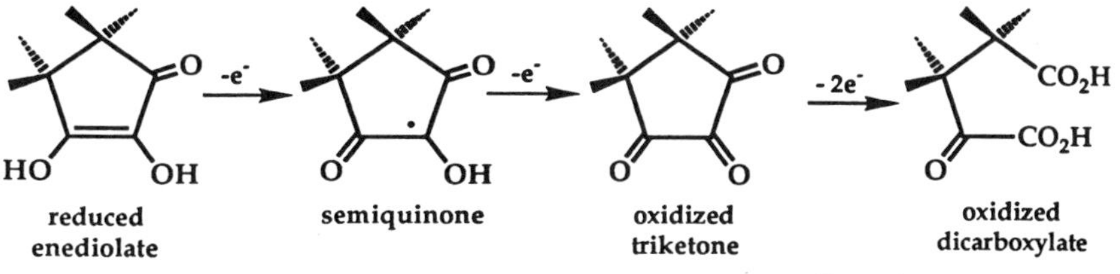

Figure 10. Redox forms of the TMRA reductant/ligand.

Table 3. Substrates and Oxidation Products in the 1/O$_2$ Catalytic Hydroxylation System[36]

Substrate	Reductant	Product (% total yield)
cyclohexane	TMRA	cyclohexanol (98), cyclohexanone (2)
methylcyclohexane	TMRA	1-Me-cyclohexanol (18), 2-Me-cyclohexanol (11) 3-Me-cyclohexanol and 4-Me-cyclohexanol (71)
ethylbenzene	TMRA	1-phenylethanol (65), other product (9) acetophenone (26)
cyclohexene	AA	2-cyclohexen-1-one (8), 2-cyclohexen-1-ol (77) cyclohexene oxide (15)

scavenger, 2,6 di-*tert*-butyl-4-methylphenol (BHT), was used to probe the nature of the active species. The overall yield and product distribution were similar with or without BHT, although the initial reaction rate was slightly decreased in the presence of this radical scavenger. These results indicate that free radicals are not involved in the mechanism, although the existence of a caged radical intermediate can not be ruled out.

Perdeuteriocyclohexane was employed to look for a possible kinetic isotope effect. With TMRA as reductant, a k_H/k_D ratio of 2.5 ± 0.1 was observed.[40] This ratio is significantly different from that of the Fenton system ($k_H/k_D \sim 1$)[50] and porphyrin system with PhIO as the oxygen source ($k_H/k_D \sim 12$-16).[51] It is similar to that of the *Gif* system ($k_H/k_D = 2.5$)[52] and comparable to that of MMO ($k_H/k_D \sim 4$-5).[53,54] These results indicate that C-H bond cleavage is important in the rate determining step(s).

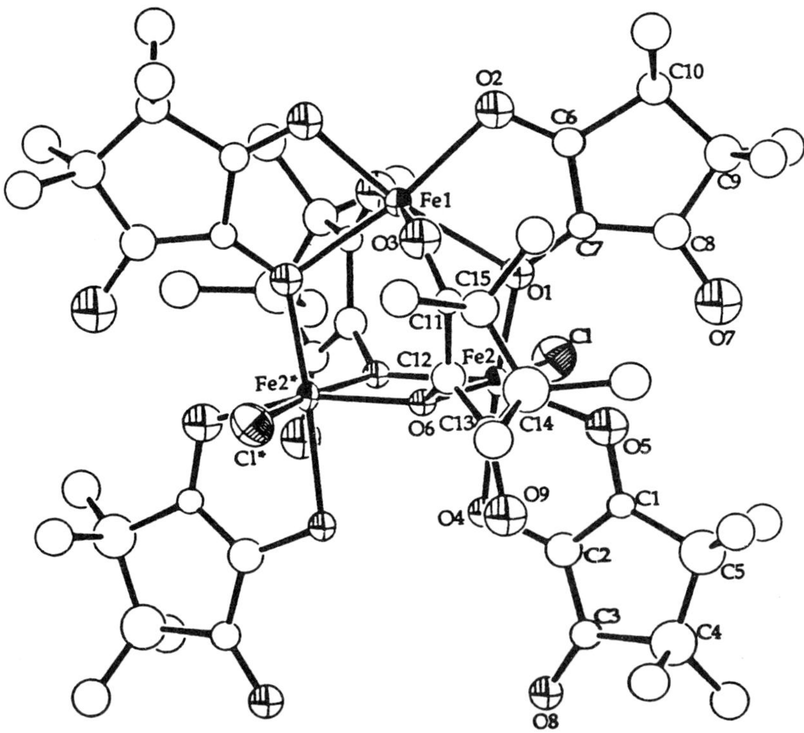

Figure 11. ORTEP drawing of **2** show the 50% probability thermal ellipsoids and atom labels for the iron, chlorine, and selected oxygen and carbon atoms. Methyl carbon atoms are represented as small spheres and hydrogen atoms are omitted for clarity. Selected interatomic distances (Å) are as follows: Fe(1)-O(1), 2.24 (3); Fe(1)-O(2), 2.12 (3); Fe(1)-O(3), 2.06 (3), Fe(2)-Cl, 2.26 (1); Fe(2)-O(1), 1.93 (3); Fe(2)-O(4), 2.16 (2); Fe(2)-O(5), 2.04 (3); Fe(2)-O(6), 2.11 (3); Fe(2)-O(6), 2.01 (3); O(1)-C(7), 1.27 (4); O(2)-C(6), 1.30 (4); O(3)-C(11), 1.24 (4); O(4)-C(2), 1.45 (5); O(5)-C(1), 1.31 (4); O(6)-C(12), 1.38 (4); O(7)-C(8), 1.37 (5); O(8)-C(3), 1.25 (4); O(9)-C(13), 1.38 (5); C(1)-C(2), 1.29 (5); C(6)-C(7), 1.36 (5); C(11)-C(12), 1.38 (5).

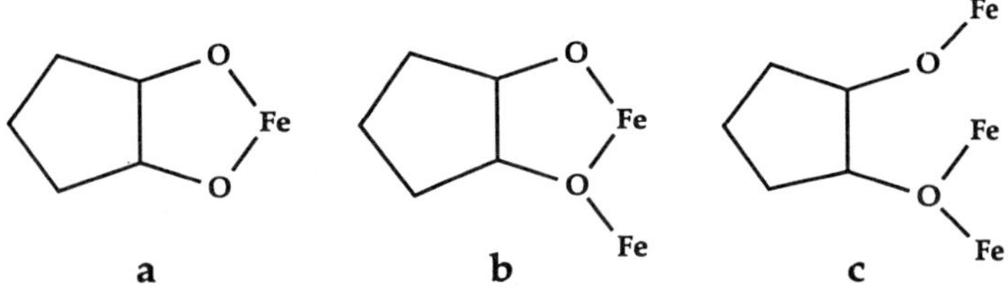

Figure 12. Representation of the three modes of TMRA coordination in **2**.

Structural Characterization of a Precursor of the Active Species in the Catalytic System. The purple complex formed between **1** and TMRA in the catalytic system described above, [Fe$_3$Cl$_2$(TMRA)$_2$(TMRASQ)$_4$] (**2**), was isolated and characterized by X-ray crystallography.[40] The molecular geometry is depicted in Fig. 11. The three iron atoms form an isosceles triangle with a crystallographically required two-fold axis passing through Fe(1) and the center of the Fe(2)···Fe(2*) edge. The longer edge of the triangle (Fe(1··· Fe(2) is 3.49 (1) Å and the shorter edge (Fe(2)···Fe(2*)) is 3.24 (1) Å. There are three types of ligand attachment modes to the iron atoms (Fig. 12). Two terminal chelating TMRA ligands coordinate to the two Fe(2) centers (Fig. 12a), two perform both chelating and bridging functions (Fig. 12b), and another two link all three iron atoms (Fig. 12c). Similar binding patterns have been observed in a tetranuclear iron complex with 3,5-di-*t*-butyl-1,2-benzoquinone as ligand.[55] Two chloride ligands on the two Fe(2) centers are coordinated in the *anti* configuration (Fig. 11), providing possible oxygen binding sites on the catalyst.

Mössbauer spectra of a solid sample of **2** indicated that one of the iron atoms is in the Fe(II) oxidation state while the other two are ferric ions. Thus the six TMRA derived ligands must bear a total of six negative charges which, together with two chloride ligands, accounts for the charge neutrality of the complex. This requirement can be satisfied by allowing the proper distribution among the three possible redox and attendant deprotonation states among the ligands (Fig. 13). Taking the symmetry and charge requirements of the molecule into consideration, there are two possible combinations. One assigns all six ligands as monoanions and the other has two neutral, two monoanionic

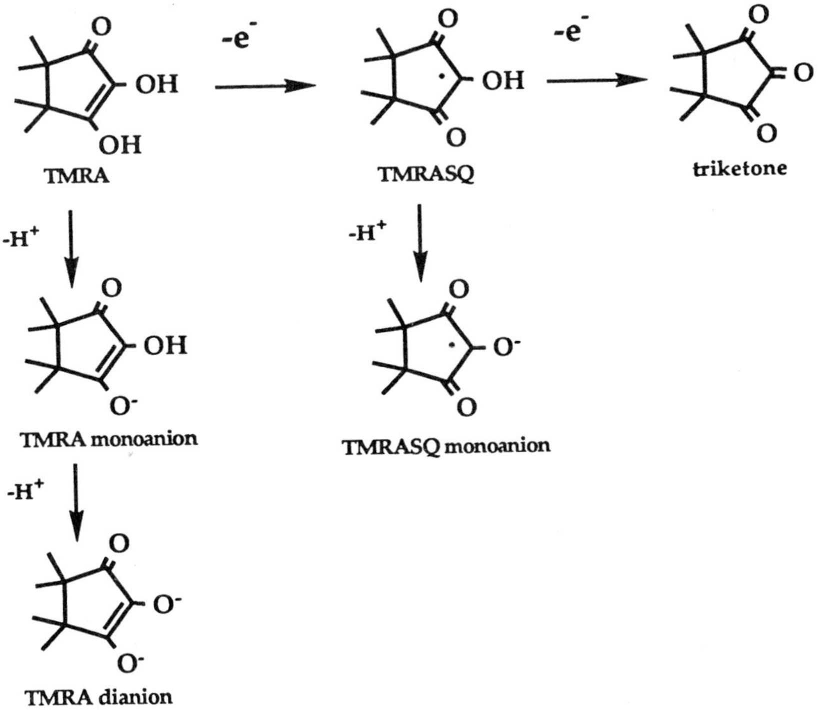

Figure 13. Representation of the redox and deprotonation states of TMRA.

and two dianionic ligands. The assignment of ligand oxidation states in such a case can sometimes be made by a detailed analysis of bond lengths, as in the case of some metal catechol complexes.[55] In the fully reduced form of TMRA, the C=C bond in the enediol fragment should be shorter than the corresponding bond in the semiquinone form, reflecting the different distributions of electron density in these two redox states. Close examination of the bond lengths (Fig. 11) reveals that the C(1)-C(2) distance of 1.29 Å in the ligand chelated to Fe(2) is much shorter than the corresponding C=C bonds in the other two types of ligands, C(6)-C(7) = 1.36 Å and C(11)-C(12) = 1.38 Å. A comparison of the C-O bond lengths in the enediol fragments of the ligands should also reveal the oxidation states of the ligands. This comparison, however, is very difficult in the complex because of their different coordination environments. Based on this analysis, we tentatively assign the two ligands chelated to Fe2 and Fe2* as TMRA monoanions and the remaining four ligands as TMRASQ monoanions.

The optical spectrum of 2 in acetonitrile exhibits a broad peak at 554 nm, a shoulder at 358 nm, and a peak at 280 nm. No band was observed in the near IR region. The Raman spectrum of 2 has a feature at 314 cm^{-1} which can be assigned to the Fe-Cl symmetric stretching mode. This feature has been reported for other dinuclear iron complexes with terminal chloride ligands.[56] Assuming that this chloride site will be the site of oxygen binding during the catalytic cycle, the kinetics of the oxygen binding process can be monitored by measuring the change of Fe-Cl resonance on the proper time scale.

It is of interest to examine whether 2 is kinetically competent to serve as an intermediate in the hydroxylation reaction. Equation 3 presents a postulated single turnover reaction stoichiometry

$$[Fe_3Cl_2(TMRA)_2(TMRASQ)_4] + 7H^+ + \frac{7}{2}RH + \frac{7}{2}O_2 \longrightarrow \frac{7}{2}ROH + \frac{7}{2}H_2O + 6 \text{ triketone} + 3Fe^{3+} + 2Cl^- \quad (3)$$

assuming that 2 is the active species in the catalytic air oxidation of alkanes. To test this hypothesis, the catalytic activity of 2 in the air oxidation of cyclohexane was examined using (i) only 2 and O_2; (ii) 2, extra proton sources, and O_2; or (iii) 2, extra TMRA, and O_2. In the system where only complex 2 serves as catalyst, the conversion of cyclohexane to cyclohexanol is very slow with an overall yield of about 40% and an alcohol/ketone ratio of 2/1. The addition of excess H^+ accelerated the oxidation reaction slightly and gave a total yield of 52%. With extra TMRA, complex 2 catalyzed the oxidation reaction with a rate and overall yield comparable to those found for the *in situ* system. A change in the product distribution pattern was observed, however. The ratio of cyclohexanol to cyclohexanone was 7/1, much lower than that observed in the *in situ* system.

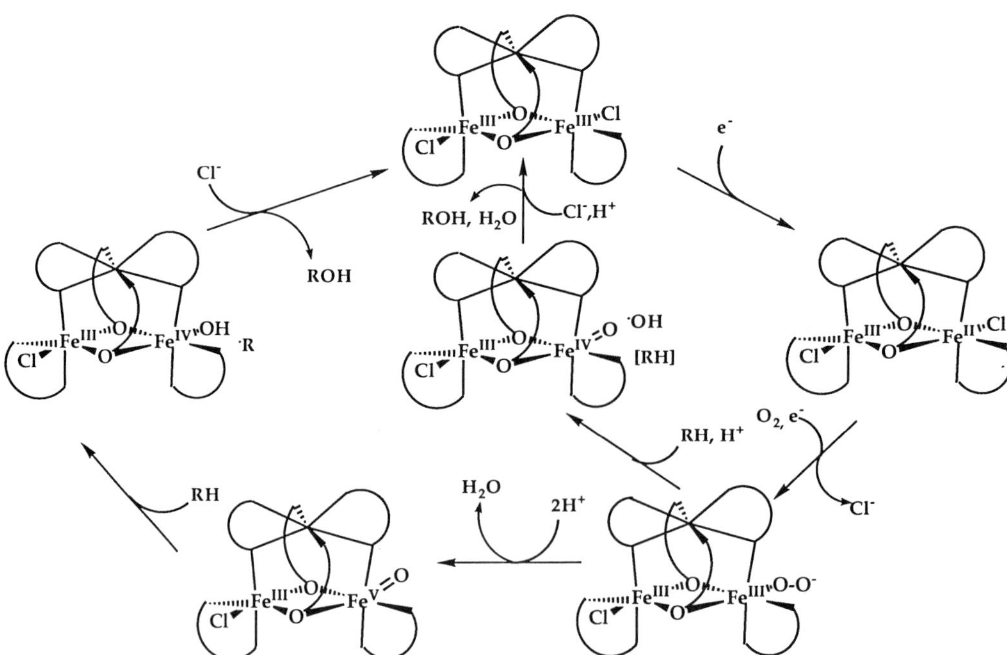

Figure 14. Proposed catalytic cycle.

Mechanistic Considerations and Conclusions. There is currently no experimental information available about the catalytic mechanism for **2**. By analogy to the cytochrome P-450 reaction, one can propose a catalytic cycle, as shown in Figure 14. This scheme, which serves as our current working hypothesis, has several key intermediates, the presence of which can be tested thorough time resolved spectroscopic and structural studies that are currently in progress. In this mechanism, the reduced, diiron(II) center in the catalyst reacts with dioxygen ultimately to form a peroxo intermediate of the kind discussed above for $[Fe_2(O_2CH)_4(BIPhMe)_2]$ and related model complexes. In particular, the postulated participation of ferric peroxide, high valent iron oxo, and caged radical intermediates all afford the potential for spectroscopic verification. It will be interesting to see which of any of these possibilities prove to be involved in the catalytic mechanism.

ACKNOWLEDGMENTS

This work was supported by grants from the National Institute of General Medical Sciences and the National Science Foundation. We are grateful to Drs. Mary E. Roth and James G. Bentsen for initiating the model and MMO studies, respectively, in our laboratories, and to our various other co-workers and collaborators whose contributions to our cited research are very much appreciated.

REFERENCES

1. H. Dalton, Oxidation of hydrocarbons by methane monooxygenases from a variety of microbes, *Adv. Appl. Microbiol.*, 26:71 (1980).
2. C. Anthony, "The biochemistry of methylotrophs," Academic Press, London (1982).
3. J. Colby and H. Dalton, Resolution of the methane monooxygenase of *Methylococcus capsulatus* (Bath) into three components. Purification and properties of component C, a flavoprotein, *Biochem. J.*, 171:461 (1978).
4. B. G. Fox, W. A. Froland, J. E. Dege, and J. D. Lipscomb, Methane monooxygenase from *Methylosinus trichosporium* OB3b: purification and properties of a three component system with high specific activity from a Type II methanotroph, *J. Biol. Chem.*, 264: 10023 (1989).
5. B. G. Fox, Y. Liu, J. E. Dege, and J. D. Lipscomb, Complex formation between the protein components of methane monooxygenase from *Methylosinus trichosporium* OB3b, *J. Biol. Chem.*, 266:540 (1991).
6. A. C. Stainthorpe, J. C. Murrell, G. P. C. Salmond, and H. Dalton, Molecular analysis of methane monooxygenase from *Methylococcus capsulatus* (Bath), *Arch. Microbiol.*, 152:154 (1989).
7. J. G. DeWitt, J. G. Bentsen, A. C. Rosenzweig, B. Hedman, J. Green, S. Pilkington, G. C. Papaefthymiou, H. Dalton, K. O. Hodgson, and S. J. Lippard, X-ray absorption, Mössbauer, and EPR studies of the dinuclear iron center in the hydroxylase component of methane monooxygenase, submitted for publication.
8. B. G. Fox and J. D. Lipscomb, Purification of a high specific activity methane monooxygenase hydroxylase component from a Type II methanotroph, *Biochem. Biophys. Res. Commun.*, 154:165 (1989).
9. J. B. Vincent, G. L. Olivier-Lilley, and B. A. Averill, Proteins containing oxo-bridged dinucleae iron centers: a bioinorganic perspective, *Chem. Rev.*, 90:1447 (1990).
10. J. Green and H. Dalton, Protein B of soluble methane monooxygenase from *Methylococcus capsulatus* (Bath). A novel protein of enzyme activity, *J. Biol. Chem.*, 260:15795 (1985).
11. J. Lund and H. Dalton, Further characterisation of the FAD and Fe_2S_2 redox centres of component C, the NADH: acceptor reductase of the soluble methane monooxygenase of *Methylococcus capsulatus* (Bath), *Eur. J. Biochem.*, 147:291 (1985).
12. W. E. Wu and S. J. Lippard, unpublished results.
13. S. J. Pilkington, G. P. C. Salmond, J. C. Murrell, and H. Dalton, Identification of the gene encoding the regulatory protein B of soluble methane monooxygenase, *FEMS Microbiol. Lett.*, 72:345 (1990).
14. W. E. Wu and S. J. Lippard, unpublished results.
15. A. C. Rosenzweig and S. J. Lippard, unpublished results.
16. A. C. Stainthorpe, V. Lees, G. P. C. Salmond, H. Dalton, and J. C. Murrell, The methane monooxygenase gene cluster of *Methylococcus capsulatus* (Bath), *Gene*, 91:27 (1990).
17. A. Ericson, B. Hedman, K. O. Hodgson, J. Green, H. Dalton, J. G. Bentsen, R. H. Beer, and

S. J. Lippard, Structural characterization by EXAFS spectroscopy of the binuclear iron center in protein A of methane monooxygenase from *Methylococcus capsulatus* (Bath), *J. Am. Chem. Soc.*, 110:2330 (1988).

18. W. H. Armstrong and S. J. Lippard, Reversible protonation of the oxo bridge in a hemerythrin model compound. Synthesis, structure, and properties of (μ-hydroxo) bis(μ-acetato)-bis[hydrotris(1-pyrazolyl)borato]diiron(III), [(HB(pz)$_3$)Fe(OH)(O$_2$CCH$_3$)$_2$Fe(HB(pz)$_3$)]$^+$, *J. Am. Chem. Soc.*, 106:4632 (1984).

19. W. H. Armstrong, A. Spool, G. C. Papaefthymiou, R. B. Frankel, and S. J. Lippard, Assembly and characterization of an accurate model for the diiron center in hemerythrin, *J. Am. Chem. Soc.*, 106:3653 (1984)

20. B. Hedman, M. S. Co, W. H. Armstrong, K. O. Hodgson, and S. J. Lippard, EXAFS studies of dinuclear iron complexes as models for hemerythrin and related proteins, *Inorg. Chem.*, 25:3708 (1986).

21. R. C. Prince, G. N. George, J. C. Savas, S. P. Cramer, and R. N. Patel, Spectroscopic properties of the hydroxylase of methane monooxygenase, *Biochim. Biophys. Acta*, 952:220 (1988).

22. X. Feng and S. J. Lippard, unpublished results.

23. P. J. Marini, K. S. Murray, and B. O. West, Iron complexes of N-substituted thiosalicylideneimines. Part 1. Synthesis and reactions with oxygen and carbon monoxide. *J. Chem. Soc., Dalton Trans.*, 143 (1983).

24. B. P. Murch, F. C. Bradley, and L. Que, Jr., A dinuclear iron peroxide complex capable of olefin epoxidation, *J. Am. Chem. Soc.*, 108:5027 (1986).

25. Q. Chen, J. B. Lynch, P. Gomez-Romero, A. Ben-Hussein, G. B. Jameson, C. J. O'Connor, and L. Que, Jr., Iron oxo aggregates. Dinuclear and tetranuclear complexes of N, N, N', N'-tetrakis(2-benzimidazolylmethyl)-2-hydroxy-1,3-diaminopropanol, *Inorg. Chem.*, 27:2673 (1988).

26. S. Yan, D. D. Cox, L. L. Pearce, C. Juarez-Garcia, L. Que, Jr., J. H. Zhange, and C. J. O'Connor, A (μ-oxo)(μ-carboxylato)diiron(III) complex with distinct iron sites, *Inorg. Chem.*, 28:2507 (1989).

27. F. Arena, C. Floriani, A. Chiesi-Villa, C. Guastini, A mixed valence μ-oxo iron(III)-iron(III) complex: a polynuclear iron-sodium-oxo aggregate from the chemical reduction of a μ-oxo diiron(III) complex, *J. Chem. Soc., Chem. Commun.*, 1369 (1986).

28. W. B. Tolman, A. Bino, and S. J. Lippard, Self-assembly and dioxygen reactivity of an asymmetric, triply bridged diiron(II) complex with imidazole ligands and an open coordination site, *J. Am. Chem. Soc.*, 111:8522 (1989).

29. D. M. Kurtz, Oxo- and hydroxo-bridged diiron complexes: a chemical perspective on a biological unit, *Chem. Rev.*, 90:585 (1990).

30. A. C. Rosenzweig, C. Bender, J. Peisach, and S. J. Lippard, unpublished results.

31. W. B. Tolman, S. Liu, J. G. Bentsen, and S. J. Lippard, Models of the reduced forms of polyiron oxo proteins: an asymmetric, triply carboxylate bridged diiron(II) complex and its reaction with dioxygen, *J. Am. Chem. Soc.*, 113:152 (1991).

32. A. S. Borovik and L. Que, Jr., Models for the Fe^{II}Fe^{III} and Fe^{II}Fe^{II} forms of iron-oxo proteins, *J. Am. Chem. Soc.*, 110:2345 (1988).

33. K. E. Liu and S. J. Lippard, Redox properties of the hydroxylase component of methane monooxygenase from *Methylococcus capsulatus* (Bath) - effects of protein B, reductase, and substrate, *J. Biol. Chem.*, in press.

34. A. Stassinopoulos, G. Schulte, G. C. Papaefthymiou, and J. P. Caradonna, Synthesis, structure, and electronic characterization of reactive diiron(II) 1,2-bis-(2-hydroxybenzamido)benzene complexes as models for methane monooxygenase, submitted for publication.

35. P. Cofré, S. A. Richert, A. Sobkowjak, and D. T. Sawyer, Redox chemistry of iron picolinate complexes and of their hydrogen peroxide and dioxygen adducts, *Inorg. Chem.*, 29:2645 (1990).

36. X. Feng and S. J. Lippard, unpublished results.

37. W. Micklitz, S. G. Bott, J. G. Bentsen, and S. J. Lippard, Characterization of a novel μ_4-peroxide tetrairon unit of possible relevance to intermediates in metal-catalyzed oxidations of water to dioxygen, *J. Am. Chem. Soc.*, 111:372 (1989).

38. N. Kitajima, H. Fukui, Y. Moro-oka, Y. Mizutani and T. Kitagawa, Synthetic model for dioxygen binding sites of non-heme iron proteins: X-ray structure of Fe(OBz)(MeCN)(HB(3,5-iPr$_2$pz)$_3$ and resonance Raman evidence for reversible formation of peroxo adduct, *J. Am. Chem. Soc.*, 112:6402 (1990).

39. S. Menage, B. A. Brennan, C. Juarez-Garcia, E. Münck, and L. Que, Jr., Models for iron-oxo proteins: Dioxygen binding to a diferrous complex, *J. Am. Chem. Soc.*, 112:6423 (1990).

40. X. Feng, M. E. Roth, D. P. Bancroft, and S. J. Lippard, manuscript to be submitted.

41. N. Kitajima, H. Fukui, and Y. Moro-oka, A model for methane monooxygenase: Dioxygen oxidation of alkanes by use of a μ-oxo dinuclear iron complex, *J. Chem. Soc., Chem Commun.*, 485 (1988).

42. J. B. Vincent, J. C. Huffman, G. Christou, Q. Li, M. A. Nanny, D. N. Hendrickson, R. H. Fong, and R. H. Fish, Modeling the dinuclear sites of iron biomolecules: Synthesis and properties of $Fe_2O(OAc)_2Cl_2(bipy)_2$ and its use as an alkane activation catalyst, *J. Am. Chem. Soc.*,110:6898 (1988).

43. G. Balavoine, D. H. R. Barton, J. Boivin, A. Gref, P. L. Coupanec, N. Ozbalik, J. A. X. Pestana, and H. Riviere, Functionalization of saturated hydrocarbons. Part X. A comparative study of chemical and electrochemical processes (GIF and GIF-Orsay systems) in pyridine, in acetone and in pyridine-co-solvent mixtures, *Tetrahedron*, 44:1091 (1988).

44. S. Inbar, A. Ehret, and K. Norland, Oxidation of tetramethyl reductic acid by silver halide, *Abstracts of Papers, Natl. Meet. Soc. Photogr. Sci.*, Minneapolis, MN, USA (1987).

45. G. A. Hamilton, R. J. Workman, and L. Woo, Oxidation by molecular oxygen. I. Reaction of a possible model system for mixed-function oxidases, *J. Am. Chem. Soc.*, 86:3390 (1964).

46. G. A. Russel, Reactivity, selectivity, and polar effects in hydrogen atom transfer reaction, *in* "Free Radicals", J. K. Kochi Ed., Wiley: New York, Vol. I:pp. 275 (1973).

47. J. Green and H. Dalton, Substrate specificity of soluble methane monooxygenase, *J. Biol. Chem.*, 264: 17698 (1989).

48. J. T. Groves, Mechanisms of metal-catalysed oxygen insertion, *in* "Metal Ion Activation of Dioxygen", T. G. Spiro ed., Wiley: New York, pp. 125 (1980).

49. J. T. Groves, and D. V. Subramanian, Hydroxylation by cytochrome P-450 and metalloporphyrin models. Evidence for allylic rearrangement, *J. Am. Chem. Soc.*, 106:2177 (1984).

50. C. R. E. Jefcoate, J. R. L. Smith, and R. O. C. Norman, Hydroxylation. Part IV. Oxidation of some benzenoid compounds by Fenton's reagent and the ultraviolet irradiation of hydrogen peroxide. *J. Chem. Soc. B* :1013 (1969).

51. J. T. Groves and T. E. Nemo, Aliphatic hydroxylation catalyzed by iron porphyrin complexes. *J. Am. Chem. Soc.*, 105:6243 (1983).

52. D. H. R. Barton, J. Boivin, N. Ozbalik and K. M. Schwartzentruber, On the mechanism of the Gif system for the oxidation of saturated hydrocarbons, *Tetrahedron Lett.*, 26:447 (1985).

53. S. G. Jezequel and I. J. Higgins, Mechanistic aspects of biotransformations by the monooxygenase system of *M. trichosporium OB3b*, *J. Chem. Tech. Biotechnol.*, 33B:139 (1983).

54. H. Dalton and D. J. Leak, Mechanistic studies on the mode of action of methane monooxygenase, *in* "Gas Enzymology", H. Degn, R. P. Cox, and H. Toftlund eds., Reidel:Dordrecht, Holland, pp. 169 (1985).

55. S. R. Boone, G. H. Purser, H. R. Chang, M. D. Lowery, D. N. Hendrickson, and C. G. Pierpont, Magnetic exchange interactions in semiquinone complexes of iron. Structural and magnetic properties of Tris(3,5-di-*tert*-butylsemiquinonato)tetrakis(3,5-di-*tert*-butylcatecholato) tetrairon (III), *J. Am. Chem. Soc.*, 111:2292 (1989).

56. R. M. Solbrig, L. L. Duff, D. F. Shriver, and I. M. Klotz, Raman and infrared spectroscopy of the oxo-bridged iron (III) complex, $[Cl_3Fe\text{-}O\text{-}FeCl_3]^{2-}$ as a spectroscopic model for the oxo bridge in hemerythrin and ribonucleotide reductase, *J. Inorg. Biochem.*, 17:69 (1982).

RELEVANCE OF GIF CHEMISTRY TO ENZYME MECHANISMS

Derek H. R. Barton and Darío Doller

Department of Chemistry
Texas A and M University
College Station, TX 77843, USA

Abstract: The Gif family of systems for the selective oxidation of saturated hydrocarbons is briefly described. The mechanism of the reaction is analysed in terms of two intermediates. The first is well characterised as an ^{V}Fe σ-carbon bond species and the second has been fully identified, at least in the case of cyclohexane, as a hydroperoxide. The utility of dynamic ^{13}C N.M.R. spectroscopy in iron containing systems is demonstrated. Gif type chemistry is closely related to the unusual enzyme methane monooxygenase.

Part I

The invention of chemical reactions which will selectively functionalise saturated hydrocarbons under mild conditions (for example at room temperature and at neutral pH) represents a noble challenge for the chemists working in the last two decades of this century[1]. Already the porphyrin based iron containing enzymes, such as the P_{450} enzymes, have received much attention[2]. Their reactivity, at least in model systems, is radical like and considered to involve an Fe^{IV} oxenoid species liganded to a porphyrin radical cation. From the practical point of view the most promising reaction is the epoxidation of olefins by model systems[3]. A long known reaction which satisfies in part our definition is the Fenton hydrogen peroxide Fe^{II} oxidation system[4]. However, this generates hydroxyl radicals, which are very reactive and thus unselective.

When we began to study the selective oxidation of saturated hydrocarbons in 1980, we were aware of a short communication from the late Prof. Tabushi[5], who was attempting a biomimetic type oxidation of adamantane **1**. In order to solubilise this hydrocarbon he used pyridine. Oxidation with oxygen, a thiol and an Fe^{II} salt, considered a surrogate for a P_{450} enzyme, gave, like other such models, very little oxidation. However, the selectivity observed was unusual, since there was more secondary than tertiary substitution. Much radical chemistry by Tabushi and others has shown that adamantane[5], as expected, is substituted mainly at the tertiary position. We repeated and confirmed these experiments and modified the system. There was little oxidation, but the selectivity continued to be unusual.

R^2 **1.** $R^1=R^2=H$
R^1 **2.** $R^1,R^2=O$

Applications of Enzyme Biotechnology, Edited by J.W. Kelly and
T.O. Baldwin, Plenum Press, New York, 1991

Life existed on Earth under anaerobic conditions long before the blue-green algae started to make oxygen. Under reducing conditions the atmosphere was full of hydrocarbons, especially methane; there was much hydrogen sulfide, from reduction of sulfate; the abundant element iron was present as metallic iron (cf. bog iron) and in the seas as Fe^{II}. A form of life took advantage of the new aerobic conditions to oxidise the iron to Fe^{III} and deposit it as pure ferric oxide in vast mountain ranges in Australia and Brazil. It seemed to one of us, on reflection, that the new form of life would have obtained far more energy from concerting the oxidation of iron and hydrocarbons together than from just making ferric oxide. The hydrocarbon oxidation products (CO_2) would have left no geological trace. Although this reasoning was naive, it led to a simple, but fundamentally important, experiment. To the adamantane in pyridine containing hydrogen sulfide Dr. Gastiger[6] added iron powder and an equivalent amount of acetic acid (to dissolve the iron) and then stirred the suspension under air. The iron powder, aided by a surface effect of the hydrogen sulfide, dissolved and the adamantane was oxidised to mainly adamantanone **2** with a 10-, then 20- and finally 30-fold increase in yield. The first significant Gif system (Gif^{III}) had been invented. Later[7], we showed that the reaction could be made catalytic in iron (up to 2,000 turnovers) by using metallic zinc as a source of electrons (Gif^{IV}). Hydrogen sulfide was not needed for Gif^{IV}, nor for Gif^{III} if the temperature was raised to 30-40°C to start the reaction. From the beginning I did not think that carbon radicals could play a major role as they are quenched efficiently by the -S-H bond. Some recent work carried out in association with Prof. G. Balavoine, Dr. A. Gref and Mlle. I. Lellouche (Table 1) has shown that when cyclohexyl radicals are generated in presence of hydrogen sulfide they are indeed efficiently quenched by hydrogen atom transfer.

Table 1

1.75 mmol	X (mmol)	Σ(%)	%	%	%	%
	7.2	85	85	0	0	0
	4.1	78	78	0	0	0
	2.0	81	43	3	22	13
	1.0	72	18	15	18	21

Further work[8] has shown that the various Gif systems (Table 2) are based on the reaction of superoxide with Fe^{II} or of hydrogen peroxide with Fe^{III} to afford an Fe^V oxenoid species[9]. The Fe^V species is the same as in P_{450} models, but it is not reduced by electron transfer from the porphyrin ligand. It has, therefore, very different chemical reactivity.

Our nomenclature (Table 2) is based on Geography. Gif comes from Gif-sur-Ivette in France where the first experiments were carried out. GO is the electrochemical system, which was studied in collaboration with Prof. Balavoine at Orsay (Université de Paris-Sud), hence Gif-Orsay. When we moved to Texas A and M, it was logical to add Agg giving us GoAgg, a theoretically important, but not practical system, as well as $GoAgg^{II}$ and $GoAgg^{III}$, where hydrogen peroxide is the oxidant for the iron and the systems are homogeneous. Professor Sawyer has recently introduced[10] another, and potentially important, GoAgg system based on the reaction of oxygen with ferrous picolinate. Dr. Geletii has carried out important, pioneering work[11] on an analogous Cu^{II} system.

Our early work was carried out to give a maximum yield of ketone. Some earlier work on picolinic acid as a ligand[12] had been carried out using Fe^{II} and hydrogen peroxide, but did not give an improved yield of ketone. It was Professor Sawyer who informed us that

Table 2

GIF SYSTEMS

Pyridine - Acetic Acid - Room Temperature

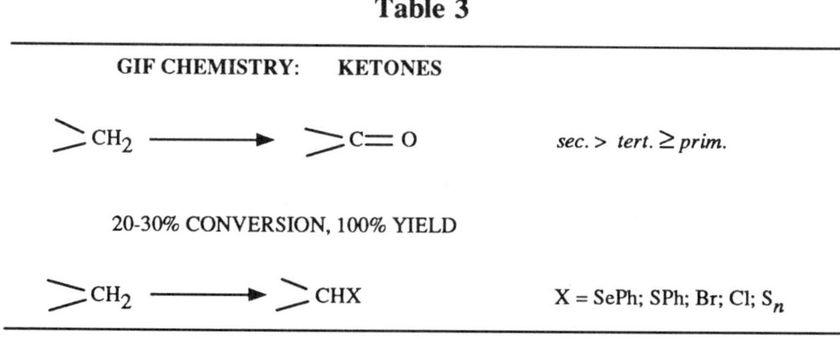

Gif^{III} $Fe^{0} - O_2$

Gif^{IV} Fe^{II} cat. - Zn^{0} - O_2 $\Big\}$ Fe^{II}

GO Fe^{II} cat. - cathode - O_2

$GoAgg^{I}$ Fe^{II} - KO_2 - Ar

$GoAgg^{II}$ Fe^{III} cat. - H_2O_2 - Ar $\Big\}$ Fe^{III}

$GoAgg^{III}$ = $GoAgg^{II}$ + ligand

$Fe^{II} + {}^-O_2 \cdot \equiv Fe^{III} + H_2O_2 \longrightarrow {}^{III}Fe-O-OH$

$\longrightarrow {}^{V}Fe=O$

picolinic acid as a ligand made the reaction more rapid[13]. We have examined a number of ligands[14] and some of them, like picolinic acid or pyrazine-2-carboxylic acid make the oxidation fifty times faster. Others, like pyridine-2-phosphonic acid, stop the reaction dead. So far changing ligands has only increased the speed of the oxidation, but has not increased the yield. Clearly, the synthesis of modified ligands is an important new objective for Gif chemistry.

In Table 3 we present a brief summary of the chemical reactivity of the Gif systems. The selective oxidation of saturated hydrocarbons to ketones is the most striking reaction and up to a reasonable conversion the reaction is quantitative. The capture of an intermediate (Table 3) is discussed in more detail later.

Table 3

GIF CHEMISTRY: KETONES

$>CH_2 \longrightarrow >C=O$ *sec. > tert. ≥ prim.*

20-30% CONVERSION, 100% YIELD

$>CH_2 \longrightarrow >CHX$ $X = SePh; SPh; Br; Cl; S_n$

A new development[15], related to the biosynthesis of penicillin and biotin, is the demonstration that Gif type chemistry in the presence of H_2S, H_2S-S or Na_2S affords both oxidation products and, especially with cyclohexane, di- and polysulfides. In the presence of picolinic acid the oxidation products are suppressed and only dicyclohexyl disulfide and related polysulfides are seen. Under $GoAgg^{III}$ conditions the sulfation process is remarkably efficient and nearly half of the sulfur in the reagent ends up in the di- and poly-sulfides. The new oxygen-ferrous dipicolinate system of Professor Sawyer[10] works using hydrogen sulfide (or other reducing agent) to reduce the Fe^{III} formed back to Fe^{II}. If this is a Gif type system then the hydrogen sulfide present should also be making polysulfides. Recent work in collaboration with the group from Orsay (Table 4) has shown that this is indeed the case. The original work[10] had overlooked the formation of these compounds.

In our original studies of the oxidation of adamantane *tert.* radicals could be detected by competitive capture by pyridine and oxygen. A careful comparative study[16] has

Table 4

$$\text{(cyclohexane)} + H_2S + \text{Catalyst} + \underset{RO_2C\quad N\quad CO_2R}{\text{(pyridine diester)}} \xrightarrow[\text{Room Temp.}]{Py,\ AcOH,\ O_2} \text{(cyclohexanone)} + \text{(cyclohexyl-}S_n\text{-cyclohexane)}$$

15 mmol	X mmol	0.28 mmol	0.56 mmol		mmol	mmol
	5.4	Fe(MeCN)$_4$(ClO$_4$)$_2$	R=Et$_4$N		0.16	0.11
	5.4	FeCl$_2$.4H$_2$O	R=H		0.18	0.07
	8.1	FeCl$_2$.4H$_2$O	R=H		0. 27	0.15
	10.8	FeCl$_2$.4H$_2$O	R=H		0.42	0.13

confirmed that secondary radicals are not involved in Gif chemistry to any significant extent. *Tert.* radicals are not normally important; adamantane is a special case.

We have recently compared[17] genuine radical bromination of saturated hydrocarbons using BrCCl$_3$ with Gif type oxidative bromination in the presence of BrCCl$_3$. By mixing pairs of hydrocarbons and running the reactions to 30% conversion, or more, a reliable quantitative rate order could be established. The results are summarised in Table 5. The radical bromination order is in agreement with limited literature data and presents no surprises. The GoAgg^{III} data are completely different. In the real radical reactions cyclohexane reacts more slowly than the other hydrocarbons. In GoAgg^{III} cyclohexane reacts faster than any other hydrocarbon examined. The data in Table 5 are normalised for the number of C-H bonds. Another remarkable difference is that the radical reaction gives, as expected, CHCl$_3$ (determined quantitatively by proton N.M.R.), whilst the GoAgg^{III} reactions produce CO$_2$ from the BrCCl$_3$ (determined quantitatively as BaCO$_3$). Clearly Gif chemistry and radical chemistry are very different and, therefore, demand different theoretical interpretations.

In Table 6 we have summarised many experimental facts. Gif chemistry produces ketones as major products. A lesser amount of the secondary alcohol is always formed. Ketone and secondary alcohol are not interconverted at a significant rate under the experimental conditions. The water content of the system has a slight effect[18] on the ketone-alcohol ratio. A comparison of Gif^{IV} with GoAgg^{II} shows[19] that the ketone-alcohol ratio is lower for the GoAgg^{II} system, where the water content (from the 30% aqueous hydrogen peroxide) is higher.

From all these facts two intermediates **A** and **B** were identified (Table 6). The first intermediate **A** is captured by Ph$_2$Se$_2$ and BrCCl$_3$ quantitatively[17,20]. It is also captured by

Table 5

Radical Bromination with BrCCl$_3$ (Normalised)

$$\text{(neopentane)} > \text{(cyclooctane)} > \text{(cyclopentane+)} > \text{(cyclopentane)} > \text{(cyclohexane)}$$

10.2 tert. only	3.3	1.9	1.3	1.0

GoAgg^{III} with BrCCl$_3$ (Normalised)

$$\text{(neopentane)} < \text{(cross)} < \text{(cyclopentane)} < \text{(cyclooctane)} < \text{(cyclohexane)}$$

tert. 0.18 prim. 0.06	0.63	0.69	0.76	1.0

Table 6

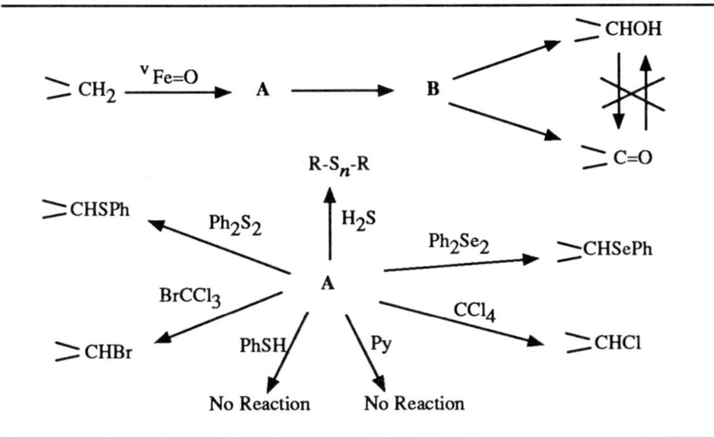

Ph_2S_2, CCl_4 and $CHBr_3$. Normally these facts would suggest radical chemistry. However **A** is not captured by pyridine under conditions where real carbon radicals are easily detected[16]. Recently we have examined carefully[19,21] the effect of thiophenol on the Gif^{VI} system. Unless stated to the contrary, Gif^{V} is Gif^{III} + picolinic acid and Gif^{VI} is Gif^{IV} + picolinic acid. As shown in Table 7 one mmol of thiophenol has a major effect on the ketone-alcohol ratio, but has no significant effect on the total oxidation. The addition of more thiophenol reduces further the ketone-alcohol ratio without a major change in the amount of oxidation. Similar results are seen when dianisyl telluride is added[22]. Again intermediate **A** cannot be a carbon radical, but the effect of thiophenol and dianisyl telluride show that a second intermediate **B** is needed.

Table 7. Effect of Thiophenol in the Gif^{IV} System [Cyclododecane (5 mmol) as Substrate]

PhSH (mmol)	Conversion (%)	ketone/alcohol
0	30.1	12.1
1.0	29.8	3.08
2.0	26.0	1.40
4.0	21.2	0.95

We have recently summarised the data in terms of the working hypothesis[23] shown in Table 8. As we have always argued, intermediate **A** contains an iron-carbon sigma bond formed by an insertion process of the Fe^{V} oxenoid species into the carbon-hydrogen bond. A related insertion process into the sigma carbon-iron bond of (say) diphenyl diselenide affords the phenylseleno derivatives.

Part II

In Table 8, **A** could fragment into alcohol and the two Fe^{III} resting state. This could explain the small water effect. To explain the effect of thiophenol and of dianisyl telluride on the ketone-alcohol ratio we argued that an intermediate **B** must be partitioned between ketone and alcohol. The latter are not interconverted under the conditions of the reaction. We decided to look for intermediate **B** using N.M.R. spectroscopy. We knew that proton studies would not be helpful due to line broadening under the influence of the paramagnetic iron. We therefore used ^{13}C N.M.R. spectroscopy and in the GoAgg^{II} system were able to observe the appearance and disappearance of an intermediate.

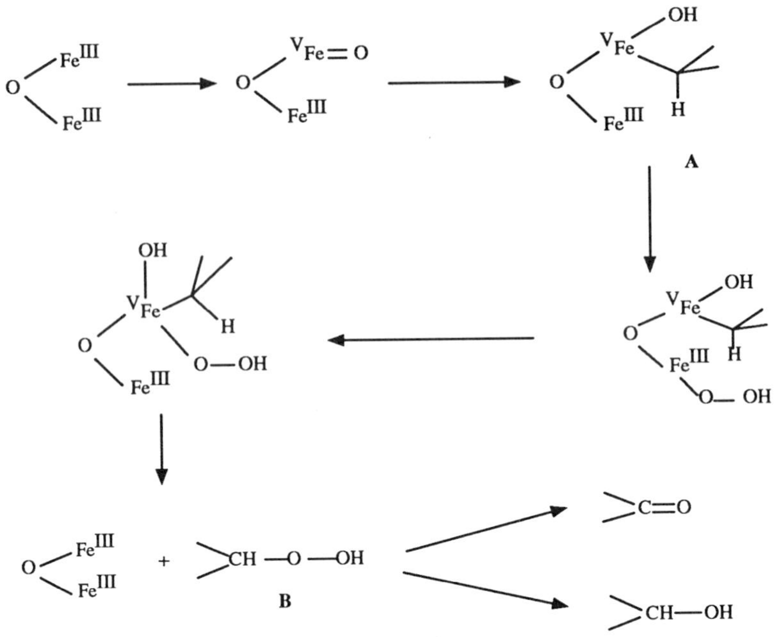

Preliminary experiments following the course of the GoAgg^{II} oxidation of cyclohexane by ^{13}C N.M.R. spectroscopy showed the convenience of using a ^{13}C-enriched substrate in order to improve the signal-to-noise ratio and increase the sensitivity of the experiment. Also, picolinic acid was not added to the reaction mixture to allow a reaction rate compatible with the N.M.R. time scale.

Figure 1 shows the series of ^{13}C-N.M.R. spectra obtained upon addition of hydrogen peroxide (30% in water) to a solution of [1-^{13}C]-cyclohexane in deuteropyridine-acetic acid containing a catalytic amount of ferric chloride. The time between consecutive spectra is about twenty minutes. The formation of an intermediate characterized by four peaks at 82.64, 30.99, 26.23, and 24.11 ppm can be seen. The intensity of these four signals increases steadily with time, reaching a maximum value at *ca.* 3.5 hours after the addition of hydrogen peroxide. The gradual disappearance of these four peaks is accompanied by the appearance of four new signals at 212.16, 42.00, 27.25, and 25.01 ppm assigned to C-1 to 4 of ^{13}C-labelled cyclohexanone, respectively[24].

We then considered the possibility of cyclohexyl hydroperoxide being the intermediate. The authentic sample was prepared[26] and its ^{13}C N.M.R. spectrum recorded. The four observed resonances (82.72, 31.32, 26.55, and 24.38 ppm) are in excellent agreement with those found in the kinetic experiment. Thus, the intermediate was characterized as cyclohexyl hydroperoxide. In Figure 2 the peak intensity for the C-2 resonances of the hydroperoxide and the ketone are compared as a function of the reaction time. It can be concluded that cyclohexyl hydroperoxide is the main source of cyclohexanone in the GoAgg^{II} oxidation of cyclohexane.

The possibility of the intermediate being a *self-assembled* iron(III) cyclohexanolate[25] was eliminated after recording the ^{13}C N.M.R. spectrum of cyclohexanol in a deuteropyridine-acetic acid-ferric chloride solution. The chemical shifts values obtained for C-1 to C-4 (69.87, 36.61, 25.02, and 26.43, respectively) did not agree with those obtained for the intermediate.

To confirm that the hydroperoxide was an intermediate in the ketonization of methylenic carbons a sample of the cyclohexylhydroperoxide was submitted to GoAgg[II] conditions (except hydrogen peroxide) and its complete transformation to cyclohexanone was verified, in agreement with previous work[27].

This enables us to complete the second part of the scheme in Table 8. After the first Fe^{III} is changed to Fe^{V} oxenoid and reacted with the C-H bond to give **A**, second mole of hydrogen peroxide reacts with the second Fe^{III} to give another hydroperoxide. Nucleophilic displacement on Fe^{III} should, of course, be much faster than on Fe^{V}. Now the

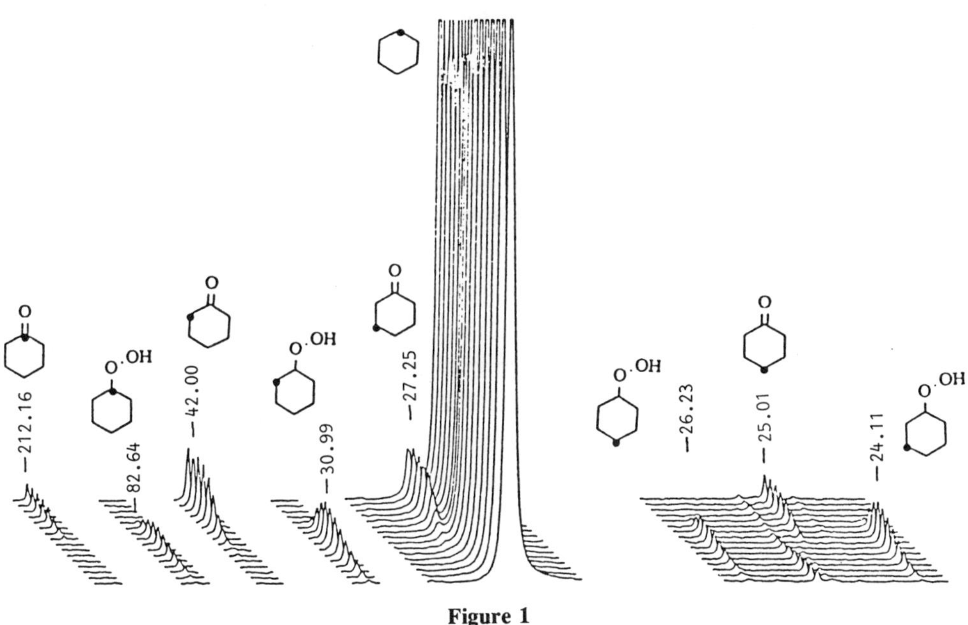

Figure 1

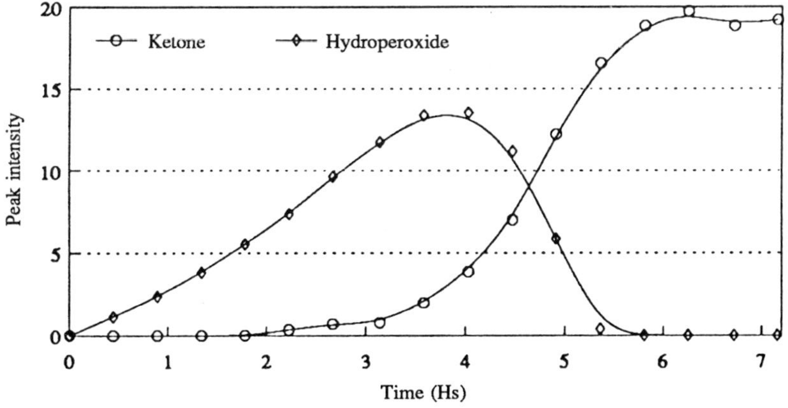

Figure 2. ^{13}C-NMR experiment: GoAgg[II] reaction.

hydroperoxide group can be transferred intramolecularly to the Fe^V species. Ligand coupling of the hydroperoxide function with the secondary carbon affords the alkyl hydroperoxide, intermediate **B**. The partitioning of **B** into ketone and alcohol then depends on the presence of reducing agents in the medium. Thiophenol and dianisyl telluride are well known to reduce hydroperoxides to alcohols. We have verified that this is a very efficient process with dianisyl telluride. However, when a larger excess of these two reagents is added the ketone-alcohol ratio decreases but so does the total amount of oxidation products. We consider that the excess of these reductants begins to reduce the first Fe^V oxenoid species back to Fe^{III}. We decided to search for a reductant which would reduce intermediate **B** but nor interfere with the Fe^V oxenoid or intermediate **A**. We found this reagent in the form of triphenylphosphine.

The working hypothesis summarised in Table 8 uses a μ-oxo dimer iron species. We make this assumption to make facile the entry of the second hydrogen peroxide. However, two ferrous irons are needed for the reaction with successive superoxides as shown in Table 9. They are also needed to explain the apparent relationship between Gif chemistry and the unusual enzyme methane monooxygenase (see further below).

Table 9

Part III
___

After defining our working hypothesis (Table 8), the next stage was the designing of experiments to test it. A key step was to find such a reagent that would give information about the reaction mechanism; i.e. react with intermediate **B** without interfering significantly with the activation process and without modifying the unusual Gif selectivity.

Of the many reagents tried, triphenylphosphine met all the requirements. As the reaction

$$PPh_3 + H_2O_2 \xrightarrow{Py,\ AcOH} OPPh_3 + H_2O$$

is very fast, this part of the work was conducted under Gif^{IV} conditions. The substrate of choice was cyclododecane (**3**), since the low volatility of this and its oxidation products, cyclododecanone (**4**) and cyclododecanol (**5**), allows excellent mass balances. The results of adding up to 3.5 mmol of PPh_3 to the Gif^{IV} oxidation of cyclododecane are shown in Figure 3. It is observed a continuous decrease in the yield of ketone, with the corresponding increment in the yield of alcohol, as the amount of PPh_3 increases. The total amount of oxidation products (**4 + 5**) remains approximately constant.

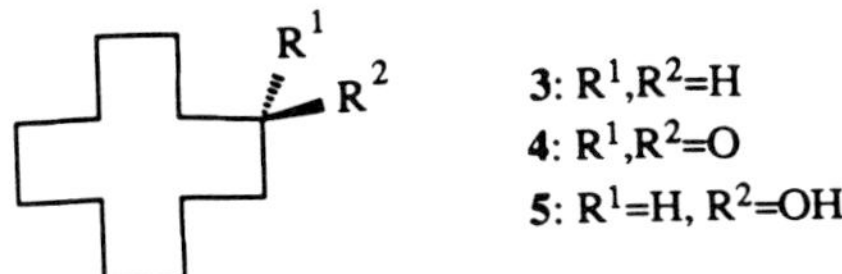

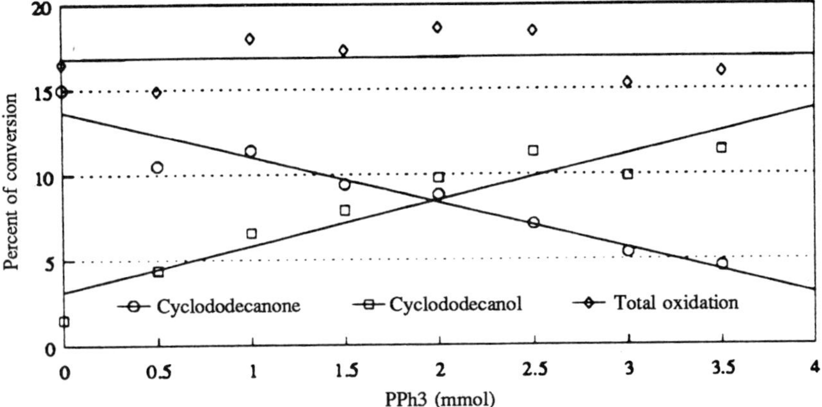

Figure 3. Effect of PPh$_3$ in Gif^{IV} reaction.

These results are explained by considering again, that the alkyl hydroperoxide (which in the absence of PPh$_3$ would have fragmented to ketone) is reduced by PPh$_3$ to alcohol. There is no effect of the phosphine on the formation of intermediate **B**. Kinetic measurements showed that PPh$_3$ has a slight effect on the reaction rate (see below). We can conclude that PPh$_3$ is not a ligand for iron in our system.

We still had to verify that the presence of PPh$_3$ in the reaction medium did not change the selectivity of the process. Thus, adamantane (**1**) was used as substrate and the effect of PPh$_3$ on the amount of products substituted at the tertiary position (C^3) and at the secondary position (C^2) was compared with the usual Gif^{IV} reaction selectivity. The ratio obtained in the presence of 2.0 mmol of PPh$_3$ (C^2/C^3 = 1.0) does not differ significantly from the value obtained for a typical Gif^{IV} oxidation (C^2/C^3 = 1.1)[28]. Therefore, the addition of PPh$_3$ to the Gif^{IV} reaction does neither change the selectivity nor the total amount of activation. Normal Gif chemistry can be performed in the presence of PPh$_3$, except that part of the expected ketone is found as alcohol.

It is very interesting that PPh$_3$ was oxidized to OPPh$_3$ almost quantitatively at the end of the reaction. Kinetic measurements (Figure 4) and blank experiments showed that the oxidation of PPh$_3$ (a very easily oxidizable compound) and the hydrocarbon (very hard to oxidize) occur at comparable rates. A competition is established for superoxide (HOO·) between Fe^{II} and PPh$_3$. If Fe^{II} wins the reaction continues with normal Gif chemistry, as shown in Table 8. Otherwise, OPPh$_3$ is produced, together with a minor amount of hydroxyl radicals, which are quenched by pyridine or afford a minor amount of autooxidation products[29].

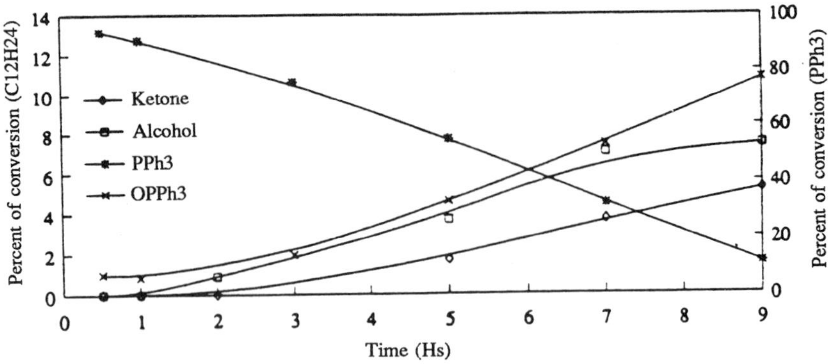

Figure 4. Gif^{IV} oxidation (2 mmol of PPh$_3$).

Part IV

Methane monooxygenase (MMO) is a remarkably interesting enzyme. Thanks especially to the work of Dalton [30,31], the enzyme has been obtained in a soluble form and characterised as a three component enzyme. The component responsible for the chemical oxidation of methane is a non-porphyrinic molecule containing a μ-oxo di-iron center[32].

Recent work by Dalton and his collaborators[33] has shown that MMO is an unusually catholic enzyme, which oxidises many hydrocarbons much larger than methane. The selectivity of an enzyme is usually a reflection of the protein associated with the active center. MMO seems to be different as it tends to give secondary alcohols[4] and attacks cyclic olefins at the allylic positions with or without a shift of the double bond. This is exactly what is seen in Gif type oxidation and has been interpreted as the participation of a π-allyl iron complex[34]. Finally, MMO oxidises adamantane with exactly the same selectivity for the secondary and tertiary positions as seen in all the Gif type systems[23]. It will be very interesting to see if methyl hydroperoxide is an intermediate in methane oxidation or not. Since the medium is water, alcohol formation may well be determined by ligand coupling in intermediate **A**. The reducing function of the enzyme could, however, also allow for methyl hydroperoxide to be reduced to methanol.

Similar considerations apply to the important enzyme prolylhydroxylase, which converts peptide bound proline to the genetically uncoded *trans*-4-hydroxyprolyl residue[35].

Part V

Gif type chemistry clearly has potential economic utility. Extensive development work has not been carried out, although the careful and precise work of Schuchardt and his collaborators[36] has shown how well cyclohexane is oxidised to cyclohexanone. The new Gif type system of Prof. Sawyer has been mentioned above[10]. His work on the selective ketonisation of saturated hydrocarbons shows the usual Gif type selectivity[13]. Complexing with picolinic acid is good support for the μ-oxo dimer of iron as the basis for Gif type reactivity. Another important article[37] shows that when an excess of Fe^{II} complex is used under GoAgg^{III} conditions, normal Fenton chemistry is indeed seen. The hydroxyl radicals attack the substrate cyclohexane to give cyclohexyl radicals which are trapped by pyridine or, if diphenyl diselenide is present, by this reagent. When all the diphenyl diselenide has been consumed then the standard GoAgg^{III} reaction is seen producing ketone.

Acknowledgments. We thank all past and present collaborators for enthusiastic efforts they have devoted to unravelling the mysteries of Gif chemistry. The present group, excluding the authors of this article, are S. Bévière, W. Chavasiri, È. Csuhai, Y. Geletii, D. Hill, H.-J. Lim, W. Liu and T. J. Weiss. We also thank the organisms that have supported this work, including the N.S.F., the N.I.H., Merck Sharp and Dohme and Quest Internatl.

References

1. *Inter alia* A. E. Shilov, *'Activation of Saturated Hydrocarbons by Transition Metal Complexes'* D. Reidel Publishing Co., Dordrecht, 1984. R. H. Crabtree , *Chem. Rev.* **1985**, *85*, 245. B. Meunier, *Bull. Soc. Chim. Fr.* **1986**, 576. *'Activation and Functionalisation of Alkanes'* Ed. C. L. Hill, J. Wiley and Sons, New York, 1989. M. B. Sponsler, B. H. Weiller, P. O. Stoutland, R. G. Bergman, *J. Am. Chem. Soc.* **1989**, *111*, 6841, and references there cited.

2. T. J. Mc Murry, J. T. Groves in *'Cytochrome* P_{450}. *Structure, Mechanism and Biochemistry'*. Ed. P. Ortiz de Montellano, Plenum Press, New York, 1985, Chap. 1. D. Mansuy, *Pure and Appl. Chem.* **1987**, *59*, 759.

3. J. P. Collman, X. Zhang, R. T. Hembre, J. J. Braumann, *J. Am. Chem. Soc.* **1990**, *112*, 5357. J. P. Collman, P. D. Hampton, J. J. Braumann, *ibid.* **1990**, *112*, 2977, 2986, and references there cited.

4. C. Walling. *Acc. Chem. Res.* **1975**, *8*, 125.

5. I. Tabushi, T. Nakajima and K. Seto, *Tetrahedron Lett.* **1980**, *21*, 2565.

6. D. H. R. Barton, M. J. Gastiger, W. B. Motherwell. *J. Chem. Soc., Chem. Commun.* **1983**, 41.

7. *Idem, Ibid,* **1983**, 731.

8. D. H. R. Barton, F. Halley, N. Ozbalik, E. Young, G. Balavoine, A Gref, J. Boivin. *New J. Chem.* **1989**, *13*, 177.

9. M. L. Kremer. *Int. J. Chem. Kin.* **1985**, *17*, 1299.

10. C. Sheu, A. Sobkowiak, S. Jeon, D. T. Sawyer. *J. Am. Chem. Soc.* **1990**, *112*, 879.

11. Y. V. Geletii, V. V. Lavrushko, G. V. Lubimova. *J. Chem. Soc., Chem. Commun.* **1988**, 936.

12. G. Balavoine, D. H. R. Barton, J. Boivin, A. Gref. *Tetrahedron Lett.* **1990**, *31*, 659.

13. See C. Sheu, S. A. Richert, P. Cofré, B. Ross, A. Sobkowiak, D. T. Sawyer, J. R. Kanofsky. *J Am. Chem. Soc.* **1990**, *112*, 1936.

14. E. About-Jaudet, D. H. R. Barton, È. Csuhai, N. Ozbalik. *Tetrahedron Lett.* **1990**, *31*, 1657.

15. G. Balavoine, D. H. R. Barton, A. Gref, I. Lellouche. *Tetrahedron Lett.* **1990**, in press.

16. D. H. R. Barton, F. Halley, N. Ozbalik, M. Schmitt, E. Young, G. Balavoine. *J. Am. Chem. Soc.* **1989**, *111*, 7144.

17. D. H. R. Barton, È. Csuhai, D. Doller, N. Ozbalik, N. Senglet. *Tetrahedron Lett.* **1990**, *31*, 3097.

18. D. H. R. Barton, È. Csuhai, D. Doller. In preparation.

19. D. H. R. Barton, È. Csuhai, N. Ozbalik. *Tetrahedron* **1990**, *46*, 3743.

20. G. Balavoine, D. H. R. Barton, J. Boivin, P. LeCoupanec, P. Lelandais. *New J. Chem.* **1989**, *13*, 691.

21. D. H. R. Barton, È. Csuhai, D. Doller. In preparation.

22. D. H. R. Barton, È. Csuhai, N. Ozbalik. *Tetrahedron Lett.* **1990**, *31*, 2817.

23. D. H. R. Barton, È. Csuhai, D. Doller, N. Ozbalik, G. Balavoine. *Proc. Natl. Acad. Sci.* **1990**, *87*, 3401.

24. E. Pretsch, J. Seibl, W. Simon, T. Clerc. *'Spectral Data for Structure Determination of Organic Compounds'*. Springer-Verlag, Berlin, **1983**.

25. S. M. Gorun, G. C. Papaefthymiou, R. B. Frankel, S. J. Lippard. *J. Am. Chem. Soc.* **1987**, *109*, 3337. R. W. Saalfrank, A. Stark, M. Bremer, H. U. Hummel. *Angew. Chem. Int. Ed. Engl.* **1990**, *29*, 311.

26. H. R. Williams, H. S. Mosher. *J. Am. Chem. Soc.* **1954**, *76*, 2984, 2987.

27. D. H. R. Barton, É. Csuhai, unpublished observations.

28. G. Balavoine, D. H. R. Barton, J. Boivin, P. Lecoupanec, P. Lelandais. *New J. Chem.* **1989**, *13*, 691.

29. D. H. R. Barton, S. Bévière and D. Doller. In preparation.

30. H. Dalton. *Adv. Appl. Microbiol.* **1980**, *26*, 71.

31. J. Colby, H. Dalton. *Biochem. J.* **1978**, *171*, 461.

32. R. C. Prince, G. N. Geoge, J. C. Savas, S. P. Cramer, R. N. Patel. *Biochim. Biophys. Acta* **1988**, *952*, 220. A. Ericson, B. Hedman, K. O. Hodgson, J. Green, H. Dalton, J. G. Bentsen, R. H. Beer, Lippard, S. J. *J. Am. Chem. Soc.* **1988**, *110*, 2330. B. G. Fox, K. K. Sureus, E. Münck, J. D. Lipscomb. *J. Biol. Chem.* **1988**, *263*, 10553.

33. J. Green, H. Dalton. *J. Biol. Chem.* **1989**, *264*, 17698. D. J. Leak, H. Dalton. *Biocatalysis* **1987**, *1*, 23.

34. D. H. R. Barton, K. W. Lee, W. Mehl, N. Ozbalik, L. Zhang. *Tetrahedron* **1990**, *46*, 3753.

35. H. M. Hanauske-Abel, V. Günzler. *J. Theor. Biol.* **1982**, *94*, 421.

36. U. Schuchardt, V. Mano. *Preprints First World Congress and Second European Workshop on New Developments in Selective Oxidation*, **1989**, *D6*, 01. U. Schuchardt, E. V. Spinacé, V. Mano. *4th International Symposium on Activation of Dioxygen and Homogeneous Catalytic Oxidation*, Balatonfüred, Hungary, **1990**.

37. C. Sheu, A. Sobkowiak, L. Zhang, N. Ozbalik, D. H. R. Barton, D. T. Sawyer. *J. Am. Chem. Soc.* **1989**, *112*, 879.

TRANSTHYRETIN ACID INDUCED DENATURATION IS REQUIRED FOR AMYLOID
FIBRIL FORMATION *IN VITRO*

Wilfredo Colon and Jeffery W. Kelly *

Department of Chemistry
Texas A&M University
College Station, Texas 77843-3255

ABSTRACT: The human plasma protein transthyretin (TTR), implicated as the causative agent in Familial Amyloid Polyneuropathy and Senile Systemic Amyloidosis, was transformed into amyloid fibrils *in vitro* under conditions that mimic the environment of a lysosome (pH 4.5). During the course of acid (HCl) induced denaturation, a folding intermediate associates to form amyloid fibrils. This procedure for making amyloid fibrils appears to be physiologically relevant and general in that IgG2 λ, another amyloidogenic protein which is associated with primary amyloidosis, was also transformed into amyloid fibrils at pH 4.5. This preliminary work suggests that denaturation is an integral part of the amyloid fibril formation mechanism *in vivo*.

INTRODUCTION

The transthyretin human plasma protein, also known as thyroxine-binding prealbumin in the older literature, is composed of four identical 127 residue subunits (MW 54,980). (1, 2) Transthyretin is encoded by a single copy gene on chromosome 18. Human TTR is synthesized in the liver hepatocytes and secreted into the plasma where it plays a major role in the transport of thyroxine and retinol, the latter via a TTR-retinol binding protein complex.(3, 4, 5, 6) Transthyretin is turned over rapidly ($t_{1/2} \approx 220$ minutes) in plasma, presumably by a receptor on the liver hepatocytes that transports TTR to a lysosome (ca. pH 4.5), where it is degraded by proteases to the constituent amino acids. The nucleotide and amino acid sequences have been reported and the cDNA has been cloned into a high expression plasmid by Sakaki and coworkers.(7)

---

*Dedicated to the late Professor E.T. Kaiser

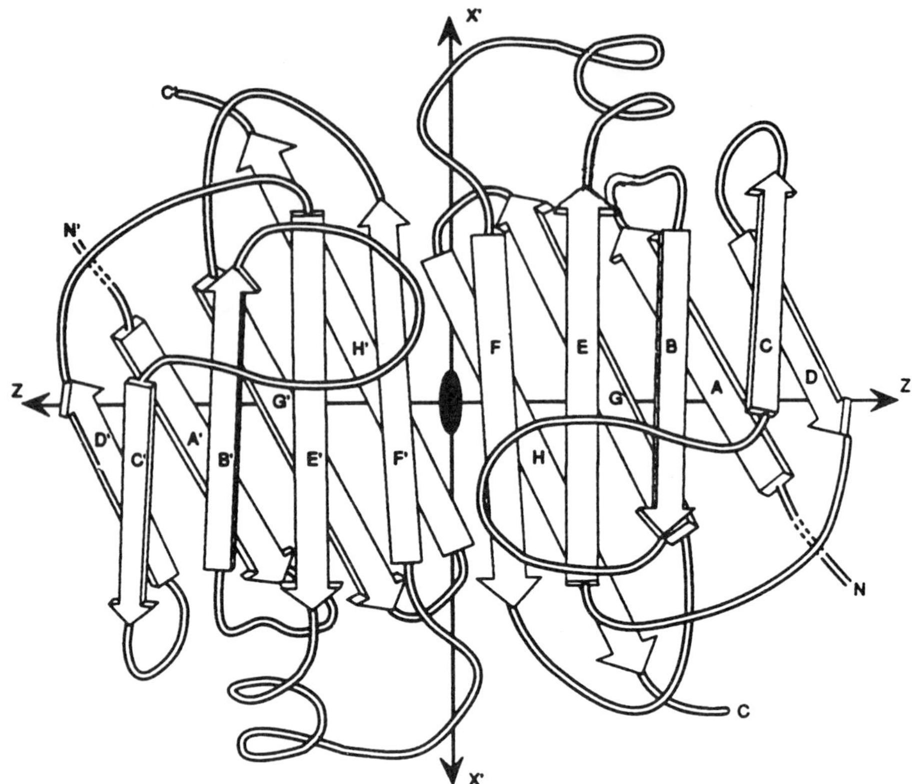

FIGURE 1   A schematic ribbon diagram showing the main-chain conformation in the prealbumin dimer. The arrows indicate β-sheet strands labeled A to H in one monomer and A' to H' in the other. (Figure based on data in References 1 and 2)

The 1.8Å X-ray crystal structure of transthyretin demonstrates that the monomer tertiary structure is an antiparallel β-sheet sandwich, <u>Fig 1</u>.(1, 2) The four stranded β-sheet sandwich dimerizes via hydrogen bonding to form an eight stranded antiparallel β-sheet sandwich which associates in a face to face fashion to form the tetramer. The hydrophobic faces of the dimer are prevented from forming a bilayer type structure by four loops which separate the hydrophobic faces affording a channel between the dimers that serves as the binding site for thyroxine.

BACKGROUND

The transthyretin protein has been implicated as the causative agent in Familial Amyloid Polyneuropathy and Senile Systemic Amyloidosis. Amyloidosis is a descriptive term that refers to the deposition of proteinacious macromolecular β-sheet fibrils in various tissues. Familial Amyloid Polyneuropathy (FAP) is a classical late onset genetic disease; symptoms appear at age thirty and the patients typically die by the fifth decade. To date, fourteen different mutations in the TTR gene have been discovered which are associated with FAP. (8) The majority of patients are heterozygotes. Differential gene expression does not seem to play a role in the disease.

Clinical symptoms include organ dysfunction and peripheral nerve damage. The amyloid fibrils in FAP patients are composed almost exclusively of the <u>variant</u> form of TTR.

Normal transthyretin is also found in amyloid fibrils in a disease prevalent in the elderly called Senile Systemic Amyloidosis, which affects 25% of the population over 80 years of age. In order to be classified as amyloid fibrils, the transthyretin macromolecular assembly must be insoluble under physiological conditions and bind congo red. The congo red-amyloid complex must exhibit green birefringence in a polarizing microscope and the UV spectrum of congo red bound to amyloid fibrils must show a significant red shift when compared to congo red in solution. Amyloid is further characterized by its distinctive, but environmentally dependent fibrilar ultra-structure observed under the electron microscope and the cross β-pattern from X-ray diffraction.(9)

Transthyretin is just one of many β-sheet proteins found in the serum which have a propensity to form amyloid fibrils.(10)   Other serum proteins capable of amyloidosis include the immunoglobulin κ and λ light chains, the Bence Jones proteins or fragments thereof, insulin, and the SAA protein, which is produced by humans in response to inflammation or cell necrosis. Interestingly, the majority of these proteins have a similar β-sheet sandwich structure in solution.

Several *in vitro* models for amyloidosis exist. Glenner demonstrated as early as 1971 that amyloid was produced *in vitro* from selected Bence Jones proteins (10mg / ml) by pepsin treatment at pH 3.5, 37°C. He recognized that amyloid was closely associated with macrophages, which have significant numbers of lysosomal organelles, and proposed that the pathogenic mechanism involves intralysosomal proteolysis.(11) It was more recently shown that Bence Jones light chains were converted into amyloid by treatment with kidney lysosomal extracts further supporting lysosomal involvement.(12) Another method employs a 1% solution of insulin (pH 2) which was converted into "amyloid fibrils" by heating to 80-90°C until a clear gel formed. The sample was then frozen at -78°C and taken through this heat / freeze cycle several times. The insulin fibrils produced have a cross-β structure as determined by X-ray diffraction.(13) Lateral fibril aggregation, although frequently noted in native amyloid fibril concentrates, was not a feature of these *in vitro* amyloid preparations.

METHODS AND PROCEDURES

Transthyretin from human plasma was obtained from Calbiochem; Insulin, IgG2 λ and congo red was purchased from Sigma. All other reagents were ultrapure from USB. Double distilled water was used for all aqueous experiments.

Transthyretin Denaturation. A 0.2 mg/ml solution of TTR (0.05M Tris, 0.1M KCl, pH 7.5) was denatured by titrating the solution with HCl to afford a final pH of 2.5. Aggregation was observed during denaturation at pH 6.0 and pH 4.5. The aggregate at pH 6.0 was formed much slower than the aggregate at pH 4.5. The aggregates were different in their appearance and in their structure, see below.

Transthyretin Amyloid Fibrils. Aggregated TTR formed at pH 4.5 was demonstrated to have the amyloid structure using a congo red binding assay, which binds amyloid specifically. Approximately 7 uM of aggregated TTR was added to a 10 uM solution of congo red (0.05M phosphate buffer, 0.1M KCl, pH 7.5) ; a red shifted congo red absorbance was observed in the UV spectra. The congo red-TTR fibril suspension was centrifuged at 12,000 rpm for 10 min. to afford a red pellet which was resuspended in water and recentrifuged. The resulting pellet remained red in color and exhibited a characteristic red shifted congo red absorbance. The aggregate formed at pH 6.0 as well as TTR in solution did not bind congo red as evidenced by the lack of a red shift in the congo red spectrum. The pellet formed at pH = 6 did not exhibit congo red absorbance after isolating it from the solution of congo red.

Microscopy. A drop of the resuspended TTR-congo red amyloid fibrils was placed on a microscope slide, covered with a cover slip, and observed in a polarizing microscope. The oriented fibers exhibited an apple green birefringence which contrasts with the black background.

Rate of TTR Amyloid Fibril Formation. Solutions of TTR (0.075 mg/ml), 0.05M phosphate buffer, 0.1M KCl were adjusted to pH 2.5. The pH was then rapidly increased to a pH between 4 and 5 to initiate refolding to a putative folding intermediate that was capable of forming amyloid. The UV absorbance at 330nm, (where proteins do not absorb) was monitored at 30 second intervals to measure the light scattering of the medium as fibril formation took place. Amyloid fibrils were also formed by pH denaturation, i.e. by lowering the pH from 7.5 (native protein) to between 4 and 5.

RESULTS

In our studies a solution of 0.2mg / ml solution of transthyretin (TTR) purchased from Calbiochem (0.1M KCl, 0.05M Tris, pH 7.5) was converted into amyloid fibrils by titrating the solution rapidly to a pH between 4.0 and 5.0, using HCl. Amyloid fibrils were also produced from a pH 2.0 solution of TTR (unfolded) that was refolded by rapidly increasing the pH so that the final pH was between 4 and 5. The rate of amyloid fibril formation as a function of pH for the denaturation and the renaturation mode is shown in Fig 2. Interestingly, the heat / freeze conditions used in the insulin model for amyloidosis are not useful for converting TTR or IgG2 λ into amyloid fibrils, indicating this mode of amyloid formation is protein dependent.

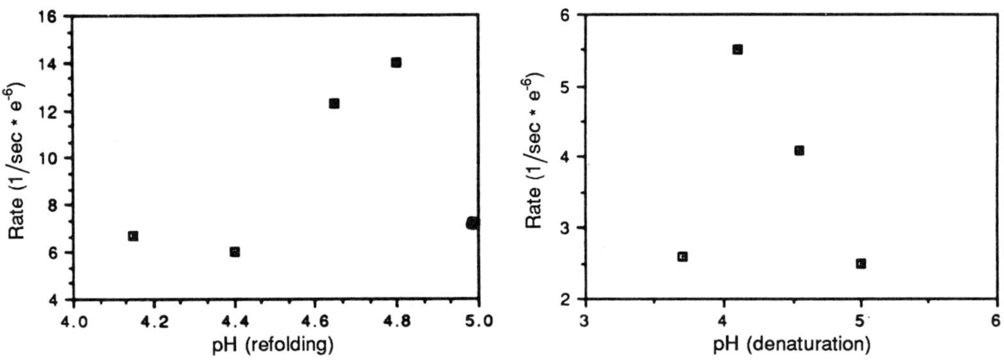

FIGURE 2. The Rate of Amyloid Fibril Formation as a Function of pH.

The affinity of congo red for amyloid fibrils has been utilized to characterize amyloid fibrils produced both *in vivo* and *in vitro*.(14) Congo red binding to amyloid fibrils is associated with a 5 to 15nm absorbance shift to longer wavelengths, (Fig 3). (15) The TTR amyloid fibrils formed at pH 4.5 have a distinct fibrilar appearance as observed in the optical microscope, bind congo red, and after repeated resuspensions in buffer exhibited a red shifted congo red absorbance. Soluble TTR and the aggregated form of TTR formed at pH 6 - 6.5 (not amyloid) do not bind congo red as determined by absence of the expected red shift observed in the congo red

UV spectrum and the lack of congo red absorption for the isolated aggregate suspension. (15, 16) Congo red appears to recognize the amyloid fibril structure specifically in that it binds to amyloid fibrils having the cross-β conformation, regardless of the protein composing the fibrils. The mechanism of congo red binding is not well characterized, however; preliminary results from our laboratory suggest that a hydrophobic interaction is most likely with specificity arising from shape complementarity.(15, 17)

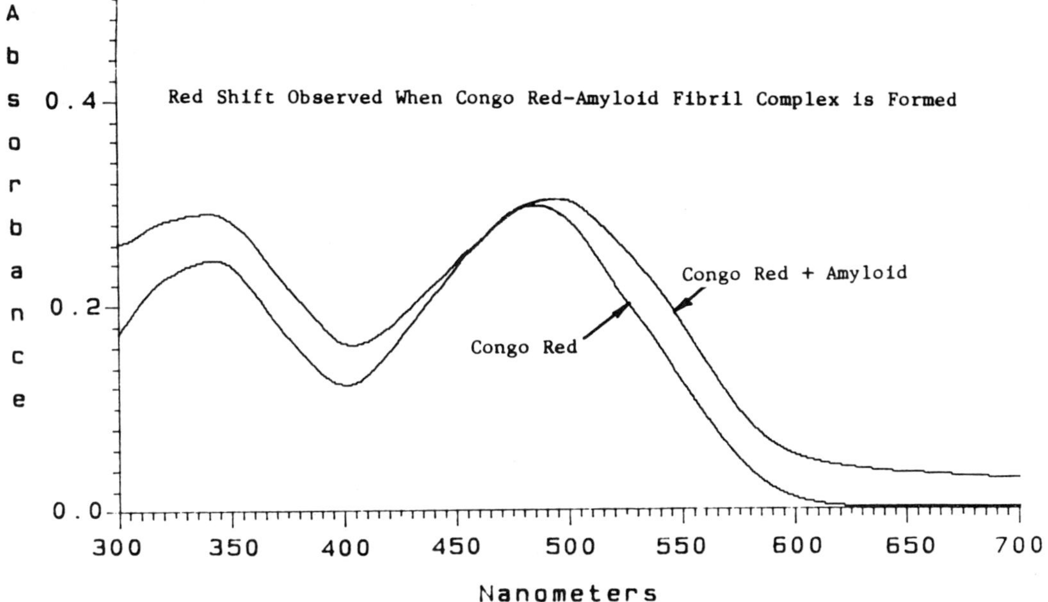

FIGURE 3. Congo Red Binding to Transthyretin Amyloid Fibrils

When viewed in a light microscope having crossed polarizers, amyloid fibrils exhibit a stunning apple green birefringence.(18, 19, 20) The birefringence is in striking contrast with the black background. Linear birefringence is caused by the difference in the refractive index of the fiber along different molecular axes. This birefringence will appear to rotate the plane of polarized light when the fiber is in the proper orientation with respect to the polarizers. When the TTR amyloid samples prepared as described above were rotated on the stage of a microscope, the color of the fibril changed a from stunning green to dark red or vice versa, depending on the orientation of the fiber. If low power magnification was used (10X), the green birefringence of various fibers disappeared and reappeared as the sample was reoriented with respect to the polarizing filters. We have compared our microscopic results to the high resolution color photographs of the green birefringence observed in Senile Systemic Amyloid patients (amyloidosis of the myocardium). The photograph of the birefringence observed from myocardium amyloid (made up of wild type TTR) samples is virtually identical to what we have seen in the case of TTR amyloid prepared by the denaturation method *in vitro*.(20) That the presence of tissue has little effect on the appearance of the fibrils has been noted previously. (21) Interestingly, the TTR fibrils produced by acid denaturation appear to undergo lateral aggregation

103

based on preliminary electron micrographs. The other *in vitro* models of amyloidosis, described above, do not produce laterally aggregated fibrils.

To demonstrate that our method for preparing amyloid is physiologically relevant and not protein dependent, we examined amyloid fibril formation from IgG2 λ light chain, which is known to be amyloidogenic and associated with primary amyloidosis in humans. The IgG2 λ light chain was transformed into amyloid fibrils under conditions identical to those described above for TTR. The conditions used in the insulin model of amyloidosis are not useful for converting the IgG2 λ light chain into amyloid fibrils. Evidence for the formation of amyloid fibrils includes congo red binding and birefringence observed in the polarizing microscope. X-ray diffraction and electron microscopy studies on the fibrils are in progress and will be reported later.

We have discovered that transthyretin amyloid fibril formation was almost completely inhibited by adding a detergent. Of the several detergents investigated, only N-tetradecyl-N,N-dimethyl-3-ammonio-1-propane sulfonate (Z 3-14) (1.0mg / ml; CMC = 0.14 mg / ml) proved to inhibit amyloid fibril formation without adversely affecting the reversible folding transition. This result suggests that the detergent micelle reversibly binds to the TTR intermediate competent to form amyloid fibrils, implying a significant hydrophobic interaction is required for amyloid fibril formation. Determining the mechanism by which micelles inhibit amyloid fibril requires further investigation.

DISCUSSION

We have demonstrated that denaturation plays an integral role in amyloid fibril formation *in vitro*. We propose that the partial denaturation of β-sheet amyloidogenic proteins in lysosomes that are either overloaded, protease deficient, or otherwise compromised, may result in amyloid fibril formation due to aggregation of the partially unfolded protein intermediates *in vivo*. That the immunoglobulin light chain amyloid has been found in macrophage lysosomes in the spleen suggests that this proposal is worthy of consideration.(22) In support of this hypothesis, Cohen has recently published strong evidence that macrophage lysosomes convert the SAA protein into AA amyloid.(12) The Bence Jones light chains also were converted into amyloid by treatment with kidney lysosomal extracts.(12) Recently, the Alzheimer's precursor protein was found in high concentrations in the lysosome. (23) Further, we report direct evidence here for the conversion of the plasma protein transthyretin and the immunoglobulin light chain (Ig G2 λ) into amyloid fibrils *in vitro* , under conditions which mimic the environment of a liver lysosome (pH 4.5).

The hypothesis that a partially folded intermediate serves as the precursor for amyloid formation is consistent with Glenner's demonstration that amyloid was produced *in vitro* from selected Bence Jones proteins (10mg / ml) by pepsin treatment at pH 3.5, 37°C.(11) The occurrence of proteolytic degradation has been clearly established in several systems, e.g. the conversion of the plasma protein SAA into AA amyloid, but it is unclear if this proteolytic processing is required for amyloid fibril formation. Proteolysis is clearly not important in the transthyretin based amyloid diseases FAP or SSA, as the predominant component of the TTR amyloid fibrils is full length TTR. However, varying amounts of proteolytic fragments are found in patients indicating proteolysis most likely occurs after fibril formation. Our data suggests that the presence of the acid environment is required for partially unfolding the β-sheet amyloidogenic precursors and dictates that further experiments are necessary concerning the role of proteolysis.

Several amyloid subunit proteins exist and are involved in different types systemic amyloidosis <u>Table 1</u>.(10) Since these precursor β-sheet sandwich proteins are transformed into fibrils which are of analogous structure according to the accepted criteria, including X-ray diffraction studies, it is likely that they utilize a common pathway to form amyloid fibrils. Based on these observations it is appropriate to use transthyretin as a model system for investigating the chemical mechanism of amyloid fibril formation.

Table 1. Amyloid Diseases and Their Component Protein

Type	Subunit Protein	Location of Amyloid
Primary Amyloidosis	Bence Jones IgG light chains	serum, urine, cerebrospinal fluid
Secondary Amyloidosis	SAA protein,circulates as part of the HDL3 complex	various extracellular locations
Senile Systemic Amyloidosis	transthyretin	heart and various other organs
Familial Amyloid Polneuropathies	variants of transthyretin	various organs and systems
Alzheimer's Disease	β-protein (A4)	Brain

Fink has shown that the acid induced denaturation of proteins is often incomplete, especially for those proteins having high β-sheet content.(24, 25, 26) Furthermore, the presence of salt facilitates anionic screening of the positive charges, minimizing charge-charge repulsion in the partly unfolded protein. This diminution in the effective charge between groups causes the protein to refold minimizing hydrophobic side chain exposure affording an unnatural structure having high secondary structure content called a "molten globule".(24, 25, 26) From pH 4 - 5, where TTR amyloid formation is fastest, approximately half of the Asp and Glu side chains are in their protonated form. The removal of the negative charge undoubtedly causes the protein to unfold due to charge repulsions between the remaining ammonium groups. Presumably, this change in conformation affords the putative folding intermediate which associates to form amyloid fibrils.

Jaenicke, King and others have shown that aggregation often results from the assembly of protein folding intermediates in an off-pathway process that competes with the normal reconstitution pathway. (27, 28, 29) King has isolated more than twenty temperature sensitive mutants of the P22 tailspike protein that aggregate at restrictive temperatures, yet are capable of folding at permissive temperatures.(30, 31) These mutations presumably increase the concentration of the folding intermediate capable of aggregation at the restrictive temperature. The mutation may induce this change in the folding pathway either by increasing the stability of the aggregation prone intermediate or by raising the activation barrier for the conversion of this intermediate to the desired product. Utilizing structural predictions, these mutations are thought to be in or near β-turn regions of the protein. Interestingly, the most common transthyretin single site mutations (amyloidogenic mutants) which cause Familial Amyloid Polyneuropathy are located at the ends of the β-strands near the β-turn regions <u>Table 2.</u>

The amyloidogenic TTR mutants are virtually identical to the wild type protein in regards to their tertiary and quaternary structure as well as their ability to bind thyroxine and retinol binding protein. However, the mutants appear to be more amyloidogenic than wild type protein as fibrils in FAP patients become a significant problem around the age of thirty. Preliminary kinetic evidence suggests that the amyloidogenic mutants form amyloid fibrils at least an order of magnitude faster than does the wild type protein under the same conditions. These results are termed preliminary because the laser desorption mass spectrum of wild type transthyretin, isolated from pooled human plasma, showed heterogeneity. The higher mass peaks largely disappeared after transthyretin was reduced, suggesting the presence of disulfides formed

Table 2. Amyloidogenic Mutations and Their Location in Figure 1

Amino Acid Substitution	Position	Location
Val to Met	30	end of β-strand B
Phe to Ile	33	in β-strand B
Thr to Ala	60	loop joining strands D & E
Ser to Tyr	77	loop joining strands E & F
Ile to Ser	84	loop joining strands E & F
Leu to Met	111	end of β-strand G
Val to Ile	122	end of β-strand H
Tyr to Val	116	in β-strand H
Thr to Gly	49	end of β-strand C

between Cys-10 and a variety of small SH containing molecules. Glutathione appeared to be present, along with Cys, and Cys-Gly. According to Petterson, nine components are released when TTR is reduced.(32, 33) We are currently characterizing all the covalently bound components by high resolution mass spectrometry.

The laser desorption mass spectrum for the wild type protein isolated from human plasma and the recombinant wild type protein are similar, but not identical . The correct mass for the wild type TTR monomer (free SH) is 13,762. This parent peak was observed in both the wild type and recombinant TTR spectra. The recombinant protein is more homogeneous than the plasma protein; however, the distribution of mixed disulfides is different. Once the disulfide distribution is understood we can be more confident concerning our initial kinetic data for the recombinant amyloidogenic FAP mutants.

Based on our preliminary data, the amyloidogenic TTR mutations appear to operate in much the same way as the tailspike temperature sensitive mutants in that the stability of the amyloidogenic precursor is increased, increasing its concentration and rate of fibril formation.

Preliminary TTR reconstitution experiments suggest the working model shown below. This is a working model which is consistent with our initial results, but most likely will change as more experimental data becomes available.

ACID INDUCED UNFOLDING / AGGREGATION MODEL

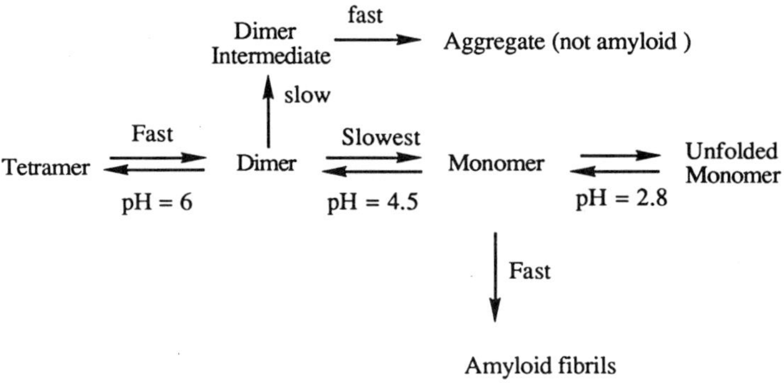

CONCLUSION

This study demonstrates that the mechanism of acid induced amyloid fibril formation *in vitro* requires the partial denaturation of transthyretin. These results suggest the participation of the lysosome in amyloid fibril formation *in vivo*. The role that the SH group plays in amyloidosis may be important and is currently being investigated by mass spectrometry.

Acknowledgments: Financial support from the Robert A. Welch Foundation and the National Institutes of Health Biomedical Research Support Grant Program is gratefully acknowledged.

REFERENCES

1. C. C. F. Blake, M. J. Geisow, I. D. A. Swan, C. Rerat, B. Rerat, *J. Mol. Biol.* 88, 1-12 (1974).

2. C. C. F. Blake, M. J. Geisow, S. J. Oatley, *J. Mol. Biol.* 121, 339-356 (1978).

3. P. P. V. Jaarsveld, H. Edelhoch, D. S. Goodman, J. Robbins, *J. Biol. Chem.* 248, 4698-4705 (1973).

4. S. F. Nilsson , L. Rask, P. A. Peterson, *J. Biol. Chem.* 250, 8554-8563 (1975).

5. A. Raz, T. Shiratori, D. S. Goodman, *J. Biol. Chem.* 245, 1903-1912 (1970).

6. D. S. Goodman, R. B. Leslie, *Biochim. Biophys. Acta.* 260, 670-678 (1972).

7. H. Furuya, *et al.*, *Biochemistry* 30, 2415-2421 (1991).

8.   M. D. Benson, *TIBS* 12, 88-92 (1989).

9.   A. S. Cohen, E. Calkins, *Nature* 183, 1202-1203 (1959).

10.   E. M. Castano, B. Frangione, *Laboratory Investigation* 58, 122-132 (1988).

11.   G. G. Glenner, *et al.*, *Science* 174, 712-714 (1971).

12.   A. S. Cohen, et.al., *Laboratory Investigation* 48, 1-4 (1983).

13.   M. J. Burke, M. A. Roughvie, *Biochemistry* 11, (1972).

14.   G. G. Glenner, E. D. Eanes, H. A. Bladen, R. P. Linke, J. D. Termine, *J. Histochem. Cytochem.* 22, 1141-1158 (1974).

15.   W. E. Klunk, J. W. Pettegrew, D. J. Abraham, *The Journal of Histochem. and Cytochem.* 37, 1273-81 (1989).

16.   H. Puchtler, F. Sweat, M. Levine, *J. Histochem. Cytochem.* 10, 355 (1962).

17.   G. G. Glenner, E. D. Eanes, D. L. Page, *J. Biochem. Cytochem.* 20, 821-826 (1972).

18.   R. D. Lillie *Histopathological Technic and Practical Histochemistry* pp 1 - 21. McGraw Hill NY (1976) 4th ed.

19.   H. R. Allcock, F. W. Lampe, *Contemporary Polymer Chemistry* (Prentice-Hall, Englewood Cliffs, New Jersey, 1981).

20.   C. Thomas, *Sandritters's Colr Atlas and Textbook of Histopathology (pp. 61)* (Year Book Medical Publishers, Inc., Chicago, 1979).

21.   R. A. Delellis, G. G. Glenner, J. Sri Ram, *J. Histochem. and Cytochem.* 16, 663-665 (1968).

22.   T. Shirahama, A. S. Cohen, *Am J. Path.* 81, 101 (1975).

23.   L. I. Benowitz, e. al., *Experimental Neurology* 105, 237-250 (1989).

24.   Y. Goto, L. J. Calciano, A. L. Fink, *Proc. Natl. Acad. Sci.* 87, 573-577 (1990).

25.   Y. Goto, N. Takahashi, A. L. Fink, *Biochemistry* 29, 3480-88 (1990).

26.   Y. Goto, A. L. Fink, *Biochemistry* 28, 945-952 (1989).

27.   R. Jaenicke, *Prog. Biophys. Molec. Biol.* 49, 117-237 (1987).

28.   R. Jaenicke, R. Rudolph, in *Methods in Enzymology* C. H. W. Hirs, S. N. Timasheff, Eds. (Academic Press, New York, 1986), pp. 218-250.

29.   G. Zettlmeissl, R. Rudolf, R. Jaenicke, *Biochemistry* 21, 3946-50 (1982).

30.   J. King, *Bio/Technology* 4, 297-303 (1986).

31.   R. Villafane, J. King, *J. Mol. Biol.* 204, 607-619 (1988).

32.   T. Pettersson, A. Carlstrom, H. Jornvall, *Biochemistry* 26, 4572-4583 (1987).

33.   T. Petterssom, A. Carlstrom, A. Ehrenberg, H. Jornvall, *Biochem. Biophys. Res. Comm.* 158, 341-347 (1989).

ISOLATION AND CHARACTERIZATION OF NATURAL

AND RECOMBINANT CYCLOPHILINS

T.F.Holzman[1], S.W.Fesik[1], C.Park[1], and J.L.Kofron[2]

[1] Drug Design and Delivery, Discovery Research,
Pharmaceutical Products Division, Abbott
Laboratories, Abbott Park, IL. 60064. [2]School of
Pharmacy, University of Wisconsin-Madison.

Introduction

Cyclophilins (Cyp's) are members of a class of proteins
which bind the pharmaceutically important immunosuppressive
agent cyclosporin A (CsA) and its analogs (1-5). In addition
to selectively binding CsA, these proteins have proven to be
enzymes which accelerate the *cis-trans* isomerization of
proline containing peptide bonds *in vitro* (8,9) and the rate
of protein folding *in vitro* (8-13). Recently. another member
of the peptidyl-prolyl isomerase family (PPIases) has been
identified, the FK binding proteins (FKBP's) (52,53). The
FKBP's have very little sequence homology with the Cyp's and
were initially identified as the binding proteins for the
potent immunosuppressant FK-506 and its analogs (52-54).
Although Cyp and FKBP are both PPIases, they apparently do not
have the same specificity for peptide substrates (16) and Cyp
will not bind FK-506 nor will FKBP bind CsA (52,53).

In a similar fashion, the *cis-trans* isomerization
activities of both Cyp and FKBP are inhibited by the binding
of their specific immunosuppressive ligands, CsA and FK-506.
However, a prokaryotic homolog of Cyp recently described in *E.
coli* does not exhibit significant inhibition by CsA in the
ligand concentration ranges normally observed to inhibit the
human or bovine enzymes (14). The comparative sequences of
some related Cyp's are presented in Figure 1. From the
alignments it is evident that the Cyp sequences from the
divergent sources are generally conserved. In particular, the
human and bovine sequences are highly conserved with
differences at only three positions. Although recent evidence
suggests the existence of a single message for human Cyp (40)
two isoelectric forms of the human splenocyte protein have
been described (2), one alkaline (pI> 9.0) the other neutral
(pI~7.4).

Both an alkaline and neutral isoform of human Cyp have
been cloned and expressed in *E. coli* (7,37). The alkaline
isoform was cloned from Jurkat cells (7) while the neutral

isoform (rhCyp) was cloned from a human renal adenocarcinoma cell line (37). Initial chemical modification data using p-hydroxymercuribenzoate (PHMB), a sulfhydryl modification reagent, suggested that Cyp isomerase activity was sulfhydryl dependent (10,15). Although the enzyme does require reducing agents like DTT or 2-mercaptoethanol for stability during isolation and storage and its activity is sensitive to metal ions (23) and PHMB (8,23), it was not inactivated by the less bulky sulfhdryl-specific reagents N-ethylmaleimide (NEM) (23) or methyl methanthiolsulfonate (MMTS) (41).

Based on comparison of the aligned sequences in Figure 1 the effects of the different chemical modification reagents on enzyme activity can be rationalized. Specifically, there are no strictly conserved sulfhydryl residues in the enzymes from these different organisms, suggesting no common conserved active-center enzyme sequence. Site-specific modification of the human enzyme Cys residues to Ala (7) did not eliminate activity, indicating these residues are not essential to

Human	.... Human Cyclophilin[29]	
Bovine	.... Bovine Cyclophilin[2]	
Yeast	.... Yeast PPIase[38]	
NC	.... Neurospora crassa PPIase[39]	
Human	MVNPTVFFDIAVDGEPL	17
Bovine	MVNPTVFFDIAVDGEPL	17
Yeast	MSQVYFDVEADGQPI	15
NC	SKVFFDLEWEG PVLGPNNK	19
Human	GRVSFELFADKVP	30
Bovine	GRVSFELFADKVP	30
Yeast	GRVVFKLYNDIVP	28
NC	PTSEIKAGRINFTLYDDVVPQS	41
Human	KTAENFRALSTGEKGFGYKGSCFHRI	56
Bovine	KTAENFRALSTGEKGFGYKGSCFHRI	56
Yeast	KTAENFRALCTGEKGFGYAGSPFHRV	54
NC	KTARNFKELCTGQNGFGYKGSSFHRI	67
Human	IPGFMCQGGDFTRHNGTGGKSIYGEKFED	85
Bovine	IPGFMCQGGDFTRHNGTGGKSIYGGKFDD	85
Yeast	IPDFMLQGGDFTAGNGTGGKSIYGGKFPD	83
NC	IPEFMLQGGDFTRGNGTGGKSIYGEKFAD	96
Human	ENFILKHTGPGILSMANAGPNTNGSQFFI	114
Bovine	ENFILKHTGPGILSMANAGPNTNGSQFFI	114
Yeast	ENFKKHHDRPGLLSMANAGPNTNGSQFFI	112
NC	ENFKKHHDRPGLLSMANAGPNTNGSQFFI	125
Human	CTAKTEWLDGKHVVFGKVKE GMNIVE	140
Bovine	CTAKTEWLDGKHVVFGKVKE GMNIVE	140
Yeast	TTVPCPWLDGKHVVFGEVVD GYDIVK	138
NC	TTVPTSWLDGRHVVFGEVADDESMKVVK	153
Human	AMERFGSRNG KTSKKITIADCGQLE	165
Bovine	AMERFGSRNG KTSKKITIADCGQI	164
Yeast	KVESLGSPSGA TKAR IVVAKSGEL	163
NC	ALEATGSSSGAIRYSKKPTIVDCGAL	180

Figure 1.   Comparative sequences of some Cyp's.   Potential N-linked glycosylation sites for human Cyp are underlined as is the site coding for potential C-terminal prenylation of bovine thymus Cyp and Trp[121] in the human enzyme.   The sequence differences between the human and bovine enzymes are double underlined.

catalysis. Additionally, steady-state kinetics, pH and temperature dependencies, and solvent and secondary deuterium isotope effects do not appear to support a simple Cys-related nucleophilic catalytic event (16). Thus, the PHMB sensitivity of Cyp (8) is likely due to distortion of the enzyme structure or steric hindrance of the enzyme active center.

At the present time there is little information on the structural details of the enzyme itself. Although predictions of secondary structure have very limited utility, the 165 residue human protein (1,2) is projected to have a high level of ß-sheet and little helical content (Fig. 2). One of the few probes of protein structure available is a single protein Trp[121] residue which undergoes a change in fluorescence quenching upon binding of CsA. This effect has been exploited to develop a useful non-isotopic binding assay for CsA to Cyp (1).

Given our interest in Cyp's as targets for the structure-based design of novel immunosuppressive agents we describe here some details of our recent studies of these enzymes (17,37). In particular, the methods employed in, and results from, purification and physical characterization of natural and recombinant Cyp's including: the kinetic assay for Cyp and its temperature dependence, sedimentation equilibrium analysis of rhCyp, isotope-edited NMR of [U-^{13}C^{9,10}]CsA bound to rhCyp, and crystallization of Cyp complexed to CsA. In addition, using 2-dimensional chromatographic analysis of natural Cyp's from bovine thymus, we present the first evidence for the existence of a family of CsA binding proteins which may be resolved and isolated using standard chromatographic techniques.

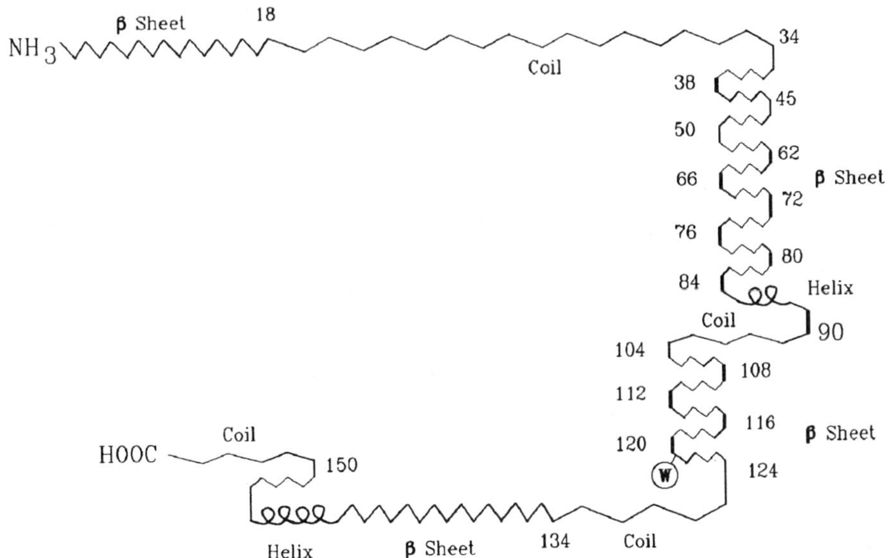

Figure 2. Schematic secondary structure of human Cyp predicted by Chou-Fasman analysis using the Wisconsin Genetics Computer Group Software package (19). The single Trp residue (#121) has been shown to interact with CsA (17).

Experimental Protocols

   Enzymes.   Natural bovine Cyp was prepared as previously described (17) (see Fig. 3).   The neutral isoform of human Cyp was cloned and expressed by Dr. R. Simmer and colleagues at Abbott Laboratories and purified as previously described (37) (Fig. 3).   Samples of protein were electrophoresed and stained with Coomaisse blue using standard SDS and IEF gel protocols from Pharmacia (37).   SDS gels were 8-25% gradient Pharmacia Phast gels, IEF gels were 5% IEF Pharmacia Phast gels.

   Enzyme Assays.   Enzyme assays were accomplished using the rapid manual mixing method developed by Fischer et al., (23) and stopped-flow measurements were accomplished as previously described (37) using a Hitech

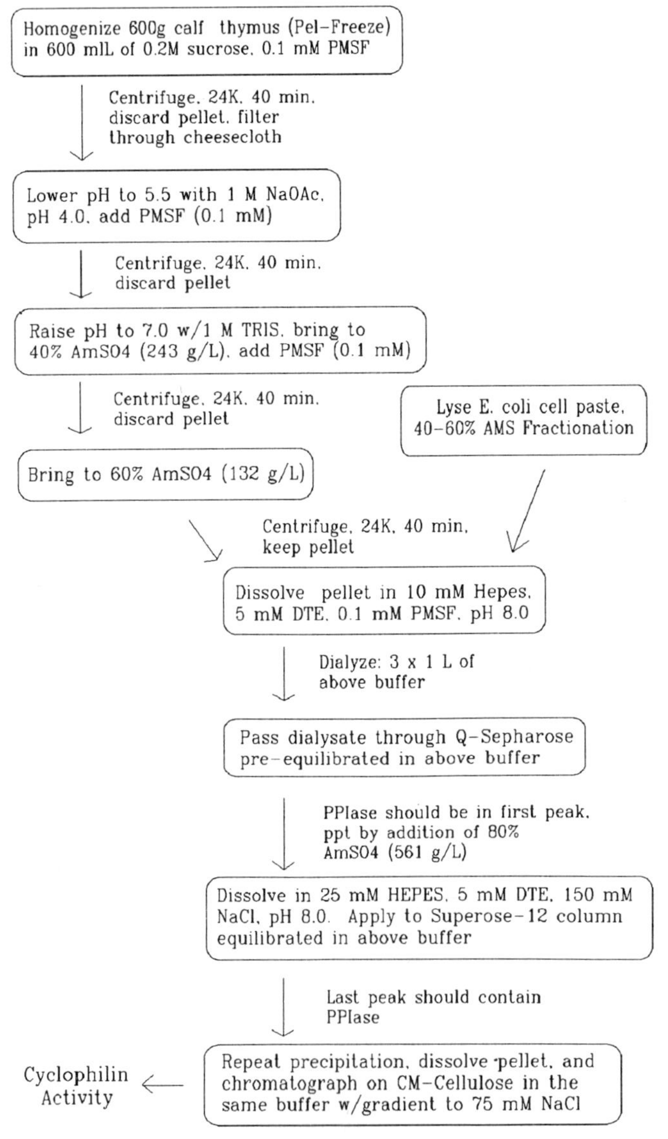

Figure 3.   Schematic diagram for Cyp purification.

model SF-51 instrument equipped with variable-ratio mixing, high intensity Xe/Hg source, and thermostated reaction cell. The system was interfaced to a Hewlett-Packard QS-20 computer using a Data Translation model 2801 high-speed A/D converter. The system dead-time is sub-millisecond and permits routine measurement of rates in excess of 2000^{-1}. Kinetic data were analyzed and rate constants were fitted using a Hitech kinetic analysis software package.

<u>Analytical Reverse-Phase HPLC</u>. Separations were accomplished with a Vydac C_4 10 μm, 300 A°, reverse phase column (4.6 x 300 mm) at a flow rate of 1.0 mL/min of 20% acetonitrile, 0.1% TFA on a Beckman HPLC system with a diode array detector. After isocratic development for five minutes a 1% gradient to 30% acetonitrile, 0.1% TFA was run followed by a separation gradient of 0.2%/min to 50% acetonitrile, 0.1% TFA after which the column was washed with 100% acetonitrile, 0.1% TFA and re-equilibrated to starting conditions. The elution of protein was detected by absorbance at 214 nm.

<u>Analytical Ultracentrifugation</u>. Sedimentation equilibrium analytical ultracentrifugation was accomplished using a Beckman Model E analytical ultracentrifuge equipped with absorption optics and photoelectric scanner. A PC-based data acquisition system interfaced to the photoelectric scanner was used to collect and analyze data (42).

<u>NMR Spectroscopy</u>. NMR samples were prepared by concentrating dilute solutions of both the rhCyp and natural bovine Cyp in Amicon centriprep cartridges (10,000 MWCO) followed by exchange into D_2O buffer (99.9%) with 50 mM potassium phosphate (pre-exchanged with D_2O), 100 mM NaCl, and 2 mM DTT at pD 6.5. Protein concentrations were estimated by absorbance at 280 nm after concentration. The $[U-^{13}C9,10]$CsA/Cyp complex was prepared as recently described (27) by gently rocking a mixture of a 1.5 molar excess of $[U-^{13}C9,10]$CsA with Cyp at 6°C for 12 hours under argon. The excess CsA was removed by centrifugation. Protein samples prepared in this fashion were found to be stable (as judged by unaltered spectral linewidths and intensities) for at least 2 months at 20°C. The NMR spectra for Figure 17 were acquired with a Bruker AM 500-MHz NMR spectrometer equipped with a 451 IF receiver and an inverse 5 mm probe. All ^{1}H NMR spectra were collected at 20°C using a spectral width of 7812.5 Hz in 32 transients using time-shared presaturation (43) of the residual H_2O signal for 1.2 sec between scans. The data were processed with 5 Hz exponential line broadening. The 2-D NOE data presented in Figure 18 were acquired on the same spectrometer as previously described (17).

Isolation of rhCyp from E. coli and Cyp's from Bovine Thymus

The Superose-12 size exclusion column profile of purified rhCyp from *E. coli* from the penultimate step in the purification procedure outlined in Figure 3 is presented in Figure 4. The buffer system was 5 mM Hepes, 5 mM DTT, 150 mM NaCl, pH 8.0. The column flow rate was 0.95 mL/min with fraction volumes of 19 mL. Enzyme activity was observed to elute in a single peak. The peak was identified by both PPIase activity and CsA binding assays and is marked with a horizontal bar. The figure inset presents SDS and IEF gels of fraction #56. Based on the stained IEF gel bands the fraction is a single species with a pI of ~7.8 (37). SDS gel electrophoresis indicated the protein was essentially homogeneous after this chromatographic step. As described previously (37) the rhCyp cloned in *E. coli* is well tolerated and is conservatively estimated to yield ~10 mg/L of purified protein. Samples of rhCyp from this step in the procedure are

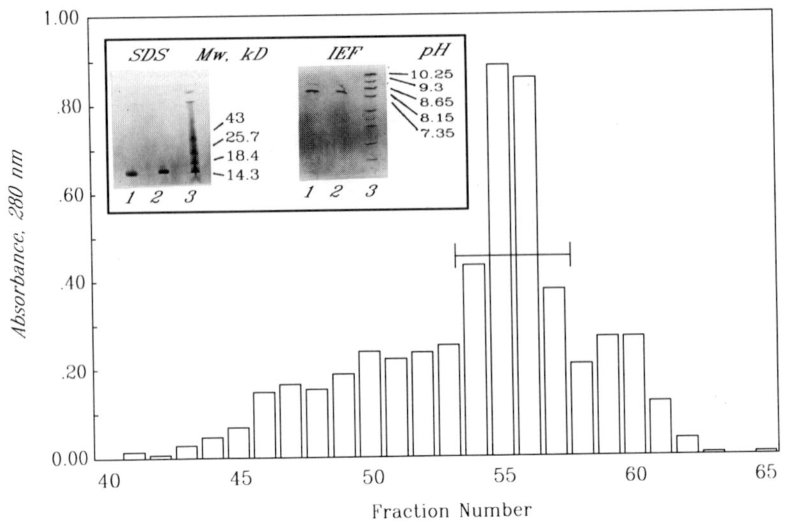

Figure 4.   Chromatography of rhCyp on Superose 12   (from 37).

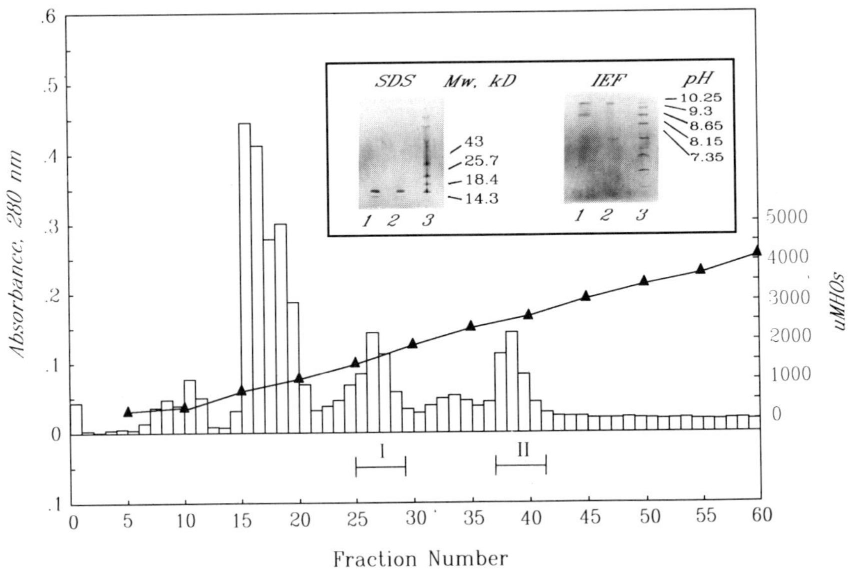

Figure 5.   Chromatography of Cyp activity from bovine thymus on CM-cellulose (from 37).

typically used for kinetic analyses, ligand binding studies,
tryptic mapping, N-terminal sequencing, amino acid analysis,
ultracentrifugation, NMR spectroscopy, etc.   However, for
samples of protein suitable for crystallization it may be
necessary to include the final CM-cellulose chromatography
step outlined in Figure 3 (41).   Recombinant human Cyp which

has been purified using this last step has been crystallized at Abbott Laboratories (below) (51).

The pI value observed for the cloned human enzyme (pI~7.8) was clearly lower than the value reported previously (pI >9.0) for the cloned protein from Jurkat cells (7). The pI of the rhCyp that we describe (37) is more similar to that of the pI 7.4 neutral isoform from human splenocytes (2). The first thirty residues of the N-terminal sequence of the neutral rhCyp isoform were sequenced and gave the expected human sequence (Fig. 2) (2,29) consistent with the sequence deduced from cDNA with no evidence of N-terminal heterogeneity (37). Analysis of CsA binding to the rhCyp described here indicates the binding process fits a simple model for a 1:1 complex and yields a dissociation constant of ~360 nM (37). Although this values is higher than has been reported for the alkaline isoform cloned from Jurkat cells (7), the binding data for the alkaline isoform does not appear to fit a simple binding process (37).

In Figure 5 a sample containing the thymus Cyp's was first subjected to anion exchange and size exclusion chromatographies (see Fig. 3) (37) and was applied to a 3x45 cm CM-column in 5 mM Hepes, 5 mM DTT, pH 8.0. The column flow rate was 0.6 mL/min with a gradient to 75 mM NaCl (closed triangles, conductivity measurements). Fraction volumes were 12 ml. Enzyme activity was observed to elute reproducibly in two distinct peaks separated by >60 mL. The peaks of enzymatic activity (marked I and II) were identified by both PPIase activity and CsA binding assays. The figure inset presents SDS and IEF gels of pooled fractions of Peaks I and II. Details of electrophoretic separations are described under Materials and Methods. SDS gel lanes: 1) peak II; 2) peak I; 3) M_W standards. IEF analysis of peak I indicated a single stained band with a pI in the range of 9.3. Based on the stained IEF gel bands we estimate peak II is separated into two species staining with almost equal intensity, one with a pI~9.3 and another more acidic form with a pI~7.8 (37).

Based on IEF gel analysis, the chromatographic profile of the two major peaks of bovine thymus Cyp is unexpected (37). Two observations are noteworthy. First, both peak I and peak II Cyp's are apparently composed of a species with an isoelectric point of ~9.3. Second, the more acidic band found in peak II elutes *after* the basic band from peak I in the salt gradient. Since the pH of the ion-exchange chromatography is only slightly *above* the isoelectric point of the acidic band of peak II, its elution more than 60 mL *after* the basic band of peak I is unusual. This behavior suggests that peak II may represent an associated complex with a pI sufficiently high to permit separation from the peak I protein or that there may be some unusual non-ionic interactions between both the peak II isoforms and the chromatographic support.

In order to further explore this peculiar chromatographic behavior, the CM-cellulose isolation of another preparation of bovine thymus Cyp's was analyzed for CsA binding across the entire chromatographic profile (Fig. 6). The results indicated the presence of multiple peaks of CsA binding activity. The two previously identified peaks of Cyp were observed to bind CsA, and peak II (Fraction ~55) was clearly

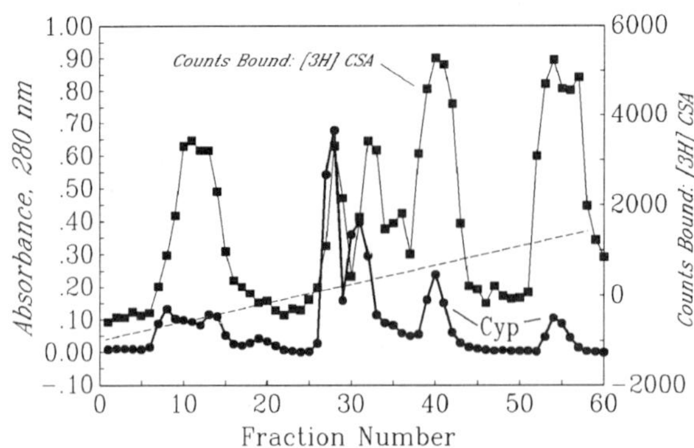

Figure 6. Chromatography of bovine Cyp's on CM-cellulose with the complete profile of ^{3}H-CsA binding activity. Elution conditions are equivalent to those described in Figure 5.

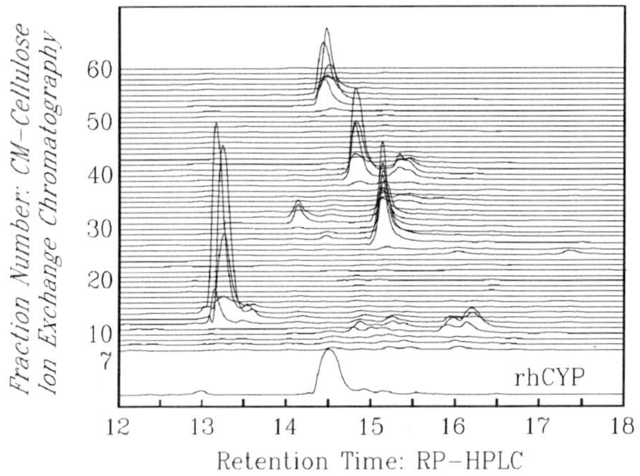

Figure 7. Two-dimensional RP-HPLC analysis of column fractions from Figure 6. HPLC separations were monitored at 214 nm and isolation conditions were similar to those described previously (37).

split into a doublet based on CsA binding. In addition, a number of other peaks of CsA binding activity were observed (Fig. 6). In an attempt to further resolve the peaks of CsA binding activity, all the fractions from the CM-cellulose of Figure 6 were subjected to RP-HPLC chromatography (Fig. 7). The resulting HPLC absorbance profiles were then sliced along the CM-cellulose isolation dimension at the HPLC retention

times corresponding to the peak I and II CM-cellulose
fractions (Fig. 8). The retention profile of rhCyp is
presented for comparison (Fig. 8).

These second dimension slices through the HPLC domain
along the CM-cellulose domain demonstrate that the peak II Cyp
is composed of two closely eluting species (Fig. 8, upper
panel) and that peak I Cyp has a single major form eluting at
a retention time of ~14.8 min. However, it is evident from
the 2D plot (Fig. 7) that the peak I Cyp (~fraction 40 in CM-
cellulose domain) also contains additional species eluting
with retention times between ~15.2 and ~15.6 minutes along the
HPLC domain. The vertical arrows in Figure 8 indicate the
absorbance peaks in HPLC domain which appear to correspond
with the CsA binding activity profile overlaid from Figure 6.

We note, however, that due to the inability to measure
accurate CsA binding activity in the HPLC-resolved profiles,
some of the additional absorbance peaks observed in the 2D
plot (Fig. 6) (notably those with Rt's at ~13.2, 14.2, and
15.2 min) may also correspond to species binding CsA. Thus
for the CM-cellulose column, the CsA binding data alone
indicates the presence of at least seven different binding
components. Taken together these data suggest that the Cyp's

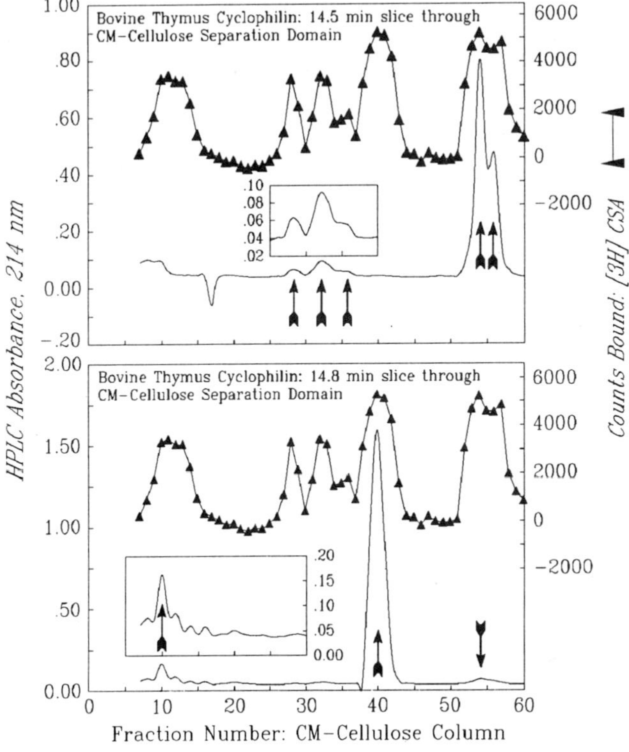

Figure 8. Second-dimension (HPLC) resolved bovine thymus Cyp's at
14.5 and 14.8 minute slices through the CM-cellulose separation
domain. The profile of CsA binding (Fig. 6) is overlaid for
comparison.

from bovine thymus can be isolated as a family of (chromatographically) related proteins.

Isomerase Reaction, Enzyme Assay, and Sedimentation Analysis

The peptidyl-prolyl isomerization reaction catalyzed by Cyp is illustrated in Figure 9 (upper panel). To date, no specific natural substrates acted upon by Cyp have been identified. It is attractive to propose that Cyp's and other PPIases are the biological response to the proline hypothesis (55). This hypothesis proposes that certain slow steps in the proper folding of proteins are the result of the need for X-pro bonds to isomerize to the appropriate cis or trans configuration.

Presently, it is not known if Cyp functions *in vivo* during protein folding to accelerate *cis-trans* isomerizations. However, the enzyme has been shown to be capable of accelerating proper folding *in vitro*, for certain protein folding reactions which are dependent upon *cis-trans* isomerization (11,12). One possibility, related to a mechanism initially discussed for Cyp (8), is outlined in

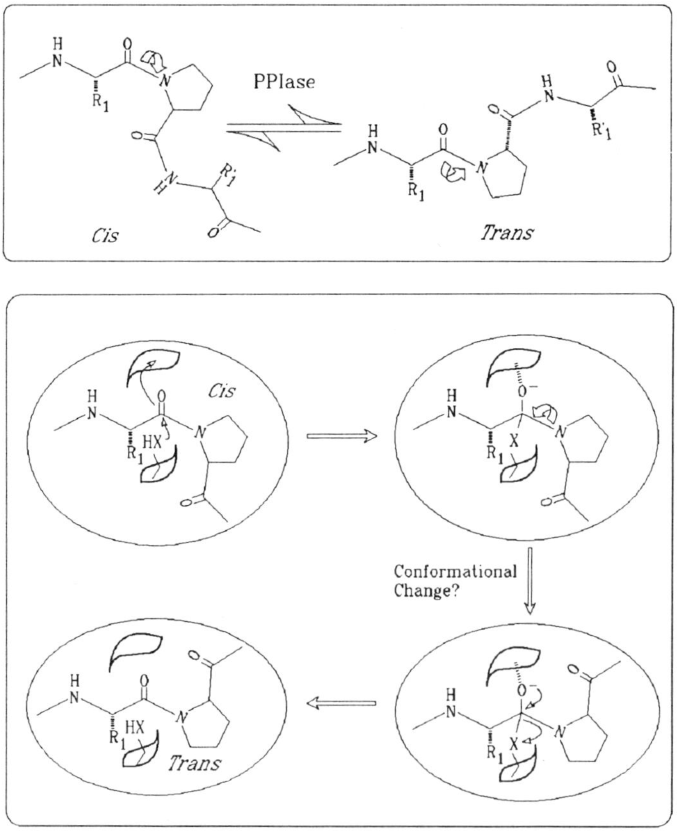

Figure 9. The peptidyl-prolyl isomerase reaction and one possible mechanism for *cis-trans* isomerization by Cyp.

Figure 9 (lower panel). In this hypothetical example the peptide (protein) is initially bound in the cis conformation. Upon binding, a nucleophilic side-chain from the protein, denoted as HX (the deprotonated form is shown reacting), attacks the peptide carbonyl. The carbonyl oxygen is somehow stabilized by another protein side-chain and/or possibly protonated. As the carbonyl looses its double bond character, a quasi-stable tetrahedral intermediate forms in which the prolyl residue can rotate into a trans conformation. For larger protein and peptide substrates, it is not unreasonable to infer that rotation may be accompanied by a more global conformational change in the enzyme. After rotation, the tetrahedral intermediate collapses, producing the final trans peptide configuration. In the initial description of the sensitivity of Cyp to PHMB (8), an enzyme bound intermediate was proposed in which the thiol group of a protein Cys residue acted as a nucleophile during catalysis to produce a hemithioorthoamide linkage to the prolyl peptide moiety. Based on the non-essential nature of the protein thiols (described above) the involvement of a protein Cys in this process is clearly ruled out. In addition, the absence of solvent and secondary deuterium isotope effects in catalysis, the pH independence of the reaction over a broad range, and the activation energy parameters exibited by Cyp, also appear to rule out a classical enzyme-bound tetrahedral intermediate mechanism in which an enzyme residue serves as a nucleophile (16). The present evidence suggests that Cyp binds and stabilizes a form of the substrate in which a partial rotation has occured about the X-Pro C-N amide (16). It is argued that the energy required to destroy the resonance stabilization of the amide is provided by binding of the substrate in a partially rotated conformation (16).

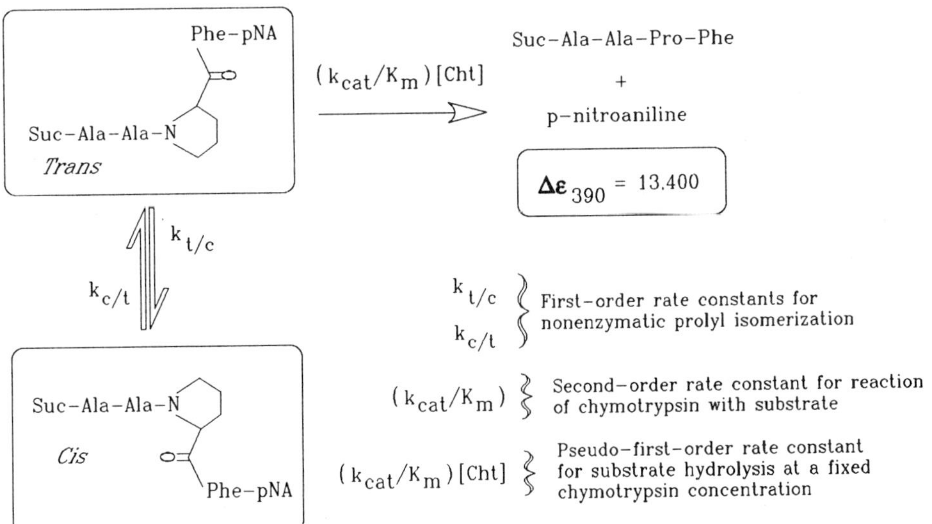

Figure 10. Overview of the coupled enzymatic assay for the peptidyl-prolyl isomerase Cyp.

A schematic diagram of the current coupled enzymatic
assay for Cyp (23) is presented in Figure 10. The assay
relies on two factors. First, the cis and trans forms of the
substrate N-succinyl-Ala-Ala-Pro-Phe-pNA are in equilibrium in
solution with a ratio of ~89% trans:~11% cis (56). Second,
the isomerization reaction is coupled to the selective
cleavage by chymotrypsin of the trans isomer. Cleavage yields
the tetrapeptide and p-nitroaniline and is readily monitored
by the increase in absorbance at 390 nm (23,56). A
fluorescence version of this assay has also been described in
which the substrate has a methylcoumaryl amide substituted for
the pNA derivative, and the reaction is monitored by the
increase in fluorescence at a wavelength above 430 nm upon the
generation of methylcoumaryl amine (11). Thus in the absence
of Cyp, addition of chymotrypsin to the substrate induces
hydrolysis of the trans conformer, leaving the cis conformer
uncleaved until it spontaneously isomerizes to a trans
conformer. An example of the absorbance-based assay is
presented in Figure 11. In the presence of sufficiently high
concentrations of chymotrypsin, a burst phase of signal
generation is observed (Fig. 11, upper panel). The kinetic
profile in the absence of Cyp is noted as "thermal".

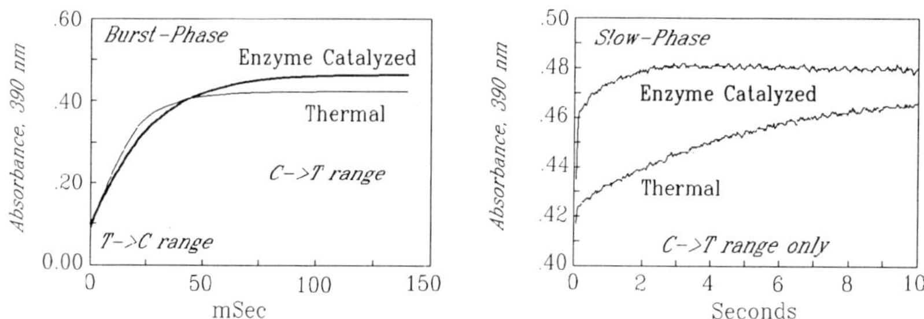

Figure 11. Typical stopped-flow burst-phase and slow-phase
kinetics observed in the coupled kinetic assay for Cyp using the
Nsuc-AAPFpNA substrate. Assay conditions: 5 mg/ml α-chymotrypsin
(Worthington) with or without 30 nM rhCyp, 40 μM substrate, pH 8.0,
37°C (37).

Once the trans isomer is exhausted, further conversion of
substrate by chymotrypsin is rate-limited by the thermal rate
of spontaneous isomerization of the remaining cis isomer to
the trans form (Fig 11, upper panel: C->T range). As
expected, in the presence of Cyp the burst-phase rate is
actually *decelerated* as Cyp catalyzes the conversion from the
trans isomer to the cis isomer (Fig. 11 upper panel: T->C
range) while the *cis->trans* conversion is accelerated (Fig 11,

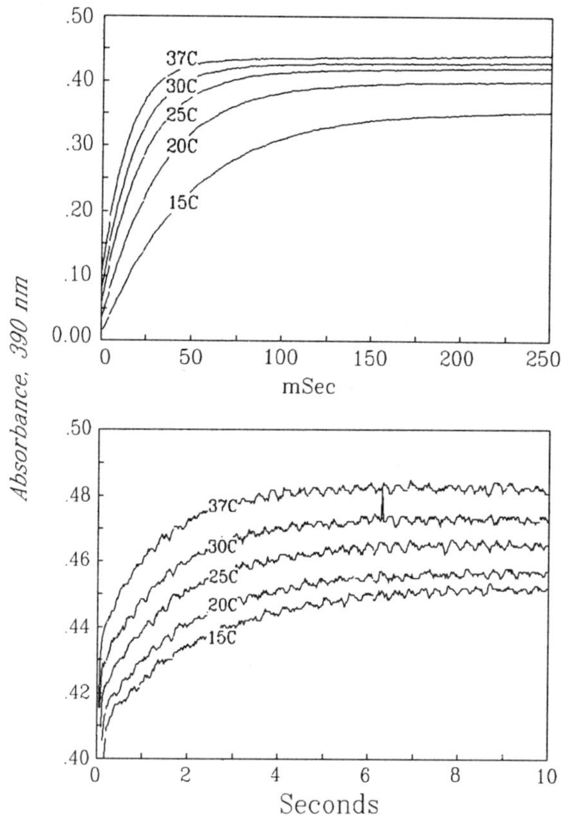

Figure 12.   Effect of temperature on the time-course of the coupled isomerase assay.   Upper panel: rhCyp catalyzed *trans-cis* isomerization during the burst-phase reaction with chymotrypsin cleavage of the *trans* substrate; Lower panel: rhCyp catalyzed *cis-trans* isomerization during the slow-phase reaction with chymotrypsin cleavage of the *trans* substrate.

upper panel: C->T range).   After the burst phase is complete the reaction time course enters a slow-phase in which the *cis->trans* isomerization process predominates (Fig. 11, lower panel).   In this slow phase the effect of Cyp on the acceleration of isomerization is readily evident.   The effect of temperature on the Cyp catalyzed burst and slow phases are presented in Figure 12.   The activation enthalpies and entropies for the thermal and Cyp catalyzed slow-phase *cis->trans* isomerization are presented in Figure 13.   In this plot k is the pseudo-first-order rate constant for *cis->trans* isomerzation, k* is Boltzmann's constant, h is Plank's constant and temperature is in °K.   The slope and intercept for the nonenzymatic reaction yield activation enthalpy and entropy values of 19.3 kcal/mole and 0.37 cal/°K-mole, respectively.

Using standard state conditions of 33 nM enzyme, the enzyme catalyzed reaction was found to give an activation enthalpy of +3.69 kcal/mole and an activation entropy of -47.3

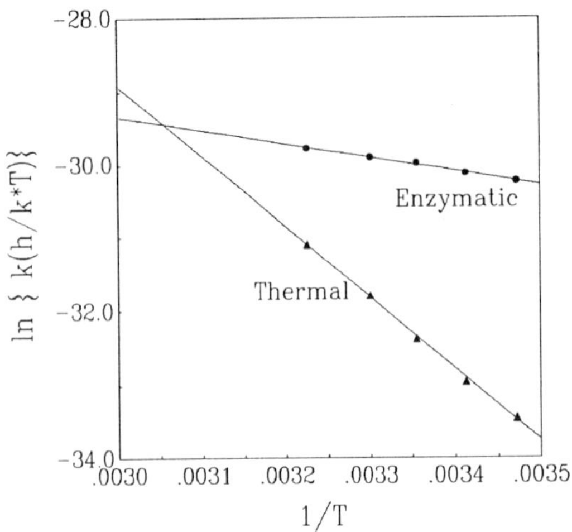

Figure 13. Eyring plot for activation enthalpies and entropies for thermal and rhCyp catalyzed *cis->trans* isomerization (from 37). Assays were performed by stopped-flow as previously described (37).

cal/°K-mole. A previous examination of the energetics of a cyclophilin from bovine thymus indicated an activation enthalpy of +3.87 kcal/mole and an activation entropy of -47.3 cal/°K-mole (16). The values for rhCyp are essentially identical to those observed for natural bovine Cyp (37) and suggest heterologous expression of human Cyp in *E. coli* results in an enzyme with normal catalytic energetics.

In order to establish the hydrodyamic form of Cyp, samples were subjected to analysis by sedimentation equilibrium analytical ultracentrifugation. Samples were prepared in 10 mM HEPES, ph 7.9, with 50 mM NaCl, 0.1 mM DTT

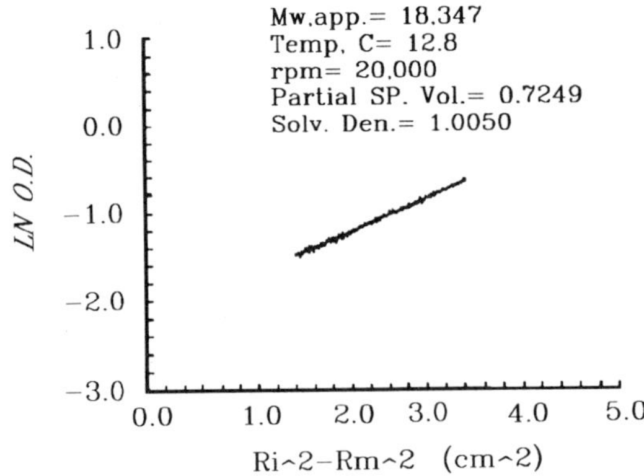

Figure 14. Analytical ultracentrifugation of rhCyp at ~0.1 mg/mL. The rhCyp partial specific volume was estimated from its amino acid composition and the solvent density was determined using an ultrasonic densitometer.

and placed in cells equipped with Yphantis multichannel equilibrium centerpieces. The samples were then subjected to an overspeed run at 36,000 rpm for ~six hours and then were sedimented to equilibrium at 20,000 rpm for 24hr. The samples were maintained at 12.8 °C during centrifugation. After equilibrium was established the samples were scanned using the photoelectric scanner and cell multiplex unit, and data were analyzed and plotted. A representative sedimentantion equilibrium profile for rhCyp is presented in Figure 14. The plot includes both the observed data and a least-squares linear fit of the data. The absence of curvature indicates the absence of sample polydispersity. The observed weight-average molecular weight (Fig. 14) is within about 2.5% of the molecular weight calculated based on amino acid composition of 17,888. Therefore, we conclude that in the absence of substrate the enzyme is a monomer with no discernable self-association at the concentrations examined.

Crystallization of Cyp's

For crystallization of bovine Cyp (peak I, Fig. 5), a sample of protein was dialyzed overnight against 20 mM Tris, pH 7.5, 5 mM 2-mercaptoethanol. About a 10-fold molar excess of CsA was dissolved in a minimum volume of acetonitrile and added to the dialyzed protein solution. The resulting CsA precipitate was removed by filtration through a 0.22 μm filter. The Cyp:CsA complex was concentrated by ultrafiltration to several different protein concentrations up to ~1 mg/mL. The precipitant solutions were seven different concentrations of $(NH_4)_2SO_4$ in 0.1M Tris, pH 7.5 spanning the

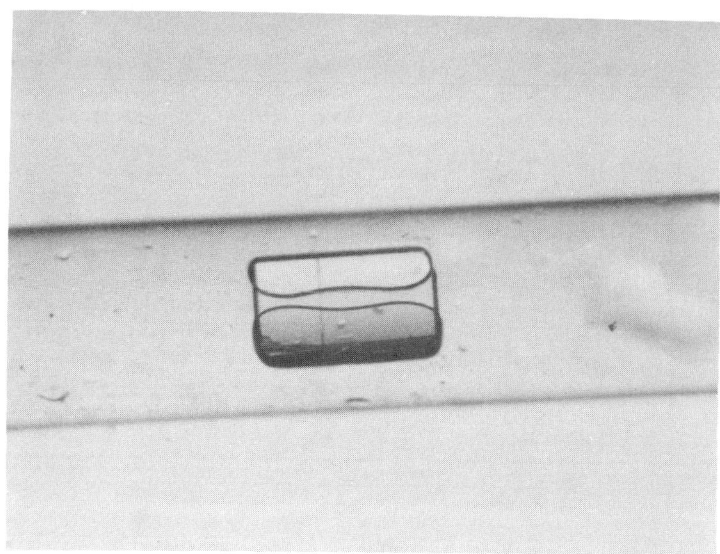

Figure 15. A Single tetragonal crystal of rhCyp complexed with CsA grown in the presence of 1% 2-methyl-2,4-pentandiol. Using this crystal a 3.5 °A resolution native data set was collected at the Midwest Area Detector facility in the Argonne National Laboratory.

range from 55% saturation to 65% saturation with intervals of 2.5%. Crystallizations were performed using the hanging drop vapor diffusion method with Linbro tissue culture plates. Hanging drops were prepared on each cover slip by mixing 5 μL of protein solution and 5 μL of precipitant solution. The volume of the precipitant solution in each well was 1 mL. Protein crystals appeared within one week of incubation at room temperature. The thickness of the plate crystals was less than 0.01 mm and the other two dimensions were typically 0.2 x 0.1 mm. Drops with precipitant concentrations higher than 65% saturation gave rise to rosettes of crystal plates. Variations in protein concentration did not give appreciable differences in crystallization. Due to the small sizes of the crystals produced by this method, no further characterization was attempted.

The conditions developed for bovine Cyp crystallization were used as a starting point for the crystallization of rhCyp. Hexagonal-bipyramidal crystals (0.2 x 0.2 x 0.4 mm) were obtained when the rhCyp was 3 mg/mL and the precipitant solution was 52.5% saturation $(NH_4)_2SO_4$ in 1.0 M Tris, pH 7.3. The space group of these crystals was $P6_1$ 2 2 (or its enantiomer) and the unit cell dimensions were: a=b= 94.6 °A, c= 432.0 °A. When 1.5% of 2-methyl-2,4-pentandiol or 2,5-hexanediol was added in the precipitant well solution, rectangular crystals of a tetragonal system could be grown (Fig. 15). The space group was $P4_1$ 2_1 2 (or its enantiomer) and the cell dimensions were a=b= 94.2 °A, c= 276.9 °A. Addition of 10% 2-pentanediol to the well solution also produced the same space group and the same cell dimensions. Even though the crystal system changed from hexagonal to tetragonal, the volume of the asymmetric unit in the unit cell remained about the same (2.8 x 10^5 °A^3 for hexagonal and 3.1 x 10^5 °A^3 for tetragonal). At present we consider the tetragonal crystal form to be a useful starting point for three-dimensional structural analysis of the Cyp:CsA complex.

Proton NMR and Isotope-Edited NMR Studies of Bovine Cyp and rhCyp and the CsA:Cyp Complex

Proton NMR spectra of natural bovine Cyp and rhCyp are presented in Figure 16A,B. As expected, due to the few differences in sequence between the bovine and human proteins, the proton NMR spectrum of rhCyp and natural bovine Cyp are very similar (Fig. 16). Each of the spectra contain proton resonances with similar line widths (~23 Hz for methyl protons at 293°K) and display characteristic upfield shifted resonances. The similarity of these spectra to that previously published (36) indicate that the purification procedures employed here yield protein comparable by proton NMR to the bovine protein isolated by more lengthy methods (2).

Upon addition of CsA to bovine Cyp some of the protein signals were shown to undergo a characteristic change in chemical shift (27,36). As shown in Figure 16C, the NMR spectrum indicates that CsA binds to the protein, and the changes that occur in the spectrum upon binding to the human protein are very similar to those observed previously (36) in

NMR spectra of the bovine Cyp/CsA complex. Taken together, these data indicated that the recombinant protein prepared as described was an excellent candidate for the study of the three-dimensional structure of the CsA/Cyp complex using isotope-edited 2-dimensional NMR methods (44-47) and heteronuclear 3-dimensional NMR techniques (48-50).

In order to simplify the complicated proton NMR spectrum of the CsA:Cyp complex, isotope-edited NMR experiments were conducted (17) using a CsA analog $[U-^{13}C^{9,10}]$CsA, in which the MeLeu residues in the 9 and 10 positions were uniformly labeled (>95%) with ^{13}C. An isotope-edited 2D NOE spectrum of $[U-^{13}C^{9,10}]$CsA bound to rhCyp is shown in Figure 17. In the $omega_1$ dimension, only those protons attached to the ^{13}C-labeled nuclei of CsA are detected. In the $omega_2$ dimension, NOE cross-peaks between these "labeled" protons and other nearby protons of CsA and Cyp are observed. The proton NMR assignments of the $[^{13}C]$CsA residues, made from analysis of ^{1}H-^{13}C correlation experiments (17), are given to the left of the spectrum. In the isotope-edited NMR spectrum (Fig. 18) a large NOE was observed between the MeLeu^{10} NCH$_3$ and the MeLeu^{9} H$^{\alpha}$ protons of CsA. This NOE and the lack of an NOE between the H$^{\alpha}$ protons of CsA residues 9 and 10 indicates that CsA adopts a trans 9,10 amide bond when bound to rhCyp, in contrast to the cis 9,10 amide bond found in the crystalline and solution conformations of CsA (17).

In addition to NOEs between CsA protons (which define the bound conformation), NOEs were observed between MeLeu^{9} of CsA and the zeta$_3$ and epsilon$_3$ protons of Trp^{121} and Phe protons of Cyp. These NOE's indicate the close proximity of these

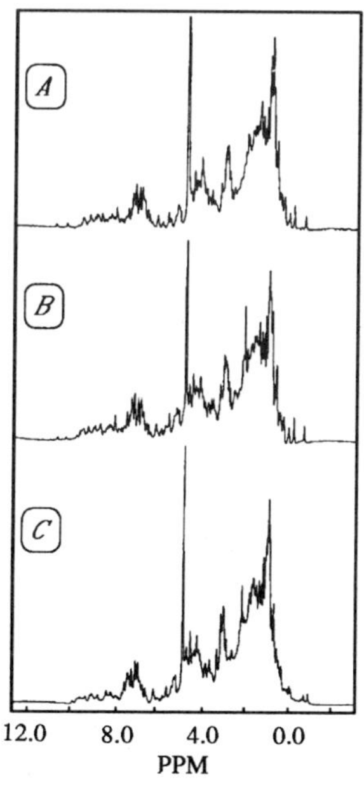

Figure 16. 500 MHz proton NMR spectra of A) natural bovine Cyp (peak II, CM-cellulose column, Fig. 10), B) rhCyp, and C) rhCyp at 1.3 mM complexed with $[U-^{13}C^{9,10}]$CsA.

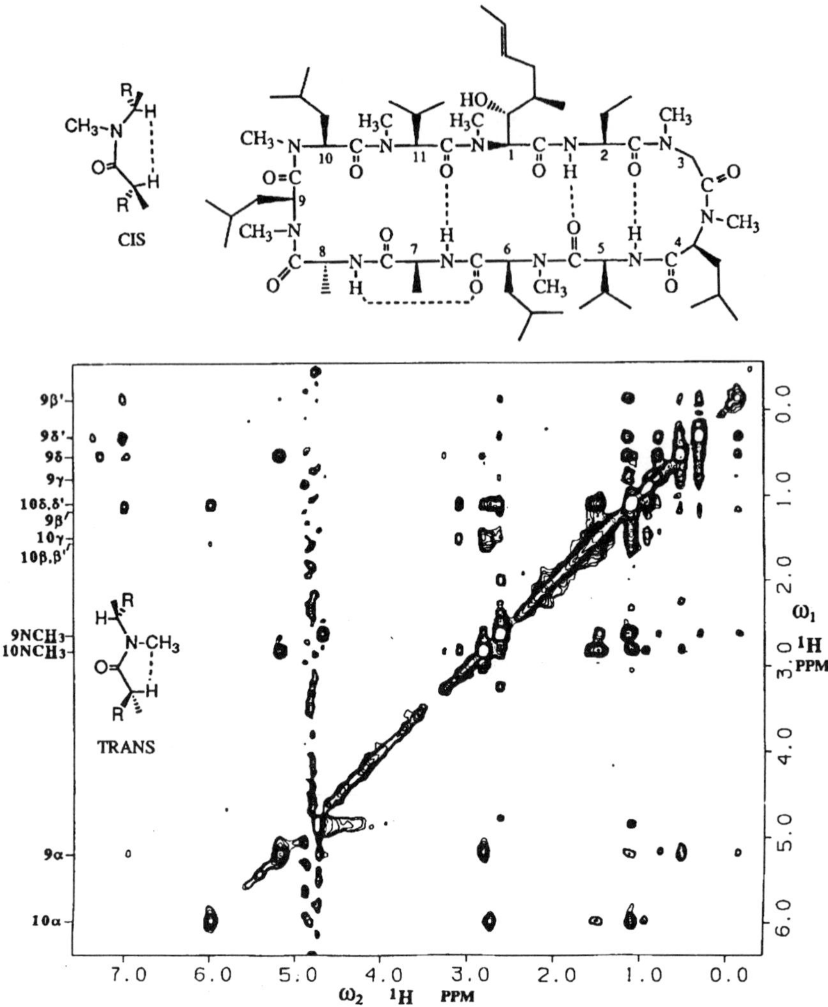

Figure 17.   Isotope-edited 2-D NOE contour plot of a 1:1 complex of rhCyp and [U-^{13}C-MeLeu^{9,10}]CsA (from 17).   The sample was prepared and analyzed as described under Materials and Methods.   The MeLeu^9 and MeLeu^{10} proton assignments are presented on the left of the contour plot.

residues in the CsA:Cyp complex.   More recently. we have applied a variety of multi-dimensional NMR methods in the study of [U-^{13}C]CsA bound to cyclophilin (58).   The ^{1}H and ^{13}C NMR signals of CsA in the bound state were assigned, and the complete bound conformation of CsA was determined from an analysis of 3D NOE data.   In addition, from CsA-Cyp NOE's we have identified those portions of CsA which interact with Cyp during binding (58).

Acknowledgements

We thank Dr. Thomas J. Perun and Dr. George W. Carter for their support of this work, Dr. Neal S. Burres for binding assays, R. Gampe for NMR spectra, D. Egan and R. Edalji for protein characterizations and Dr. J. Luly and Drs. D. Rich and V. Kishore for providing the $[U-^{13}C^{9,10}]$CsA.

References

1) Handschumacher,R.E., Harding,M.W., Rice,J., Drugge, R.J., & Speicher, D.W. (1984) Science 226, 544-547.
2) Harding,M.W., Handschumacher,R.E., & Speicher,D.W. (1986) J.Biol.Chem. 261, 8547-8555.
3) Stiller,C.R., & Keown,P.A. (1984) Progress in Transplantation, Morris, P.J. ed., 1, 11.
4) Wenger,R.M. (1985) Angew. Chem. 24, 77-85.
5) Wenger,R.M., Payne,T.G. & Schreier,M.H. (1986) Prog. Clin. Biochem. Med. 3, 157-191.
6) Emmel,E.A., Verweij,C.L., Durand,D.B., Hoggins,K.M., Lacey,E., & Crabtree,G.R. Science (1990) 246, 1617-1620.
7) Liu,J., Abers,M.W., Chen,C.M., Schrieber,S.L., Walsh,C.T. (1990) Proc. Natl. Acad. Sci. USA 87, 2304-2308.
8) Fischer,G., Wittman-Liebold,B., Lang,K., Kiefhaber,T., & Schmid,F.X. (1989) Nature 337, 476-478.
9) Takahashi,N., Hayano,T., & Suzuki,M (1989) Nature 337, 473-475.
10) Fischer,G., & Bang,H. (1985) Biochem. Biophys. Acta 828, 39-42.
11) Bächinger,H.P. (1987) J. Biol. Chem. (1987) 262, 17144-17148.
12) Lang,K. Schmid,F.X., & Fischer,G. (1987) Nature 329, 268-270.
13) Lang,K., & Schmid,F.X. (1988) Nature 331, 453-455.
14) Liu,J., & Walsh,C.T. (1990) Proc. Natl. Acad. Sci. USA 87, 4028-4032.
15) Fischer,G., Berger,E., & Bang,H. (1989) FEBS ,Lett. 250, 267-270.
16) Harrison,R.K., & Stein,R.L. (1990) Biochemistry 29, 1684-1689.
17) Fesik,S.W., Gampe,R.T., Holzman,T.F., Egan,D.A., Edalji,R., Luly,J., Simmer,R. Helfrich,R., Kishore,V., & Rich,D.H. (1990) Science 250, 1406-1409.
18) Maniatis,T., Fritsch,E.F., & Sambrook,J. (1982) Molecular Cloning: A Laboratory Manual (Cold Spring Harbor Lab., Cold Spring Harbor, N.Y.).
19) Devereux,J., Haeberli,P., & Smithies,O. (1984) Nuc. Acids Res. 12, 387-395.
20) Saiki,R.K., Scharf,S., Faloona,F., Mullis,K.B., Horn, G.T., Erlich.H.A., & Arnheim,N. (1985) Science 230, 1350-1354.
21) Faloona,F., & Mullis,K.B. (1987) Meth. Enzymol, 155, 335-350.
22) McFarland,J. (1907) J. Am. Med. Assoc. 49, 1176-1178.
23) Fischer,G., Bang,H., & Mech,C. (1984) Biomed. Biochim. Acta 43, 1101-1111.
24) Agarwal,R.P., Threatte,G.A., & McPherson,R.A. (1987) Clin. Chem. 33, 481-485.
25) Deranleau,D.A. (1969) J. Am. Chem. Soc. 91, 4044-4049.
26) Deranleau,D.A. (1969) J. Am. Chem. Soc. 91, 4050-4054.
27) Heald,S.L., Harding,M.W., Handschumacher,R.E., & Armitage,I.M. (1990) Biochemistry 29, 4466-4478.
28) Danielson,P.E., Forss-Peter,S., Brow,M.A., Calavetta,L.,

     Douglass,J., Milner,R.J., & Sutcliffe,J.G., (1988) DNA $\underline{7}$,
     261-267.
29) Haendler,B., Hofer-Warbinek,R., & Hofer,E. (1987)
     EMBO J. $\underline{6}$, 947-950.
30) Vieira,J., & Messing,J. (1982) Gene $\underline{19}$, 259-268.
31) DeBoer,H.A., Comstock,L.J., Vasser,M., (1983) Proc.
     Natl. Acad. Sci. USA, $\underline{80}$, 21-25.
32) Gordon,M.J., Huang,X., Pentoney,S.L., Zare,R.N.,
     (1988) Science $\underline{242}$, 224-228.
33) Deyl,Z., Rohlicek,V., & Struzinsky,R. (1989) J.Liq.
     Chrom. $\underline{12}$, 2515-2526.
34) Farnsworth,C.C., Gelb,M.H., Glomset,J.A. (1990) Science
     $\underline{247}$, 320-322.
35) Rilling, H.C., Breunger,E., Epstein, W.W., & Crain, P.F.
     (1990) Science $\underline{247}$, 318-320.
36) Dalgarno,D.C., Harding,M.W., Lazarides,A., Armitage,I.M.,
     & Handschumacher,R.E. (1986) Biochemistry $\underline{25}$, 6778-6784.
37) Holzman,T.F., Egan,D.A., Edalji,R., Simmer,R.L.,
     Helfrich,R., Taylor,A., Burres,N.S. (1991) J. Biol. Chem.
     $\underline{266}$, 2474-2479.
38) Dietmeier,K. & Tropschug,M (1991) Nuc. Acids Res. $\underline{18}$, 373.
39) Tropschug,M., Nicholson,D.W., Hartl,F.U., Kohler,H.,
     Pfarmer,N. Wachter,E., & Neupert,W. (1988) J.Biol.Chem.
     $\underline{263}$, 14433-14440.
40) Haendler,B. & Hofer,E. (1990) Eur. J. Biochem. $\underline{190}$,
     477-482.
41) Holzman,T.F. (1990) unpublished observations.
42) Holzman,T.F., Egan,D.A., Chung,C.C., Rittenhouse,J.,
     Turon,M. (1990) Biophys. J. $\underline{57}$, 378.
43) Zuiderweg,E.R.P., Hallenga,K., & Olejniczak,E.T.
     (1986) J. Mag. Res. $\underline{70}$, 336-343.
44) Otting,G., Senn,H., Wagner,G., & Wüthrich,K., J. Mag.
     Reson., $\underline{70}$, 500-505.
45) Bax,A., & Weiss,M.A. (1987) J. Mag. Reson. $\underline{71}$, 571-575.
46) Fesik,S.W., Gampe,R.T., & Rockway,T.W. (1987) J. Mag.
     Reson. $\underline{74}$, 366-371.
47) Fesik,S.W., Luly,J.R., Erickson,J.W., & Abad-Zapatero, C.
     (1988) Biochemistry $\underline{27}$, 8297-8301.
48) Fesik,S.W., & Zuiderweg,E.R.P. (1988) J. Mag. Reson. $\underline{78}$,
     588-593.
49) Marion,D., Kay,L.E., Sparks S.W., Torchia,D.A., Bax,A.
     (1989) J. Am. Soc. $\underline{111}$, 1515-1517.
50) Fesik,S.W., & Zuiderweg,E.R.P. (1990) Q. Rev. Biophys. $\underline{23}$,
     97-131.
51) Park,C. (1990) unpublished observations.
52) Sierkierka,J.J., Hung,S.H.Y., Poe M., Lin,C.S., &
     Sigal,N.H. (1989) Nature $\underline{341}$, 755-757.
53) Harding,M.W., Galat,A., Uehling,D.E., & Schreiber, S.L.
     (1989) Nature $\underline{341}$, 758-760.
54) Siekierka,J.J., Weiderrecht,G., Greulich,H., Boulton,D.,
     Hung,S.H.Y., Cryan J., Hodges,P.J., & Sigal, N.H. (1990)
     J. Biol. Chem. $\underline{265}$, 21011-21015.
55) Brandts,J.F., Halvorson,H.R., & Brennan,M. (1975)
     Biochemistry $\underline{14}$, 4953-4963.
56) Fischer,G., Bang,H., Berger,E., Schellenberger,A. (1984)
     Biochim. Biophys. Acta $\underline{791}$, 87-97.
57) Yphantis, D.A. (1964) Biochemistry $\underline{3}$, 297-317.
58) Fesik,S.W., Gampe,R.T., Eaton,H.L., Gemmecker,G.,
     Olejniczak,E.T., Neri,P., Holzman,T.F., Egan,D.A.,
     Edalji,R.H., Simmer,R., Helfrich,R., Hochlowski,J.,
     & Jackson, M. (1991) Biochemistry, in press.

MUTATIONS AFFECTING PROTEIN FOLDING AND MISFOLDING *IN VIVO*

Anna Mitraki, Ben Fane, Cameron Haase-Pettingell, and Jonathan King

Department of Biology
Massachusetts Institute of Technology
Cambridge, Massachusetts 02139

INTRODUCTION

In vivo folding and *in vitro* refolding studies of many proteins have established that the polypeptide chain does not attain the native conformation directly, but must pass through partially folded intermediates, (Creighton 1978, Kim and Baldwin 1982, Goldenberg and King 1982). It has often been assumed that such species are conformational subsets of the fully folded native state. In fact the existing data suggest that folding intermediates have properties of their own, not necessarily reflected in the native state (Creighton and Goldenberg, 1984). Numerous cases have now been described in which partially folded intermediates form transient complexes with helper proteins, (chaperonins) within the cells, (Pelham 1986, Hemmingsen et al. 1988, Goloubinoff et al. 1989a,b). The necessity of proteins to fold in physiological environments can explain why intermediates might have distinct properties from those of the native state. These chains may have specific sites and properties that can mediate recognition with molecular chaperones, membrane transport sites and other factors. For proteins destined to be exported through a membrane channel, the chain must be prevented from prematurely reaching the native conformation, (Randall and Hardy, 1988). Thus, folding intermediates must have been evolved with respect to their *in vivo* folding environments.

The search for the final native conformation requires not only the passage from the correct intermediate conformations, but also avoidance of sterically available but incorrect conformations Indeed, formation of non-native aggregated states frequently competes with folding into the native conformation during the *in vitro* refolding of polypeptide chains (Zettlemeissl et al., 1979; London et al. 1974, Mitraki et al. 1987). Aggregation is also observed for polypeptide chains synthesized within both prokaryotic and eukaryotic cells, particularly at higher temperatures. The expression of the protein product of cloned genes in foreign hosts often results in accumulation of the newly synthesized polypeptide chains in an aggregated non-native state or inclusion body; (Marston 1986, Schein 1989). The few systematic studies on this subject suggest that aggregates are off-pathway polymeric structures derived, both *in vitro* and *in vivo,* from folding intermediates in the productive pathway, (Mitraki and King 1989).

GENETIC ANALYSIS OF PROTEIN STRUCTURE, FUNCTION AND STABILITY

Within the polypeptide chain one class of residues must carry information specifying the conformational features of the native structure, as well as its biological function, once this structure is reached, (Reidhaar-Olson and Sauer, 1989). Classical genetic studies allowed the identification of this class of critical residues, through the recovery of mutations which altered particular properties of the native state (thermal stability or biological activity, for example). This approach generally requires that the mutant polypeptide chain will still be able to attain a conformation that allows conventional biochemical characterization. The residues identified are those that are critical for the activity and stability of the native state of the protein. Examples include identification of critical residues in the active site of tyrosyl t-RNA synthetase, residues causing temperature sensitivity or reduced activity in T4 lysozyme, reduced activity in staphylococcal nuclease, and reduced activity in DNA binding proteins (Trp and λ repressors). For excellent reviews on genetic analysis of function and stability of proteins, see Fersht and Leatherbarrow (1987), Goldenberg (1988), Shortle (1989), and Alber (1989).

Once mutations are in hand altering specific properties, it becomes possible to search for mutations restoring the wild-type phenotype. Such restoration could be due to change of the mutant amino acid back to the wild type residue (true revertant). Alternatively, the mutant residue might be conserved, but another residue in a completely different position in the chain can mutate and restore the wild-type property. These are referred to as pseudorevertant or second-site suppressor (Helinski and Yanofsky, 1963). Isolation of second-site suppressors provides a methodology to map interactions between amino acid residues, again at the level of the native structure.

SUPPRESSOR MUTATIONS

Second-site suppressor analysis has been carried out with several DNA binding proteins (repressors) for whom missense mutations have been isolated. Missense mutations of the lambda repressor exist, either decreasing the stability of the native protein or altering the affinity of the folded mutant protein for its operator DNA. Second site suppressor mutations for both types of primary mutations have been isolated. Of three second-site suppressors isolated, two were able to suppress more than one primary mutation (Hecht and Sauer, 1985). Proteins that bear the second-site mutations only were purified and found to have increased affinity and specificity for operator binding (Nelson and Sauer, 1985). For the *E. Coli* Trp repressor five second-site mutations were isolated (Klig et al., 1988) that are compensatory for the initial missense mutants, i.e. they restore activity to intermediate levels (not a 100% wild type activity). Of those five suppressors, three were previously isolated as "super-repressors", i.e. mutants with higher than wild type activity, (Kelley and Yanofsky, 1985). Such a super-repressor protein was purified and it was shown to have increased affinity for the operator DNA than wild type, (Klig and Yanofsky, 1988). Each suppressor was able to correct the missense phenotype of more than one starting mutation, suggesting a possible "global" mechanism of action for the correction of the missense defects. The word "global" was in fact introduced by Shortle and coworkers, to characterize suppressors of missense mutations for staphylococcal nuclease, (Shortle and Lin, 1985). Those studies suggested that the missense mutations affected the stability and/or the activity of the native form of the proteins. The proposed mechanism of global action was that the second-site mutations contribute new overall stabilizing forces to the native structure, thus counterbalancing the initial defects.

Another protein for which there is extensive genetic analysis of structure and function is the phage T4 lysozyme, (Alber and Matthews, 1987). Several temperature-sensitive mutants have been isolated and characterized, as well as activity mutants. Second-site suppressors of the later have been recently isolated, (Poteete et al., 1991).

GENETIC ANALYSIS OF *IN VIVO* FOLDING PATHWAYS

If the polypeptide chain has to form and pass through the correct folding intermediates, clearly it will not in general be possible to deduce the nature of the folding rules simply by comparing native structure to amino acid sequence. An additional class of information must be required, determining the conformation and succession of folding intermediates.

We have been involved in developing a methodology for identifying those residues and local sequences which direct the conformation of the intracellular folding intermediates. Such a study requires isolating mutations which specifically affect the folding pathway, rather than the native protein, and then determining if the mutant polypeptide chain has folded incompletely or incorrectly within the cell. For those mutants that are blocked in the chain folding pathway, mapping and sequencing them identifies critical residues. Such mutants will only exist where there is indeed an intracellular folding pathway in which intermediates are well differentiated from the native structure. However, mutants that block the folding of the chain are generally absolute lethal mutations, causing the technical problem of the absence of a final stable conformation amenable to biochemical study. This problem can be bypassed by the isolation of conditional lethal mutations, for example temperature-sensitive mutations, which are defective at restrictive (high) temperature, but not at permissive (low) temperature. There are two classes of temperature sensitive mutants: TL (thermolabile), and TSS (temperature-sensitive for synthesis) (Sadler and Novick, 1965). TL mutants render the native form of the protein thermolabile, while in TSS mutants the protein is synthesized at high temperatures, but fails to attain the native conformation. The native form of the protein is reached at permissive temperature, and stays native after shift to restrictive temperature.

THE PHAGE P22 TAILSPIKE ENDORHAMNOSIDASE AS A MODEL SYSTEM FOR PROTEIN FOLDING WITHIN THE CELL

These efforts to identify residues and sequences controlling the conformation of intermediates and perhaps also off pathway steps, have been carried out with the bacteriophage P22 tailspike protein, a homotrimer of three 666 amino acid chains (Goldenberg et al., 1982; Sauer et al. 1982) whose secondary structure is dominated by beta sheet, as revealed by Raman spectroscopy (Sargent et al. 1988, Thomas et al., 1990). The native protein is resistant to SDS, proteases and heat and there are no covalent modifications known to be required for its maturation.

Due to a number of properties of the tailspike and of phage-infected cells, the folding and aggregation pathway of this system can be studied *in vivo* (Goldenberg and King 1982; Goldenberg et al. 1983). A mutation that blocks capsid assembly can be introduced to the phage strains, so that tailspikes can be produced in soluble form inside the cell. After release from the ribosome, an early productive single chain intermediate forms which can further proceed in the productive pathway and form the protrimer, a species in which the chains are associated but not fully folded. The protrimer folds further into the native spike, with concomitant acquisition of the resistance to SDS, proteases and heat. The early step in the pathway is thermolabile. The chains can partition between the productive pathway, or form aggregates, the aggregation path being favored at high temperatures. However, chains synthesized at high temperatures can reenter the productive pathway if shifted to permissive temperature early enough (Smith and King, 1981; Goldenberg et al., 1983; Haase-Pettingell and King, 1988). This indicates that aggregation is a conformational trap for folding intermediates that can be kinetically avoided.

Purified tailspikes can also be refolded *in vitro* after denaturation with acid urea, and the *in vitro* refolding pathway is currently under investigation (Seckler et al., 1989, Fuchs et al., 1991). Off-pathway aggregation also competes with productive folding in the *in vitro* refolding pathway, especially at temperatures above 25° C. (**fig.1**)

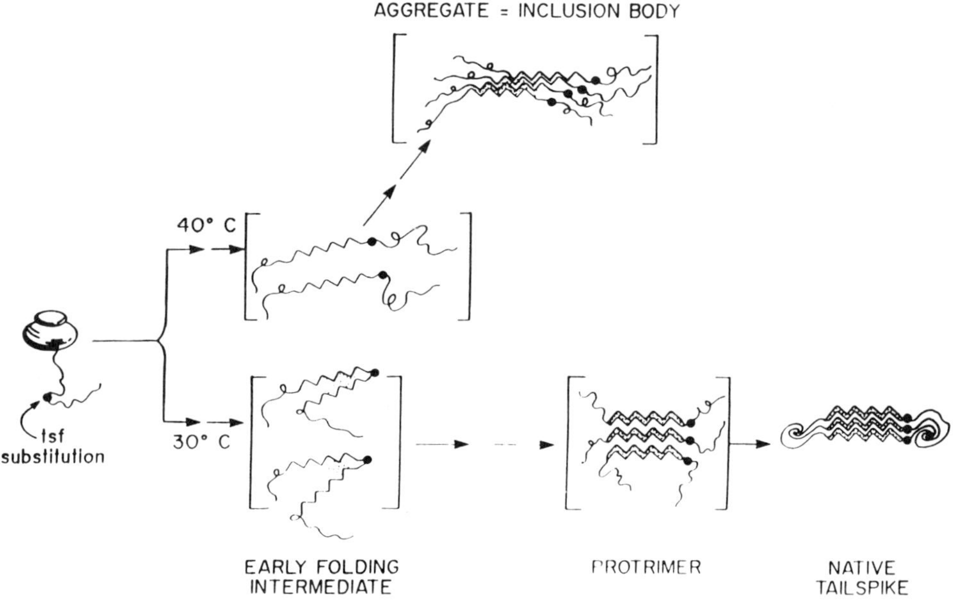

Figure 1

TEMPERATURE-SENSITIVE FOR FOLDING MUTATIONS IDENTIFY CRITICAL RESIDUES FOR THE FOLDING PATHWAY

Over 100 temperature-sensitive folding (*tsf*) mutations have been isolated in gene *9*, clustered in the central region of the gene **(Table 1)** The mutant polypeptide chains are synthesized at high temperatures, but fail to reach the native state at high temperatures (Goldenberg et al., 1983). At restrictive temperatures, the *tsf* mutant polypeptide chains accumulate as intracellular aggregates corresponding to inclusion bodies (Haase-Pettingell and King, 1988). These inclusion bodies form from the early folding intermediate, which partitions between aggregation and productive pathways depending on temperature. At low temperatures native trimeric tailspikes form, (Goldenberg and King 1981, Yu and King, 1984; Thomas et al, 1989; Sturtevant et al, 1989). The melting temperatures of the purified mutant proteins are comparable to the 88° C Tm of the wild type, as measured by differential scanning calorimetry (Sturtevant et al., 1989). Their physiological functions, such as binding to phage heads to produce infectious viral particles, and adsorption to bacterial cells are not distinguishable from wild type, (Goldenberg and King, 1981).

Some of the mutants confer altered electrophoretic mobility to the native trimer. Yu and King, after careful biochemical characterization of the purified mutant proteins, concluded that the altered mobility is not due to a conformational change in the protein. They suggested that it is due to the mutations being located at the protein surface, probably marking beta-turn positions (Yu and King, 1988; Villafane and King, 1988). These mutations may prevent correct beta-sheet formation at the level of the critical folding intermediates at restrictive temperature, but once this critical stage passed, they can be accommodated at the surface of the native form. Thus, the failure of the *tsf* mutations to reach the native state at high temperatures is not due to lowered stability or activity of the native state. Therefore, this class of mutations indeed confers information critical for the folding pathway,

Table 1. Sequences which May Kinetically Direct Turns in the Tailspike Polypeptide Chain

Mutation	Residue	Substitution	Local Sequence								
tsU9	177	Gly>Arg	Phe	Ile	Gly	Asp	<u>Gly</u>	Asn	Leu	Ile	Phe
tsH304	244	Gly>Arg	Val	Lys	Phe	Pro	<u>Gly</u>	Ile	Glu	Thr	Leu
tsH302	323	Gly>Asp	Asn	Tyr	Val	Ile	<u>Gly</u>	Gly	Arg	Thr	Ser
tsU38	435	Gly>Glu	Leu	Leu	Val	Arg	<u>Gly</u>	Ala	Leu	Gly	Val
tsH300	235	Thr>Ile	Gly	Tyr	Gln	Pro	<u>Thr</u>	Val	Ser	Asp	Tyr
tsH301	368	Thr>Ile	Thr	Trp	Gln	Gly	<u>Thr</u>	Val	Gly	Ser	Thr
tsU18	307	Thr>Ala	Asp	Gly	Ile	Ile·	<u>Thr</u>	Phe	Glu	Asn	Leu
tsU5	227	Ser>Phe	Thr	Leu	Lys	Gln	<u>Ser</u>	Lys	Thr	Asp	Gly
tsN48	333	Ser>Asn	Gly	Ser	Val	Ser	<u>Ser</u>	Ala	Gln	Phe	Leu
tsU19	285	Arg>Lys	Gly	Phe	Leu	Phe	<u>Arg</u>	Gly	Cys	His	Phe
tsU53	382	Arg>Ser	Asn	Leu	Gln	Phe	<u>Arg</u>	Asp	Ser	Val	Val
tsU57	230	Asp>Val	Glu	Ser	Lys	Thr	<u>Asp</u>	Gly	Tyr	Glu	Pro
tsmU9	309	Glu>Val	Ile	Ile	Thr	Phe	<u>Glu</u>	Asn	Leu	Ser	Gly
tsmU8	344	Glu>Lys	Asn	Gly	Gly	Phe	<u>Glu</u>	Arg	Asp	Gly	Gly
tsH303	250	Pro>Ser	Glu	Thr	Leu	Leu	<u>Pro</u>	Pro	Asn	Ala	Lys
tsU11	250	Pro>Leu	"	"	"	"	"	"	"	"	
tsU24	258	Ile>Leu	Lys	Gly	gln	Asn	<u>Ile</u>	Thr	Ser	Thr	Leu
tsRAF	270	Val>Gly	Glu	Cys	ILe	Gly	<u>Val</u>	Glu	Val	His	Arg
tsU7	311	Leu>His	Thr	Phe	Glu	Asn	<u>Leu</u>	Ser	Gly	Asp	Trp
ts9.1	334	Ala>Val	Ser	Val	Ser	Ser	<u>Ala</u>	Gln	Phe	Leu	Arg
tsR(am)A	203	Trp>Gln	Thr	Thr	Thr	Pro	<u>Trp</u>	Val	Ile	Lys	Pro
am^{ts}H1200	207	Trp>...	Val	Ile	Lys	Pro	<u>Trp</u>	Thr	Asp	Asp	Asn
am^{ts}H840	315	Trp>...	Leu	Ser	Gly	Asp	<u>Trp</u>	Gly	Lys	Gly	Asn

but not for the function or stability of the protein. This class of temperature-sensitive folding mutations, influencing folding and /or assembly but not stability or activity has also been described for the heterodimeric enzyme luciferase (Sugihara and Baldwin, 1988).

ISOLATION AND IDENTIFICATION OF SECOND-SITE SUPPRESSORS OF FOLDING MUTANTS

Since mutations affecting tailspike folding intermediates exist, it was likely that selection of suppressors for them would identify interactions in the intracellular maturation pathway (Fane and King, 1991). Fane and King searched for second-site suppressors using the following strategy: They started with amber mutations (mutations that prematurely terminate the polypeptide chain elongation by introducing a stop codon). Such amber mutations have been isolated at more than 60 sites. By growing these mutants on *Salmonella* hosts with altered transfer RNAs that insert different amino acids at the stop codon, it was possible to generate missense polypeptide chains with different amino acids at the amber site (Miller, et. al. 1979; Fane and King, 1987). In the tailspike many of these are amino acid insertions that confer temperature-sensitive phenotypes. Since the starting amino acid position is occupied by a non-wild-type residue, a search for revertants at restrictive conditions can subsequently used to identify second-site suppressors. Fane and King (1991) isolated second site suppressors of these missense proteins, that corrected the ts defects. A subset of these suppressors mapped within gene 9. Sequencing of the starting amber/suppressor mutants revealed that many of the isolated suppressors were located at two positions in the chain: Valine 331 mutated to alanine, and alanine 334 mutated to valine, (Fane et al., 1991 in press).

The repeated isolation of V331>A and A334>V with a variety of starting mutations suggested that they might have a global character, (Shortle and Lin, 1985). To test the hypothesis that the suppressors might act globally, they were crossed with a variety of well-characterized *tsf* mutations. We chose *tsf* mutations that conferred altered electrophoretic mobilities to the native trimer, in order to have a precise screening assay for putative double mutants. Since an alanine-valine interchange is not likely to confer altered electrophoretic mobility, a recombinant that maintains the starting mutation and a second-site suppressor will have the mutant electrophoretic mobility. The parental phage strains consisted of a *tsf* mutant, which maps between the original amber mutation and the suppressor, and either *su331/amber* or *su334/amber*. Recombinants were selected by plating on an restrictive host (host that does not insert an amino acid at the amber site) at 39°C. The parent carrying both the amber and suppressor mutations cannot grow on the restrictive *Salmonella* host; the *tsf* parent cannot grow at elevated temperatures. Recombinants arising from recombination events between the amber and the *tsf* mutation will carry the suppressor mutation alone and they will form plaques under these conditions. The other recombinant, arising from a recombination event between the *tsf* mutation and the suppressor, would presumably be the su/*tsf* double mutant. It will form plaques at elevated temperatures only if the suppressor can correct for the folding defects associated with the *tsf* mutation. These double mutants can be easily distinguished, since they will maintain the altered electrophoretic mobility of the starting mutant. **(Figure 2)**

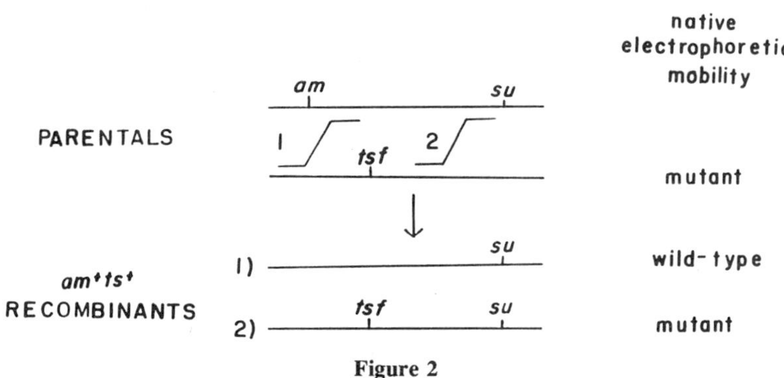

Figure 2

The *su334* mutation suppressed the defects associated with at least eight *tsf* sites while the *su331* mutation suppressed the defects associated with at least seven *tsf* sites. These sites map between residues 177 and 405. In summary, the *su331* and *su334* suppressors alleviate many *tsf* mutations mapping to the central region of gene *9*. **(Table 2)**

Table 2

Residue	Wild type amino acid	Insertion causing defect	Suppressor
45	Glutamine	Serine	val84>ala
122	Lysine	Glutamine	undetermined
156	Glutamine	Serine	undetermined
202	Tryptophan	Glutamine	ala334>val
207	Tryptophan	Glutamine	ala334>val
207	Tryptophan	Tyrosine	val331>ala
232	Tyrosine	Serine	val331>ala
365	Tryptophan	Tyrosine	val331>ala

THE TAILSPIKE SECOND-SITE SUPPRESSORS IDENTIFY RESIDUES THAT CORRECT FOLDING DEFECTS

The suppressor mutations do not affect the activity and stability of the native form of the tailspikes, by themselves or in combination with a *tsf* mutation (Mitraki et al., 1991). Thus, the tailspike suppressors do not seem to operate by conferring stabilizing forces to the native state of the protein. In this respect, those second-site suppressors are clearly different from the ones described above for the lambda and trp repressor systems. However, mutants defective in tailspike stability and head binding have also been described by Peter Berget and colleagues, (Schwartz and Berget 1989 a,b). Maurides, Schwartz and Berget (1990) isolated a second site suppressor which corrects such a defect in stability and binding.

Since the *tsf* mutations affect an early folding intermediate, and not the stability or function of the native protein, it is reasonable to think that second-site suppressors would correct the *tsf* defects by acting also in the folding pathway. The second-site suppressors have a general or global character, suppressing mutations with different amino acid substitutions. There is no obvious correlation between the starting mutations in terms of size, charge or hydrophobicity. The sequence surrounding the suppressor region is the following:

327				suala			suval				338
ser	tyr	gly	ser	val	ser	ser	ala	gln	phe	leu	arg

Position 333 is the locus of a *tsf* mutation, serine 333 to asparagine, indicating that the local conformation might be a surface beta turn. Myeong-Hee Yu and colleagues have made multiple amino acid substitutions at the 331 and 334 sites, and found that only two substitutions at each site were able to suppress *tsf* mutations, (M-H. Yu, personal communication). Given that the primary mutations span over 200 residues in the polypeptide chain, it is rather unlikely that the suppressors act through direct residue-residue interactions. It is more probable that they could operate through non-residue specific interactions, probably critically affecting intermediates in the folding pathway. Since the ts defects lie at the level of the early folding intermediate, which partitions between aggregation and productive pathways depending on temperature, it is possible that alleviation of the ts defects also operates at that level.

Investigation of the mechanism of suppression indicates that the suppressors alleviate aggregation, (Mitraki et al., 1991). They may act by stabilizing the thermolabile folding intermediate or altering the rates of the off-pathway aggregation step. Alternatively, the folding inefficiency of wild type at high temperatures might be due to a poor interaction with molecular chaperonins. GroE overproducing strains have been reported to rescue some tailspike *tsf* mutations, (Van Dyck et al, 1989). In this case the suppressor mutations may act to improve or restore a recognition site for molecular chaperonins. The possible involvement of chaperonins is presently under investigation, (S. Sather and J. King, personal communication).

Mutations alleviating inclusion body formation of human interferon have been isolated by Wetzel et al (1991) and maybe of the same character as the tailspike suppressors. Recently, a second-site suppressor was reported to alleviate many temperature sensitive-mutations in the human receptor-like protein tyrosine phosphatase (Tsai et al., 1991). This suppressor seemed also to alleviate inclusion body formation by the ts mutants. Thus, the isolation of such mutations in other proteins, besides providing an additional route of identifying critical residues in folding pathways, may be useful in elucidating the mechanism of intracellular steps leading to inclusion body formation, (Mitraki and King, 1989; Mitraki et al., 1991). The suppression of aggregation pathways by this class of mutations, without alteration of the native protein properties offers the industrially important perspective of engineering proteins for optimum production properties.

ACKNOWLEDGMENTS: The research was supported by NSF Grant DMB 8704126 and NIH Grant 17980 (to J.K.)

REFERENCES

Alber, T. 1990. Ann. Rev. Biochem. 59: 765-798

Alber, T. and Matthews, C. R. 1987. In Protein Engineering (Oxender, D. and Fox C. eds) Alan Liss Co. N.Y. pp. 289-298.

Creighton, T.E. 1978. Prog. Biophys. Mol. Biol. 33:231-298.

Creighton, T.E. and Goldenberg, D.P 1984. J Mol. Biol. 179:497-526.

Fane, B and King J., 1991. Genetics 127:263-277

Fane, B., Villafane, R,. Mitraki, A., and King. J. 1991. J. of Biol. Chem. in press.

Fane, B. and King, J., 1987. Genetics 117: 157-171.

Fersht, A., and Leatherbarrow, R. J. 1987. in: Protein Engineering, Oxender, D. and Fox, F. editors, pp 269-278. Alan Liss, New York.

Fuchs, A., Seiderer, C. and Seckler, R. 1991. Biochemistry, in press.

Goldenberg, D. P., Berget, P. B., and King, J., 1982. J. Biol. Chem. 257:7864-7871.

Goldenberg, D.P., Smith, D.H., and King, J. 1983. Proc. Nat. Acad. Sci. USA 80: 7060-7064.

Goldenberg, D. and King, J., 1982. Proc. Natl. Acad. Sci. USA 79:3403-3407.

Goldenberg, D. and King, J. 1981. J. Mol. Biol. 145:633-651

Goldenberg, D.P. 1988. Ann. Rev. Biophys. Biophys. Chem. 17: 481-507.

Goldenberg, D. and King, J., 1982. Proc. Natl. Acad. Sci. USA 79:3403-3407.

Goloubinoff, P., Christeller, J.T., Gatenby, A. A., and Lorimer, G. H. 1989. Nature 342:884-889.

Goloubinoff, B., Gatenby, A.A. and Lorimer, G. 1989a. Nature 337: 44-47

Haase-Pettingell, C. and King J., 1988. J. Biol. Chem. 263 : 4977-4983.

Hecht, M.H., and Sauer, R.T. 1985. J. Mol. Biol. 186, 53-63.

Helinski, D.R. and Yanofsky, C. 1963. J. Biol. Chem. 238: 1043-1048.

Hemmingsen, S. M., Woolford, C., van der Vies, S. M., Tilly, K., Dennis, D. T., Georgopoulos, C. P., Hendrix, R. W., and Ellis, R. J. 1988. Nature 333: 330-334.

Kelley, R. L. and Yanofsky, C. 1985. P.N.A.S. 82, 483-487.

Kim, P.S. and Baldwin, R. L. 1990. Ann. Rev. Biochem. 59 631-660

Klig, L.S., Oxender, D.L., and Yanofsky, C. 1988. Genetics 120, 651-655.

Klig, L.S. and Yanofsky, C. 1988. J. Biol. Chem. 263, 243-246.

London J., Skrzynia C., and Goldberg M. 1974. Eur. J. Biochem. 47: 409-415.

Marston, F. A. O., 1986. Biochem. J. 240:1-12.

Maurides, P.A., Schwarz, J.J., and Berget, P.B. 1990. Genetics 125: 673-681.

Miller, J.H., Coulondre, C., Hofer, M., Schmeissner, U.,Sommer, H., and Schmitz, A. 1979. J. Mol. Biol. 131: 191-222.

Mitraki A. and King J. 1989. Bio/Technology 7:690-697

Mitraki A., Betton J.-M., Desmadril M. and Yon J. 1987. Eur. J. Biochem. 163: 29-34.

Mitraki, A., Haase-Pettingell, C. and King, J. 1991. in: Protein refolding, G. Georgiou and E. de Bernardez, eds. American Chemical Society, Washington, D.C.

Mitraki, A., Fane, B. Haase-Pettingell, C., Sturtevant, J., and King, J. 1991. Science, in press.

Nelson, H.C. and Sauer, R.T. 1985. Cell 42, 549-558.

Pelham, H. R. B. 1986. Cell 46, 959-961.

Poteete, A. R., Dao-Pin, S., Nicholson, H., and Matthews, B. W. 1991. Biochemistry 30, 1425-1432.

Randall, L.L. and Hardy, S.J.S. 1988. Science 243: 1156-1159.

Reidhaar-Olson J. and Sauer, R. 1989. Science 241: 53 - 57.

Sadler, J. R. and Novick, A. 1965. J. Mol. Biol. 12:305-327.

Sargent, D., Benevides, J.M., Yu, M-h., King, J. and Thomas, Jr., G.J. 1988. <u>J. Mol. Biol.</u> 199: 491-502.

Sauer, R. T., Krovatin, W., Poteete, A. R. and Berget, P. B., 1982. <u>Biochem.</u> 21: 5811-5815.

Schein, C. 1989. <u>Bio/Technology</u> 7:1141-1149

Schwarz, J. and Berget P. 1989. <u>J. Biol. Chem</u> 264 : 20112-20119.

Seckler, R. Fuchs, A., King, J. and Jaenicke, R. 1989. <u>J. Biol. Chem.</u> 264:11750-11753

Shortle, D., and Lin, B. 1985. <u>Genetics</u> 110: 539-555.

Shortle, D. 1989. <u>J.Biol. Chem.</u> 264: 5315-5318.

Smith, D.H., Berget, P.B., and King, J., 1980. <u>Genetics</u> 96: 331-352.

Smith, D.H., and King, J., 1981. <u>J. Mol. Biol</u> 145: 653-676.

Sturtevant, J., Yu, M-h, Haase-Pettingell, C. and King, J. 1989. <u>J. Biol. Chem.</u> 264:10693- 10698

Sugihara, J., and Baldwin, T. O. 1988. <u>Biochemistry</u> 27: 2872-2880.

Thomas, G. J. Jr., Becka, R., Sargent, D., Yu, M-H.,and King J. 1990. <u>Biochemistry</u> **29**:4181-4187.

Tsai, A.Y.M., Itoh, M., Streuli, M., Thai, T. and Saito, H. 1991. <u>J. Biol. Chem.</u> 266: 10534-10543

Van Dyk, T. K., Gatenby, A. A. and LaRossa, T. A. 1989. <u>Nature</u> 342: 451-453

Villafane R. and King, J. 1988. <u>J. Mol. Biol.</u> 204:607-619

Yu, M.-H., and King, J., 1984.<u>Proc. Natl. Acad. Sci. USA</u>. 81: 6584-6588.

Yu, M.-H., and King, J., 1988.<u>J. Biol. Chem.</u> 263: 1424-1431

Wetzel, R., Perry, L.J. and Vielleux, C. 1991. <u>Bio/Technology</u>, in press.

Zettlmeissl, G., Rudolph, R., and Jaenicke, R., 1979. <u>Biochemistry</u> 18: 5567-5571.

PROTEIN FOLDING: LOCAL STRUCTURES, DOMAINS AND ASSEMBLIES

R. Jaenicke

Institut für Biophysik und Physikalische Biochemie
Universität Regensburg
D-8400 Regensburg, Federal Republic of Germany

SUMMARY

Globular proteins show the intrinsic property of acquir-
ing their spatial structure in an autonomous way, based
solely on their amino-acid sequence and their aqueous or non-
aqueous environment. In vivo folding is assumed to occur
cotranslationally; in contrast, in vitro renaturation after
preceding denaturation refers to the integral chain. Since
the final product of reconstitution is authentic with respect
to all available physicochemical and functional criteria, in
vitro experiments may be considered a sound basis for the
thermodynamic and kinetic analysis of the folding pathway.

In order to gain insight into the mechanism of folding,
the essential steps in the "hierarchical condensation" from
the nascent (unfolded) state to the native state of a given
protein have to be characterized. As taken from spectral
data, short-range interactions stabilize well-defined local
structures (α-helices, β-turns, loops) in independent segments of the polypeptide chain. In proceeding from elements
of secondary- and supersecondary structure to subdomains and
domains, the native tertiary and quaternary structure are
finally generated by the merging and docking of domains and
subunits. The kinetic analysis of reconstitution shows that
the overall mechanism of folding and association may be described by a sequential uni-bi-unimolecular scheme, where
folding and/or association may be rate-determining. The
formation of "inclusion bodies" in overexpressing strains of
bacteria may be quantitatively described by the superposition
of rate-determining folding and diffusion-controlled aggregation. The trapped protein may be "unscrambled" by denatura-
tion/renaturation; commonly, optimization leads to the recovery of pure and authentic material in high yield.

Applications of Enzyme Biotechnology, Edited by J.W. Kelly and
T.O. Baldwin, Plenum Press, New York, 1991

Globular proteins acquire their spatial structure autonomously and spontaneously, based exclusively on their amino-acid sequence and their solvent environment. Their structural integrity in solution depends on the solvent parameters. Accordingly, one would predict that protein folding is strongly influenced by the environment. However, a variety of experimental findings have proven that the solvent conditions upon translation and reconstitution are less critical than expected: in vitro folding and assembly may be accomplished in dilute buffer solution in the absence of components involved in cellular folding events; biologically active thermophilic proteins may be expressed in mesophilic hosts; cotranslational and posttranslational modifications such as glycosylation or processing do not necessarily interfere with the intrinsic capacity of the polypeptide chain to acquire its native three-dimensional structure (Jaenicke 1991a,b). Except for the influence of viscosity (Teschner et al., 1987) and specific ligands (coenzymes, substrates, ions, etc) (Jaenicke, 1987), hardly any attempts have been made to mimic the cytoplasm in folding experiments.

Folding in vivo is assumed to parallel protein biosynthesis as a "vectorial process". On the other hand, in vitro renaturation after preceding denaturation refers to the complete polypeptide chain. There is ample evidence which proves that the final product of reconstitution is authentic with respect to all available physicochemical, biochemical and biological criteria. Thus, in vitro experiments may be considered a sound basis for the thermodynamic and kinetic analysis of the mechanism of protein self-organization (Creighton, 1978, 1990; Jaenicke and Rudolph, 1989).

The fact that the nascent or refolding chain requires neither extrinsic factors nor the input of energy in order to generate the native structure has been considered sufficient evidence to postulate that the genetic code governs both translation and folding. Whether there is a unique folding code as the "second half of the genetic code" remains still to be shown (Fasman, 1989). That it cannot be colinear is trivial for the following reasons: both local next-neighbor and non-local through-space interactions are involved in the minimization of potential energy; as a consequence, identical stretches of polypeptide chain may determine different three-dimensional structures; widely differing ("homologous") sequences code for identical topologies; subdomains and domains as cooperative entities are separated by connecting peptides exhibiting anomalous configurations; extrinsic effects or effectors (not inherent in the amino-acid sequence) may play a significant role in the folding process. The latter argument has been shown to be essential in cases where cofactors or chaperones serve to stabilize intermediates of folding or assembly (Gerschitz et al., 1978; Ellis, 1990; Fischer and Schmid, 1990). Other cell-biological implications that may interfere with a general 1D $\longrightarrow$ 3D algorithm of protein

folding are: cellular compartmentalization, genome organization, transcription control, codon usage, amino-acid pools,
kinetic competition of folding and association in overexpressing hosts, discontinuity in the rate of translation, etc
(Jaenicke, 1987, 1988, 1991b).

In spite of these pitfalls, there have been numerous
attempts to forecast the three-dimensional structure of proteins or their mode of folding: Search programs for sequence
homologies have been successfully applied to correlate given
primary structures to a limited number of protein "families".
Statistical analyses of preferences for α-helices, β-strands,
turns, or random structures provide secondary structure
predictions with reliabilities of the order of 65% (Fasman,
1989). Topological considerations and docking procedures have
been developed to optimize both minimum hydrophobic surface
area and maximum packing (Wodak et al., 1987). Energy minimization and molecular dynamics calculations, as well as semi-
quantum mechanical and statistical mechanical methods proved
useful in reducing the number of possible conformations from
an astronomically high value to only few (McCammon and Harvey, 1988). They have been most valuable in characterizing
conformational changes with high precision. A combination of
all available methods in terms of knowledge-based computer-
aided structure predictions has been conceived by Blundell et
al. (1987, 1991). The result of $\approx$ 90% correct prediction
with an rms deviation <3 Å is most satisfactory from the
theoretical point of view; however, to predict the functional
state, one has to be > 99% correct, so that at present all
structure predictions have still to be taken with a grain of
salt.

THERMODYNAMICS VS KINETICS

In considering the principles of protein self-organization, two questions are of crucial importance: How do the
intrinsic physical and chemical properties of the building
blocks and their mutual interactions determine the energetics
underlying stability and flexibility, and what are the ele-
mentary processes on the kinetic pathway of structure formation. It is now widely accepted that the acquisition of the
native three-dimensional conformation of a protein is determined by a well-defined pathway rather than random search.
Based on this hypothesis, folding occurs by the fastest route
available, with a late event along the reaction coordinate as
the rate-limiting step (Creighton, 1990). It includes well-
populated intermediates and generates the native state as the
kinetically accessible state of minimum potential energy. As
shown by side reactions competing with correct folding or
domain pairing, this must not necessarily represent the
global minimum belonging to the most stable state (Wetlaufer,
1984; Jaenicke, 1987).

The energy minimum depends on the environment, relating
to both the cellular topography or compartmentalization and
the parameters of the solvent. The first refers to the "ad-

dresses" to which proteins are sent: into the cytoplasm or into the membrane, or the external medium. In this context, differences in polarity define extreme types of protein structure, in that membrane proteins are hydrophobic in their periphery, whereas the basic principle of the structure of soluble globular proteins is the minimization of water accessible hydrophobic surface area (Richards, 1977). The second is connected with the physical conditions which (beyond the limits of stability) lead to denaturation, dissociation and deactivation (Privalov and Gill, 1988; Dill, 1990; Jaenicke and Závodszky, 1990).

The free energy of stabilization, $-\Delta G_{stab}$, is typically 20-80 kJ/mol, i.e., on balance, the equivalent of a few hydrogen bonds or ion pairs (Jaenicke, 1988; 1991b). Considering the large number of hydrogen bonds which contribute to the stabilization of the secondary structure, or the stabilization of the inner core of the molecule by hydrophobic interactions, ΔG_{stab} emerges as a small difference between large numbers. Evidently, the structure of native proteins is not optimized for maximum stability. On the contrary, under physiological conditions, proteins are at the margins of their capacity to exist as native species: ΔG_{stab} is mostly not even 10kT. Thus, the optimization in the course of evolution was obviously promoted toward function, i.e. catalysis, regulation, mobility and turnover (Wetlaufer, 1980).

There is no doubt that folding starts cotranslationally: for domain proteins, "modular assembly", i.e. sequential folding from the N- to the C-terminal end of the nascent polypeptide chain has been clearly established (Bergman and Kuehl, 1979). This does not imply that the C-terminal end is structurally insignificant: in the case of ribonuclease, its cleavage blocks the folding reaction (Teschner and Rudolph, 1989). Since in many cases the N- and C-terminal ends of proteins are in close neighborhood, their (Coulomb) interaction may contribute significantly to the overall stability.

Considering the unfolding/folding equilibrium transition, single-domain proteins have been shown to obey the two-state model

$$U \rightleftharpoons N \tag{1}$$

according to which the protein exists in one out of two states only, either the (polymorphous) unfolded state (U), or the native state (N) (Kim and Baldwin, 1990). On the other hand, the kinetic analysis clearly proved intermediates (I_i) on the folding path to be populated, thus establishing a sequential scheme

$$U \longrightarrow I_i \longrightarrow N \tag{2}$$

In the model case of ribonuclease, the unfolded state has been shown to consist of a fast-folding substate in which all proline residues are in their native configuration, and a slow-folding one in which two prolines have to overcome the energy barrier of trans-cis proline isomerization (Schmid and Baldwin, 1978; Kiefhaber et al., 1990a,b). Prior to this rate-determining event on the folding path, "hierarchical condensation" occurs:

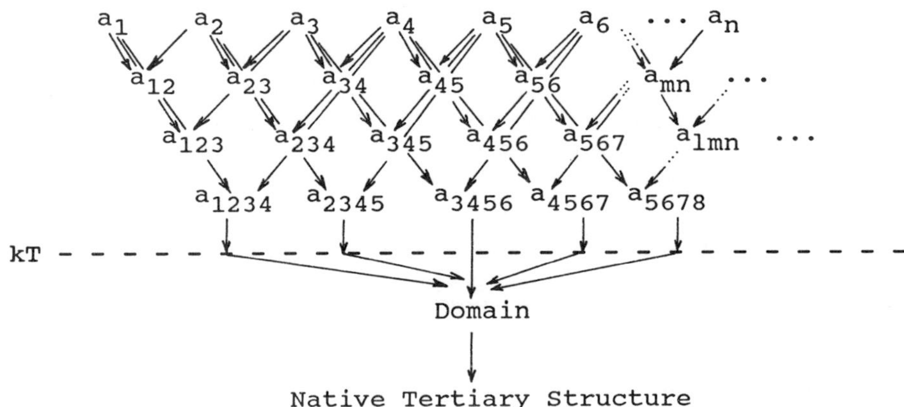

Next-neighbor interactions, in a fast reaction, generate stretches of secondary structure which are still unstable when exposed to the polar solvent. Non-local interactions, then, lead to metastable supersecondary structures (and subdomains), which, finally, collapse into the native-like "molten globule" state (Kuwajima, 1989). Subsequent shuffling to form the native tertiary structure occurs by a limited number of pathways, with the rate-limiting step as a late event. Characteristics of the "molten globule" as the most relevant intermediate on the overall pathway are summarized in Table 1.

Table 1. Characteristics of the "Molten Globule" State of Proteins.

Method [a]	Characteristics
CD, IR	Native-like secondary structure
NMR, UV-vis, CD, F_{em}	Altered environment of aromatic residues as in denatured state
Solvent perturbation	Change in exposure of trp, unaltered tyr titration
Ultracentrifugation	≈ 10% increase in Stokes radius, tendency to aggregate
DSC	Non-cooperative thermal transition
Polarization of F_{em}	Slow structural fluctuations

[a] CD, circular dichroism spectroscopy; DSC, differential scanning calorimetry; F_{em}, fluorescence emission; IR, infrared, UV-vis, ultraviolet-visible, and NMR, nuclear magnetic resonance spectroscopy, respectively.

In analyzing the folding path, two questions need consideration: (i) do local non-random conformations in the nascent or reconstituting polypeptide chain initiate folding and (ii) do elements of secondary structure or subdomains of a given protein form native-like three-dimensional structures and what is the minimum size of stable subdomains?

(i) Local non-random conformations in short peptide fragments of proteins in aqueous solution under conditions where native proteins fold have been demonstrated unambiguously, e.g. by immunological approaches and 2D NMR: Linear peptides as short as 4 or 5 residues form intramolecular hydrophobic clusters in aqueous solution (Wright et al., 1988). Due to the stabilizing effect of charged side chains interacting with the helix dipole, short α-helical peptides are found to be more stable in isolation than would be predicted by Zimm-Bragg's helix-coil parameters (Shoemaker et al., 1987). Using oligopeptides with 16 or 17 residues, the role of ion pairs on the secondary structure of synthetic peptides has been studied in detail; spacing the charged residues at different positions clearly established the stabilizing effect predicted for a helical array (Marqusee and Baldwin, 1987; Baldwin, 1990; 1991). β-turns were observed, e.g. in an immunogenic peptide fragment of the influenza virus hemagglutinin HA1 chain; upon truncation of the nonapeptide YPYDVPDYA, the N-terminal tetrapeptide still retained the reverse-turn conformation (Dyson et al., 1988).

In summarizing available data, it is obvious that local transient structures (in rapid equilibrium with the fully randomized chain) should be present in the nascent polypeptide chain. For small proteins such as ribonuclease, the "unfolded state" seems still to retain residual structure (Haas et al., 1988).

It is tempting to assume that the formation or retention of local structures in certain regions of the polypeptide chain is the initiating step in protein folding. Although such structures are only marginally stable, they may efficiently reduce the conformational space, thus directing the folding path. Clearly, the role of such "initiation sites" must be restricted to early folding steps, prior to the formation of well-populated intermediates, and much before the rate-limiting step. A number of experimental findings seem to support the idea that α-helices, β-turns and hydrophobic clusters are "seeds" of protein self-organization (Wright et al., 1988; Yu and King, 1988). However, there is some indication that the previously mentioned oligopeptides have no substantial tendency to adopt the same conformation in unrelated protein structures; also, the reverse turns observed in the small peptides seem to be absent in the known three-dimensional structures of proteins which contain these sequences. In the case of BPTI and RNase, Creighton (1988) has shown that non-random conformations in the unfolded or nascent proteins cannot be significant for the folding reaction. Thus, folding models based on well-defined initiation sites and subsequent modular assembly are still on shaky ground.

(ii) As mentioned, short α-helices and β-bends show exceptionally high stability without necessarily representing identical structures within and without their intact parent protein. In studying protein fragments obtained by limited proteolysis or semisynthesis the same problem arises, apart from the question whether a given primary structure altogether yields a stable and well-defined three-dimensional structure (Jaenicke, 1987). Using thermolysin as a model, the folding/unfolding of fragments has shown that the native-like structure persists and that fragments down to the size of subdomains may undergo reversible denaturation (Rashin, 1984; Fontana, 1989) (Table 2).

Table 2. Physicochemical characterization of thermolysin fragments[a]

Fragment	$s(S)$	M_C	M/M_C	H_{rel}	$T_m(^{\circ}C)$	ΔG_{stab}
1–316	3.18	34 227	1	100	87	55
121–316	2.36	20 904	1	96	74	47
206–316	1.70	11 829	1	95	67	31
225–316	1.54	9 560	1	92	65	26
255–316	1.7	6 630	≤ 2	101±10	64	20
297–316	n.d.	2 128	> 100	0	< 5	n.d.

a $s(S)$, sedimentation coefficient; M_C, calculated molecular mass; M/M_C, degree of association taken from sedimentation equilibrium; H_{rel}, rel. helicity; T_m, temperature of $N \longrightarrow U$ equilibrium transition; ΔG_{stab} (kJ/mol) free energy of stabilization (Vita et al., 1989).

DOMAINS

In proceeding from structural elements and subdomains to higher levels in the hierarchy of protein structure, it is important to note that the weak interactions involved in the stabilization of the native tertiary and quaternary structure show a high degree of specificity. The minimization of accessible surface area accomplished by docking and association allows domains and subunits to "recognize" their respective counterparts (Richards, 1977; Wodak et al., 1987). This holds even if the linker peptide connecting the domains is missing; for example, in the case of tryptophan synthase and lactate dehydrogenase, nicked subunits (due to the specificity of their interdomain contacts) are capable of recombining to "proteolytic dimers" (Goldberg and Zetina, 1980; Opitz et al., 1987). On the other hand, in vitro reconstitution may lead to wrong domain pairing (Zettlmeissl et al., 1984). In vivo, the vectorial character of the folding reaction, as well as chaperones are assumed to minimize side reactions. However, misfolding in vivo does occur (Mitraki and King, 1989; Pelham, 1989; Hurtley and Helenius, 1989); it leads to wrong conformers which are subsequently degraded or deposited in inclusion bodies (Rudolph, 1990).

Domains may be defined by visual inspection of crystal structures, surface area calculations, deconvolution of equilibrium transitions, limited proteolysis, sequence homology at the protein level, or gene organization at the DNA level (Jaenicke, 1987). Operationally, they are characterized by "folding-by-parts" (Wetlaufer, 1981), reflecting the independence of globular entities connected by more or less flexible linker peptides. As shown by Bergman and Kuehl (1979), domain folding occurs cotranslationally in discrete modules, this way speeding up the self-organization of large proteins by many orders of magnitude, and, at the same time, minimizing wrong intramolecular non-local interactions.

Summarizing a vast amount of data (for review, cf. Jaenicke, 1987), the overall folding reaction is characterized by multistep transitions with independent folding/unfolding kinetics involving consecutive folding and merging of the individual "lobes". In the simplest case, then, folding of a two-domain protein may be described by superimposing on one another the individual folding reactions of the constituent parts of the complex with a subsequent pairing reaction. Depending on mutual stabilization effects, the latter may or may not be significant. To give an example, Fig. 1 illustrates the equilibrium transition and the unfolding/folding kinetics for γ-II-crystallin (Rudolph et al., 1990).

Both thermodynamics and kinetics may be quantitatively described by the simple three-state model

$$N \rightleftharpoons I \rightleftharpoons U \tag{3}$$

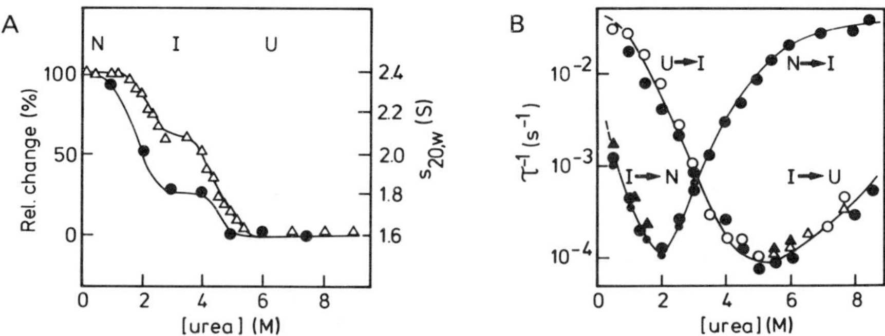

Fig.1. Unfolding/refolding of bovine γ-II-crystallin (0.1M NaCl/HCl, pH 2.0, 20°C). A. Urea induced unfolding: CD ($\triangle$); $s_{20,w}$ ($\bullet$). B. Dependence of the rate constants on urea concentration: ratio of 360nm/320nm fluorescence emission of native protein N ($\bullet$/O) and intermediate I ($\blacktriangle$,$\triangle$); FPLC gel filtration ($\bullet$).

where N indicates native γ-II-crystallin, I the intermediate
with the N-terminal domain intact and the C-terminal domain
in its random state, and U the denatured state, respectively.

In the given example, merging of the two domains remains
undetectable. For octopine dehydrogenase, a single-chain
analog of lactate dehydrogenase in molluscs, increasing
viscosity of the solvent has been shown to slow down the
folding kinetics. Obviously, in this case, the rate-limiting
process involves domain pairing rather than domain folding.
That merging of domains may lead to wrong domain interactions
follows from the incomplete reactivation which (in repetitive
cycles) yields 70%, 50% and 34%, respectively. The remaining
30% in each of the cycles represent inactive monomers with
native-like secondary structure, but perturbed fluorophores.
The overall mechanism

$$U \xrightarrow{\text{fast}} U' \xrightarrow{\text{fast}} I_N \xrightarrow{\text{slow}} N \qquad (4)$$
$$\downarrow$$
$$U''$$

again reflects the hierarchy of protein structure in that
$U \longrightarrow U'$ refers to the formation of secondary and supersecondary structure which subsequently collapse to form "structured domains" ($U' \longrightarrow I_N$), or the inactive substate (U'').
Only I_N is able to acquire the native conformation in a
(viscosity dependent) first-order reaction ($I_N \longrightarrow N$) (Zettlmeissl et al., 1984; Teschner et al., 1987). Most significant
in the given context is the occurrence of N and U'', which
seem to prove that a unique amino-acid sequence may generate
different stable conformations.

ASSEMBLY AND RECONSTITUTION

The early stages during the (re-)folding of oligomeric
proteins are expected to be identical with those involved in
the self-organization of single-chain proteins. Thus, we may
assume subunit polypeptide chains to fold first into subdomains and/or domains which will subsequently collapse to form
"structured monomers" with native-like structure. They will
then undergo association and shuffling to yield the native
quaternary structure so that the overall process may be writ-
ten as a sequence of (unimolecular) folding reactions and
(bimolecular) association steps according to

$$nM \longrightarrow nM' \longrightarrow n/2\,D \longrightarrow n/2\,D' \longrightarrow n/4\,T \longrightarrow n/4\,T' \longrightarrow .. \longrightarrow M_n$$
$$\downarrow \qquad\quad \downarrow \qquad\quad \downarrow \qquad\quad \downarrow \qquad\quad \downarrow$$
$$A \qquad\quad A \qquad\quad A \qquad\quad A \qquad\quad A \qquad\qquad (5)$$

where n is the number of subunits, M, M', D, D', T, T', ...
are monomers, dimers, tetramers, ... in different conformational states, M_n the n-mer, and A "wrong aggregates" (Jaenicke and Rudolph, 1986; Jaenicke, 1987). Since there is no
qualitative difference between interdomain and intersubunit

interactions, it is obvious that in the given hypothetical mechanism kinetic competition between folding and association may occur. In vitro the reaction leads to "wrong aggregation" (Zettlmeissl et al., 1979); in overexpressing strains of bacteria, "inclusion bodies" result from precisely the same mechanism. In fact, deposition of recombinant proteins at high expression levels can be quantitatively described by the kinetic competition of first-order folding and diffusion-controlled second-order aggregation (Rudolph, 1990).

Taking tetrameric lactate dehydrogenase (consisting of two-domain subunits with a molecular mass of 36 kDa) as an example, it becomes clear that the mechanism of reconstitution depends on the denaturation conditions. As illustrated in Fig. 2, reactivation after acid denaturation follows second-order kinetics, while reactivation after complete randomization in 6 M guanidine·HCl may be quantitatively described by a sequential uni-bimolecular mechanism, in accordance with consecutive folding/association (Jaenicke, 1987).

Optimization of in vitro reconstitution is pure alchemy, blended with some insight into the above mechanisms and the physicochemical basis of protein stability (Jaenicke and Rudolph, 1986, 1989). The full description of the mechanism of protein folding and protein association requires the determination of the structure and time-dependence of the various intermediates and transition states connecting the un-folded (nascent) state with the native and renatured states. As indicated, different denaturants yield different denatured states consisting of numerous heterogeneous conformations of closely similar energy. Commonly, maximum yields of renaturation are obtained after "complete randomization", under essentially irreversible conditions, and at exceedingly low protein concentrations (to minimize "wrong aggregation"). In order to increase the steady state concentration of the refolding protein, recycling techniques have been devised.

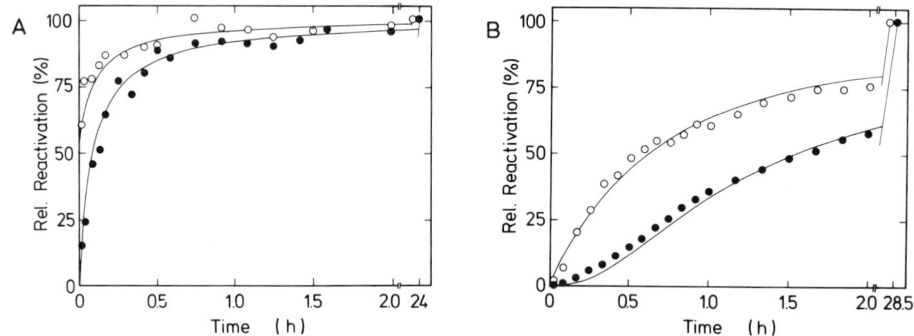

Fig. 2. Reactivation of pig LDH-M$_4$ after denaturation in 1M gly/H$_3$PO$_4$, pH 2.3 (A), and 6M guanidine·HCl (B). ●/o, activity in standard assay/in 1.5M ammoniumsulfate. Profiles calculated (A) for 2D →T with 0/50% subunit activity, k=3·10^{4}M^{-1}·s^{-1} and (B) for 4M →4M'→2D →T with 0% activity, k$_1$=8·10^{-4}s^{-1}, k$_2$=3·10^{4}M^{-1}·s^{-1}.

False intermediates, trapped under strongly native conditions, may be destabilized at optimally chosen residual denaturant concentrations, thus increasing the yield of reconstitution (Rudolph, 1990).

There have been reports that renaturation cannot be accomplished (Müller and Jaenicke, 1979; Rinas et al., 1990). Whether molecular chaperones, co- and posttranslational processing, or the vectorial character of protein folding in vivo are responsible for the negative results is unresolved (Rothman, 1989; Ellis, 1990). In connection with covalent modification, glycoproteins deserve mentioning: The carbohydrate moiety not only stabilizes the protein in its native state but also exhibits a solubilizing effect so that (wrong) aggregation is suppressed. Taking invertase from yeast as an example, the highly glycosylated extracellular form shows a dimer-tetramer-octamer dissociation equilibrium; the dimeric carbohydrate-free internal enzyme shows a strong tendency to form higher aggregates (Fig.3).

Compared to in vivo folding (which occurs within the time range of seconds), in vitro reconstitution may take exceedingly long. Mechanisms which are assumed to accelerate the reaction include: cotranslational folding instead of folding of the complete chain, nucleating effects of ligands, catalysis by specific isomerases such as protein disulfide isomerase (PDI), and peptidyl-prolyl cis-trans isomerase (PPI), or chaperones (Ellis, 1990; Fischer and Schmid, 1990; Jaenicke, 1991). Mimicking intracellular redox conditions in the case of proteins cross-linked by cystine bridges does not

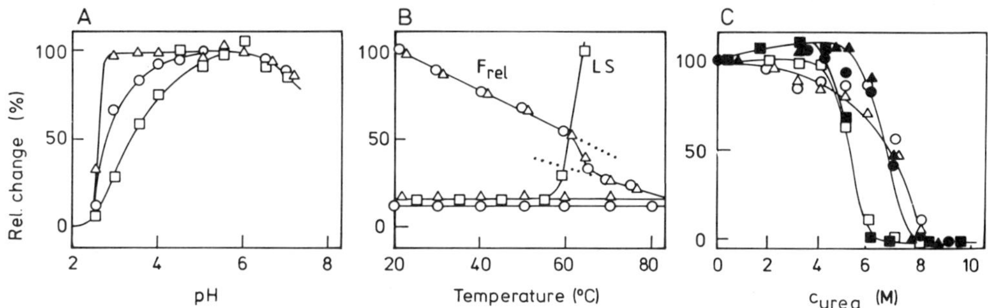

Fig. 3. Stability of internal (□), core-glycosylated (o) and external (△) invertase from yeast against pH (A), temperature (B) and urea (C).
A. Residual activity at 30°C after 42h in 50mM citrate/phosphate. B. Fluorescence emission (F_{rel}) at 325nm (λ_{exc}=280nm) and light scattering (LS) at 500nm upon heating at a rate of 0.25 K/min; 50mM Na-acetate pH 5.0. Half-times of deactivation at 65°C for the internal (non-glycosylated), core-glycosylated and external enzymes were 2.5, 30 and 32 min, respectively. C. Residual activity (open symbols) and fluorescence emission (closed symbols) at 20°C after 24h in 50mM Na-acetate pH 5.0.

Table 3. Influence of Various Proteins on the Reactivation and Aggregation of Citrate Synthase after Preceding Denaturation in 6M Guanidine·HCl [a].

Added Component	Reactivation Yield (%)
GroEL + GroES + ATP	28
GroEL + ATP	1
GroES + ATP	2
GroEL	0
addition of GroES + ATP after 45 min	28
GroEL + GroES	0
addition of ATP after 45 min	29
BiP + ATP	0
BSA + ATP	7
Lysozyme + ATP	6
ATP	3
None	1

[a] 0.1 M Tris/HCl pH 8.0, 20 mM dithioerythritol, 0.3 μM citrate synthase, 2 mM ATP; 25°C.

enhance the in vitro shuffling reaction to the rate observed in vivo (Rudolph and Fuchs, 1983).

Chaperones have the capacity to arrest partially folded or non-assembled polypeptide chains and prevent them from premature aggregation. In vitro, this "assistance" may improve the yield of reassociation dramatically (Buchner et al., 1991). How chaperones establish specificity, and how proper assembly actually takes place at the molecular and cellular level is still unknown. The competition of reactivation and aggregation in cases such as ribulose bis-phosphate carboxylase, rhodanese and citrate synthase in the presence of GroEL (Goloubinoff et al., 1989; Buchner et al., 1991) (Table 3), or the correlation of folding, association and misassembly of the tail spike protein of Salmonella phage P22 are first examples providing access to mechanistic details of in vivo folding. For the tail spike protein a large set of temperature sensitive folding mutants have been analyzed which may allow the correlation of folding patterns with specific changes in the amino-acid sequence of a given protein (King et al., 1987; Seckler et al., 1989).

ACKNOWLEDGMENTS

Work has been financed by the Deutsche Forschungsgemeinschaft (Ja 78/27-29, SFB 4 and SFB 43) and the Fonds der Chemischen Industrie. Generous support and hospitality of the John E. Fogarty International Center for Advanced Study at the NIH, Bethesda, is gratefully acknowledged.

REFERENCES

Baldwin, R.L., 1990, Pieces of the folding puzzle, <u>Nature</u>, London, 346:409.

Baldwin, R. L., 1991, Experimental studies of pathways of protein folding, <u>Ciba Foundation Symp.</u>, 161: in press.

Bergman, L. W. and Kuehl, W.M., 1979, Cotranslational modification of nascent immunoglobulin heavy and light chains, <u>J. Supramol. Struct.</u>, 11:9.

Blundell, T. L., 1991, From comparison of 3D structures to protein modelling and design, <u>Ciba Foundation Symp.</u>, 161: in press.

Blundell, T. L., Sibanda, B. L., Sternberg, M. J. E., and Thornton, J. M., 1987, Knowledge-based prediction of protein structure and the design of novel molecules, <u>Nature</u>, London, 326:347.

Buchner, J., Schmidt, M., Fuchs, M., Jaenicke, R., Rudolph, R., Schmid, F. X. and Kiefhaber, T., 1991, GroE facilitates refolding of citrate synthase by suppressing aggregation, <u>Biochemistry</u>, in press.

Creighton, T. E., 1978, Experimental studies of protein folding and unfolding, <u>Prog. Biophys. Mol. Biol.</u>, 33:231.

Creighton, T. E., 1988, On the relevance of non-random polypeptide conformation for protein folding, <u>Biophys. Chem.</u>, 31:155.

Creighton, T. E., 1990, Protein folding, <u>Biochem. J.</u>, 270:1.

Dill, K. A., 1990, Dominant forces in protein folding, <u>Biochemistry</u>, 29:7133.

Dyson, H. J., Rance, M., Houghten, R. A., Lerner, R. A. and Wright, P. E., 1988, Folding of immunogenic peptide fragments of proteins in water solution. I. Sequence requirements for the formation of a reverse turn, <u>J. Mol. Biol.</u>, 201:161.

Ellis, R.J., ed., 1990, Molecular chaperones, <u>Seminars in Cell Biol.</u>, 1:1.

Fasman, G. D. ed., 1989, "Prediction of Protein Structure and the Principles of Protein Conformation", Plenum Press, New York, London.

Fischer, G, and Schmid, F. X., 1990, The mechanism of protein folding. Implications of in vitro refolding models for de novo protein folding and translocation in the cell, <u>Biochemistry</u>, 29:2205.

Fontana, A., 1989, Structure and stability of thermophilic enzymes: Studies on thermolysin, <u>Biophys. Chem.</u>, 29:181.

Gerschitz, J., Rudolph, R. and Jaenicke, R., 1978, Refolding and reactivation of liver alcohol dehydrogenase after dissociation and denaturation in 6 M guanidine hydrochloride, <u>Eur. J. Biochem</u>, 87:591.

Goldberg, M. E. and Zetina, C. R., 1980, Importance of interdomain interactions in the structure, function and stability of the F_1 and F_2 domains isolated from the β_2 subunit of E. coli tryptophan synthase, <u>in:</u> "Protein Folding", Jaenicke, R., ed., Elsevier/North-Holland, Amsterdam: 469.

Goloubinoff, P., Christeller, J. T., Gatenby, A. A. and Lorimer, G. H., 1989, Reconstitution of active dimeric Rubisco from an unfolded state depends on two chaperone proteins and Mg-ATP, Nature, London, 342:884.

Haas, E., McWherter, C. A. and Scheraga, H. A., 1988, Conformational unfolding in the N-terminal region of ribonuclease A detected by nonradiative energy transfer: Distribution of interresidue distances in the native, denatured and reduced-denatured states, Biopolymers, 27:1.

Hurtley, S. M. and Helenius, A., 1989, Protein oligomerization in the endoplasmic reticulum, Annu. Rev. Cell Biol., 5:277.

Jaenicke, R., 1987, Folding and association of proteins, Progr. Biophys. Mol. Biol., 49:117.

Jaenicke, R., 1988, Is there a code for protein folding? , in: "Protein Structure and Protein Engineering", E. L. Winnacker and R. Huber, eds., Springer-Verlag Berlin, Heidelberg, 39. Colloquium Mosbach: 16.

Jaenicke, R., 1991a, Protein stability and protein folding, Ciba Foundation Symp., 161:in press.

Jaenicke, R., 1991b, Protein folding: Local structures, domains, subunits and assemblies, Biochemistry, 30:in press.

Jaenicke, R. and Rudolph, R., 1986, Folding and association of oligomeric proteins, Meth. Enzymol., 131:218.

Jaenicke, R. and Rudolph, R., 1989, Folding proteins, in: "Protein Structure and Function: A Practical Approach", T. E. Creighton, ed., IRL Press, Oxford: 191.

Jaenicke, R. and Závodszky, P., 1990, Proteins under extreme physical conditions, FEBS Lett., 268:344.

Kiefhaber, T., Quaas, R., Hahn, U. and Schmid. F. X., 1990a, Folding of ribonuclease T_1., Biochemistry, 29:3053, 3061,

Kiefhaber, T., Grunert, H.- P., Hahn, U. and Schmid, F. X., 1990b, Replacement of a cis-proline simplifies the mechanism of ribonuclease T_1 folding, Biochemistry, 29:6475.

Kim, P. S. and Baldwin, R. L., 1990, Intermediates in the folding reactions of proteins, Annu. Rev. Biochem., 59:631.

King, J., Haase, C. and Yu, M., 1987, Temperature-sensitive mutations affecting kinetic steps in protein-folding pathways, in: "Protein Engineering", Oxender, D. L. and Fox, C. F., eds., A. R. Liss Inc., New York: 109.

Kuwajima, K., 1989, The molten globule state as a clue for understanding the folding and cooperativity of globular proteins, Proteins: Struct., Funct. & Genetics, 6:87.

Marqusee, S. and Baldwin, R. L., 1987, Helix stabilization by Glu^-...Lys^+ salt bridges in short peptides of de novo design, Proc. Natl. Acad. Sci. U.S.A., 84:8898.

McCammon, J. A. and Harvey, S. C., 1988, "Dynamics of Proteins and Nucleic Acids", Cambridge University Press, Cambridge.

Mitraki, A. and King, J., 1989, Protein folding intermediates and inclusion body formation, Bio/Technology, 7:690.

Müller, K. and Jaenicke, R., 1980, Deanturation and renaturation of bovine liver glutamic dehydrogenase after dissociation in various denaturants, <u>Z. Naturforsch.</u>, 35c:222.

Opitz, U., Rudolph, R., Jaenicke, R., Ericsson, L. and Neurath, H., 1987, Proteolytic dimers of porcine muscle LDH. Characterization, folding and reconstitution of the truncated and nicked polypeptide chain, <u>Biochemistry</u>, 26:1399.

Pelham, H. R. B., 1989, Control of protein exit from the endoplasmic reticulum, <u>Annu. Rev. Cell Biol.</u>, 5:1.

Privalov, P. L. and Gill, S. J., 1988, Stability of protein structure and hydrophobic interaction, <u>Adv. Protein Chem.</u>, 39:193.

Rashin, A. A., 1984, Prediction of stabilities of thermolysin fragments, <u>Biochemistry</u>, 23:5518.

Richards, F. M., 1977, Areas, volumes, packing, and protein structure, <u>Annu. Rev. Biophys. Bioeng.</u>, 6:151.

Rinas, U., Risse, B., Jaenicke, R., Abel, K.- J. and Zettlmeissl, G., 1990, Denaturation renaturation of the fibrin-stabilizing factor XIIIa isolated from human placenta. Properties of the native and reconstituted protein, <u>Biol. Chem. Hoppe-Seyler</u>, 371:49.

Rothman, J. E., 1989, Polypeptide chain binding proteins: Catalysts of protein folding and related processes in cells, <u>Cell</u>, 59:591.

Rudolph, R., 1990, Renaturation of recombinant, disulfide-bonded proteins from "inclusion bodies", <u>in:</u> "Modern Methods in Protein and Nucleic Acid Research", Tschesche, H., ed., de Gruyter, Berlin: 149.

Rudolph, R. and Fuchs, I., 1983, Influence of glutathion on the reactivation of enzymes containing cysteine or cystine, <u>Hoppe-Seyler's Z. Physiol. Chem.</u>, 364:813.

Rudolph, R., Siebendritt, R., Nesslauer, G., Sharma, A. K. and Jaenicke, R., 1990, Folding of an all-β protein: Independent domain folding in γ-II-crystallin from calf eye lens, <u>Proc. Natl. Acad. Sci. U.S.A.</u> 87:4625.

Schmid, F. X. and Baldwin, R. L., 1978, Acid catalysis of the formation of the slow-folding species of RNase A: Evidence that the reaction is proline isomerization, <u>Proc. Natl. Acad. Sci. U.S.A.</u>, 75:4764.

Seckler, R., Fuchs, A., Jaenicke, R. and King, J., 1989, Reconstitution of the thermostable trimeric phage P22 tail spike protein from denatured chains in vitro, <u>J. Biol. Chem.</u>, 264:11750.

Shoemaker, R. K., Kim, P. S., York, E. J., Stewart, J. M. and Baldwin, R. L., 1987, Test of the helix dipole model for stabilization of α-helices, <u>Nature</u>, London, 326:563.

Teschner, W. and Rudolph, R., 1989, A carboxypeptidase Y pulse method to study the accessibility of the C-terminal end during the refolding of RNase A, <u>Biochem. J.</u>, 260:583.

Teschner, W., Rudolph, R. and Garel, J.- R., 1987, Intermediates on the folding pathway of octopine dehydrogenase from Pecten jacobaeus, <u>Biochemistry</u>, 26:2791.

Vita, C., Jaenicke, R. and Fontana, A., 1989, Folding of thermolysin fragments: Hydrodynamic properties of isolated domains and subdomains, <u>Eur. J. Biochem</u>, 183:513.

Wetlaufer, D. B., 1980, Practical consequences of protein folding mechanisms, <u>in:</u> "Protein Folding", Jaenicke, R., ed., Elsevier/North-Holland, Amsterdam: 323.

Wetlaufer, D. B., 1981, Folding of protein fragments, <u>Adv. Protein Chem.</u>, 34:61.

Wetlaufer, D. B., 1984, <u>in:</u> "The Protein Folding Problem", Wetlaufer, D. B. ed., Westview, Boulder: 29.

Wodak, S., de Crombrugghe, M. and Janin, J., 1987, Computer studies of interactions between macromolecules, <u>Prog. Biophys. Molec. Biol.</u>, 49:29.

Wright, P. E., Dyson, H. J. and Lerner, R. A., 1988, Conformation of peptide fragments of proteins in aqueous solution: Implications for initiation of protein folding, <u>Biochemistry</u>, 27:7167.

Yu, M. H. and King, J., 1988, Surface amino acids as sites of temperature-sensitive folding mutations in the P22 tailspike protein, <u>J. Biol. Chem.</u>, 263:1424.

Zettlmeissl, G., Rudolph, R. and Jaenicke, R., 1979, Reconstitution of lactic dehydrogenase: Non-covalent aggregations versus reactivation. I. Physical protperties and kinetics of aggregation, <u>Biochemistry</u>, 18:5567.

Zettlmeissl, G., Teschner, W., Rudolph, R. Jaenicke, R. and Gäde, G., 1984, Isolation, physicochemical properties and folding of octopine dehydrogenase from Pecten jacobaeus, <u>Eur. J. Biochem</u>, 143:401.

APPLICATIONS OF CONTROLLED PORE INERT MATERIALS AS IMMOBILIZING SURFACES FOR MICROBIAL CONSORTIA IN WASTEWATER TREATMENT

Ralph J. Portier

Institute For Environmental Studies
Louisiana State University
Baton Rouge, Louisiana

INTRODUCTION

Bioprocess technology, in the form of immobilized cells and organelles, is rapidly coming to the forefront in the third industrial revolution, as the platform upon which new "biotech" industries will be based. The number of organic compounds introduced into the environment by humans has increased dramatically in recent years (Pfaender and Bartholomew, 1982). As a consequence of this xenobiotic, i.e., man-made, pollution, the fate of these compounds, such as pesticides, in the environment is an important issue. Of particular concern is disappearance, persistence, and/or partial transformation of such compounds and their potential hazardous effect. While many are readily biodegradable, others have proven to be recalcitrant and persistent in soil and water. In recent years, a great deal of research has been done on the biochemistry and genetics of toxicant-degrading microorganisms. Both the newer literature on biotechnology, and the older literature on industrial microbiology, describe commercial processes in which microorganisms play important roles. Although some bacteria can cause adverse effects, most species are benign, and many are involved in processes of direct benefit to man.

A number of papers have been published recently on the immobilization of microbial cells for the purpose of transforming organic compounds (Chackrabarthy, 1982). Bacterial immobilization involves the entrapment of cells onto a matrix that contains a ligand recognized by bacterial surface receptors, binding the cell and making it accessible to the medium which surrounds the matrix. Immobilized bacterial cells have a number of well-established applications and considerable potential in many other areas of biotechnology.

This technology is based on the principle that natural populations of microorganisms may adapt to decompose refractory molecules. The study of techniques for the preparation of immobilized cells have only been developed during the last decade. Concomitantly, a rather broad spectrum of applications has been investigated. Applications focused on hazardous waste effluent/groundwater biotreatment, however, have not been developed. This technology, although having many advantages, also has some drawbacks, such as the difficulty of delivering a good oxygen supply to dense cell preparations, cell growth within the support, and, in some cases, changes in metabolic patterns. To partly reduce these drawbacks in the conventional immobilization method, alternative immobilization techniques have ben developed in our laboratory during the last few years. Based on this new knowledge, it is now possible to adapt new strains that are able to attack insecticides, herbicides, and other potential environmental pollutants previously thought to be quite resistant to biodegradation. Although these processes vary widely, it is only the beginning. What is clear is that there is a great deal to be done in fundamental microbiology, biotechnology, and the interface between the two.

IMMOBILIZATION OF MICROORGANISMS

There is considerable interest in the use of immobilized microbial cells in industrial biotechnology, and it represents a new trend in biochemical processing, with its full potential just being recognized (McGhee and Grant, 1982). During the last decade, investigations and new techniques for the preparation of immobilized whole cells without loss of catalytic activity have become the subject of increased investigation. A great deal of research has gone into the immobilization of commercially important enzymes, and associated problems of reactor design and scale-up procedures. Many papers have been published on enzyme immobilization procedures to transform organic compounds. Continuous enzyme reactions using immobilized microbial cells have already provided industry with major biocatalytic processes to produce pro-insulin, and the compounds L-aspartic acid, L-maleic acid and penicinallic acid, key pharmaceutical intermediates. The potential of the technique for industrial wastewater processing, is the focus of this presentation.

Most of the published techniques for enzyme immobilization have also been found to be suitable for cells and organelles. Basic cellular biocatalytic functions reside in enzymes. These enzymes are produced by, and frequently operate within, the whole living cell, where they form part of a series of interdependent multi-stage processes. However, it is worth noting that the success of certain membrane reactors and the interest in immobilized whole cells represent the latest stages in a long-standing divergence of view on whether co-factors can be regenerated artificially, or whether the mechanisms built into a living cell can be effectively exploited.

An important characteristic about adsorption of cells is the multipoint attachment binding, which makes them stick more strongly to the sorbent. Properties of the support are of utmost importance. Composition of the matrix, surface charge, surface area, and pore size are of some of the factors to be taken into account (Mattiasson, 1983). Properties of the cells, such as cell wall composition, charge, and age are also important factors in the adsorption process.

Particulate materials, such as cells, are theoretically simpler to immobilize. Adsorption to water insoluble supports, whether organic or inorganic, has been the simplest immobilization technique. However, the low catalytic intensity, sensitivity to stress, and the presence of many unwanted reactions and contaminants makes the immobilization of the live cells disadvantageous. However, it has been found that microorganisms may adhere to the surface reversibly for a short time period, and become irreversible later. The primary interaction between the cell and the solid surface must somehow induce a secondary, irreversible interaction (Mattiasson, 1983).

The advantage of using the immobilized whole cell over using intact cells in a batch reactor has been well documented (Chibata, 1979). Previous studies have demonstrated that this technique has the advantage of accelerated reaction rate owing to increased cell density. Most applications have utilized immobilized cells as biocatalysts, making possible their convenient use, like solid catalysts are used in the synthetic chemistry industry, thus providing a more cost-competitive posture.

Immobilization is seen as a technique which confines a catalytically active cell within a reactor system. In general, hydrophilic gels, such as natural and synthetic polymers, have been employed as support for cell immobilization. Techniques which utilize existing matrix structures are generally simpler to use (Mattiasson, 1983). The support must be sufficiently robust and stable to be retained by simple physical means within a reactor system, where the cells can be contacted with substrate or nutrient. It is to be expected that the biocatalytic activity of immobilized cells depends on the nature of the polymer matrix.

Microbial cells and enzymes, whether individual enzymes or multi-enzyme complexes within living cells, can be immobilized by confinement within a porous membrane or by attachment to a solid surface. The porous membrane may either be formed directly around the biocatalyst as a three-dimensional gel, or be pre-formed and employed as an integral part of the system. Alternatively, the biocatalyst can be bound to the surface of a stationary phase by a range of natural forces, varying in strength and specificity from simple non-specific adsorption to fully covalent bonding.

The concern of early investigations was to identify new methods of binding enzymes and to elucidate the underlying kinetics of heterogeneous catalysts, while more recent emphasis has shifted toward developing ideas on whole cell immobilization and attempts to identify practical outlets for more mature ideas. A large number of methods concerning immobilization of biocatalysts have been reported. Rosevear (1982) classifies these methods in three basic categories: entrapment within a support, adsorption to a support, and covalent binding to a support. For industrial wastewater biotreatment, in which a mixed waste is often encountered, the combination of entrapment and adsorption to a support, is the preferred mechanism.

<u>Entrapment within a Support</u>.This approach is usually based on the confinement of the cell in a three-dimensional structure, behind a membrane or within a gel. In principle it is one of the least disruptive methods of immobilization. The cells are inside of the structure, freely suspended in fluid and interacting with substrates that diffuse inside and outside of the cell. Living cells have the ability to divide, and eventually may contribute to the breakdown of the support. One of the advantages of this method is that the catalyst, when encased in a lattice, is protected from macromolecule inhibitors.

We have used entrapment, the most common immobilization method used in both laboratory experiments and in field pilot demonstrations. The aminopolysaccharide , chitin, in its deacetylated form, is used as an effective fixing agent (Portier and Meyers, 1984). Protan Laboratories(Redmond, Washington), produces commercially different industrial grades of chitosan (deacetylated chitin).

<u>Physical Adsorption</u> The oldest method of catalyst immobilization is that of physically adsorbing the cell or enzyme onto a polymer matrix without covalent bonding. In nature, many microorganisms adhere strongly to surfaces, particularly in nutrient-depleted environments, as found in soil. Adsorption is used here in its broadest sense to cover all those processes by which biocatalysts are held at a phase boundary by non-covalent forces usually between a solid stationary phase and an

aqueous mobile phase. The strength of the binding can range from simple Van der Waal's forces, through hydrophobic interactions, to strong ionic binding. Polymers other than cellulose were investigated as supports to which enzymes could be covalently bonded (Mosbach et al., 1966) , suggesting that, in general, the level of activity of an immobilized enzyme depends on the degree of hydration of the polymer matrix. Polysaccharides are not always the ideal support materials for whole cell immobilization, because they suffer from microbial attack. However, this "attack" of the entrapment media further induces glycocalyx formation and the establishment of a niche by the axenic culture introduced. In time, the entrapment approach is replaced by an adsorption mechanism and, finally, by sheer numbers, a mature axenic microbial community. This approach insures the generation of an adequate concentration of whole cells for effective biocatalysis while at the same instance, providing a fixed film difficult to dislodge by other microbial strains.

APPLICATIONS IN THE FIELD

Two summaries of case studies of application of these whole cell biotreatment systems in the field are presented. the first is a groundwater associated with the production of commercial pesticide products. Leaching of formulated product from tanks seeped below the facility and was recovered by a pump and treat approach coupled to an immobilized whole cell bioreactor. The groundwater consisted primarily of organophosphates, carbamates and organochlorines. This multiple carbon source ground water is typical of many streams associated with agriculture and consumer products manufacturing. The second effluent, a high sucrose centrifuge concentrate from the processing of sugarcane, was evaluated as a possible renewable resource by biological fermentation to citric acid.

Approach

Isolation of Microorganisms. The microorganisms isolated and adapted for continuous degradation of industrial effluents are discussed in detail in Portier et al., 1983. These strains were developed for bioconversion/detoxification of specific industrial streams using aquatic microcosm systems. Progenitors of these organisms are continuously modified using conventional mutation methods, and additional toxic and/or loading methods in aquatic microcosm systems, such that the resulting microbial population has been enzymatically induced and is capable of being immobilized on an appropriate substrate for continuous biotreatment efforts.

Immobilization to Controlled Pore Inert Materials Figure 1 shows the general design for the bioprocessing waste treatment module consisting of an immobilized

microbial community on an inert controlled pore support material or "surface". In this
instance, the aminopolysaccharide, chitin, processed from shrimp and crab shells, is
shown as the activated support phase. Bacterial strains are immobilized on the solid
supports mentioned following guidelines, as described by Messing et al., 1979.
Entrapment and adsorption to supports is the primary mechanism for immobilization.
Chitin and chitosan are conditioned using methods outlined by Hood, 1973, and
Portier, 1982, for the culture and growth of marine chitinoclastic populations.
Following a distilled water wash, the polysaccharide materials are suspended in water,
pH adjusted to 6.7 with either sodium hydroxide or hydrochloric acid solutions, and
allowed to stand overnight. Following the decanting, this material is applied to a
diatomaceous earth support (Manville Corporation, R630-634) washed in a 1%
glutaraldehyde solution, consolidated and sealed into the module-holding chamber.
Microorganisms are continuously introduced and recycled through the module for 24
hours by a peristaltic pump to initiate viable cell attachment. Cell loading optimum
(gram wet cells per liter wet catalyst) was determined using microbial ATP approaches
(Portier, 1986). Aliquots of post-module effluents are analyzed for viable free-cell
biomass and compared to initial biomass inocula. Additional aliquots are also removed,
washed and analyzed for ATP from support surfaces. Completion of loading rate is
determined by percent free cell washout detected post-module (5% of total microbial
ATP).

     Chromatographic Analysis. Liquid chromatography (HPLC) approaches are
routinely used to determine the kinetic efficiency (biotransformation) of the immobilized
microbial population, by comparison of influent and effluent targeted toxicant
concentrations. Rate coefficients based on Michaelis Menten kinetics are used to
predict microbial transformation rates using a multiphasic mathematical model (Lewis et
al., 1984). Final post treatment effluent concentrations will also be analyzed for the
presence of microbial metabolites indicative of parent compound biodegradation.

Groundwater Remediation

     The microorganisms isolated and adapted for continuous biodegradation of
chlorinated pesticides and organophosphates are discussed in detail elsewhere (Portier,
1987). These strains, primarily _Pseudomonas sp._ and _Arthrobacter sp._ were developed
for detoxification use from earlier aquatic microcosm studies on the environmental fate
and effect of polychlorinated biphenyls in coastal delta wetlands.

     Groundwater feeding the biotreatment facility was pumped from a sand and
gravel aquifer occurring 25 feet below ground surface. The stratigraphy of the site

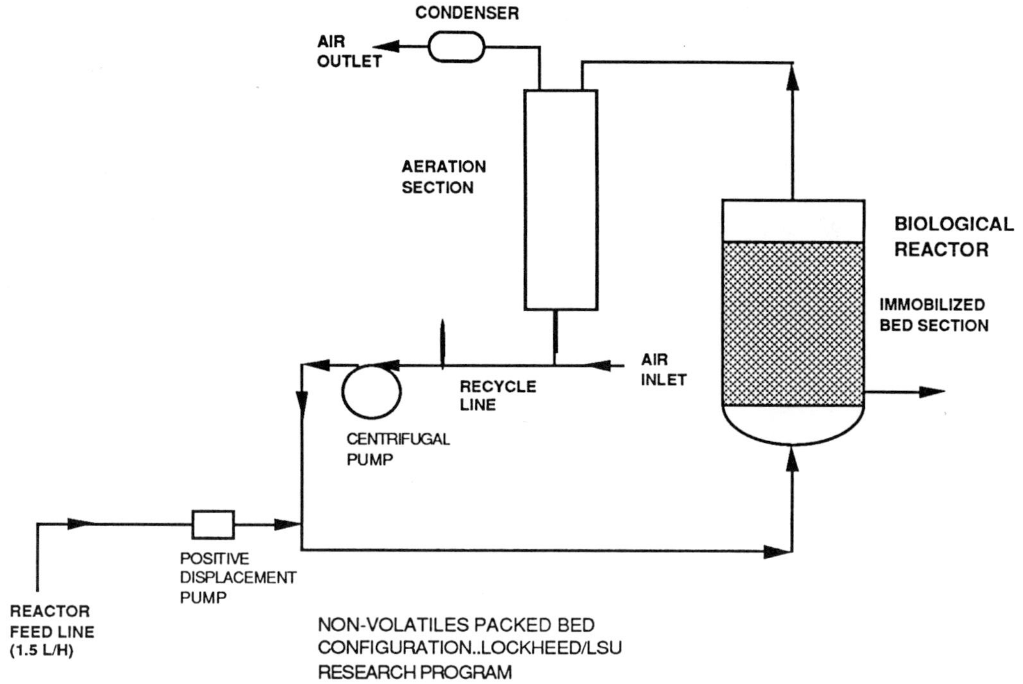

Fig. 1. A typical reactor process flow diagram for effluent biotreatment.

consists of an upper stratum consisting of 13 feet of low permeability clay. The biological tower, designated Manville Packed Bed Reactor (MPBR), was configured as a plug flow reactor containing fine bubble diffusion with a maximum aeration of 0.3 scfm. The unit was packed with 5.4 cu. ft. of carrier (Celite™ R-630) provided by Manville Corporation. A carbon column was available as a polishing unit. After initial start-up, the carbon unit was bypassed to place waste directly into the biological reactors at a rate of 80 gallons per day (gpd). Aeration rates were maintained at 0.2 scfm per tower (saturation levels for both units).

Figure 2 shows the net inputs and subsequent biotreatment system export values for two carbon sources, the organophosphate malathion and the organochlorine, aldrin. The MPBR system was effective in the reduction of malathion for all data points . On day 28 with the highest feed input of malathion, the net reduction in this toxicant concentration remained consistent with earlier export values. It is important to note that organochlorine-degrading microbial populations were used to colonize surfaces in the MPBR treatment units. Significant biotransformation of malathion may have been a consequence of the recognized nutritive values associated with this organophosphate as a nitrogen and phosphorus source. This is a very common phenomenon in the biotransformation of complex mixtures. In previous studies (Portier et al., 1986), sequential biotransformation of polynuclear aromatics occurred as a consequence of

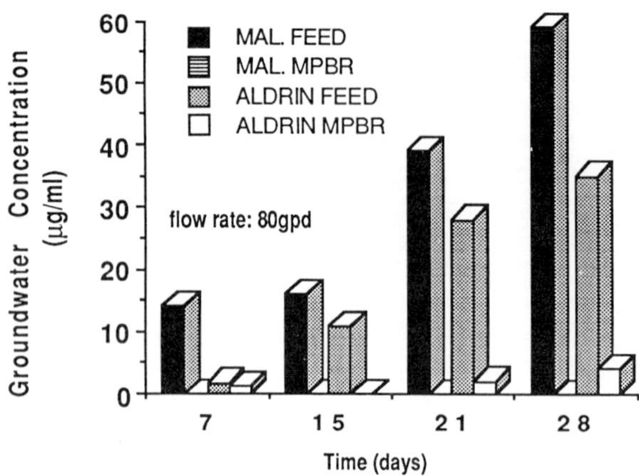

Fig.2 Continuous groundwater remediation of mixed pesticide wastes.(a) MAL, the organophosphate malathion which provide needed nitrogen and phosphates; (b) Aldrin, a common organochlorine of the dieldrin family (MPBR, Manville™ R630 carrier packed bed reactor)

nitrogen availability as well as complexity of carbon structure. In that work, the net biotransformation of all toxicant components in these occurred when appropriate retention times were used.

Feed and effluent mole fractions for the organochlorine groundwater contaminants in the groundwater stream is also shown in Figure 2. At a flow rate of 80 gpd, fluctuations in aldrin varied from trace to 35 µg/ml. The MPBR system appeared to be more effective when aldrin concentrations increased.The kinetic removal rate for malathion was 4 mg/ 1000 m^3/D. The kinetic removal rate for aldrin was 0.18 mg/1000 m^3/D.

<u>Citric Acid Fermentation</u>

Available economical substrates, such as process effluents from the preparation of soybean meal, sugar cane, molasses and other related agricultural materials , generate significant high biological oxygen demand (BOD5), point sources into local water bodies. Recognizing these streams as underutilized substrates for pharmaceutical and food additives production solves a troublesome environmental problem while generating potential new sources of income. A 5 liter pilot plant , consisting of <u>Aspergillis niger</u> immobilized to diatomaceous earth surfaces, was employed to continuously treat a sucrose laden effluent from a sugarcane mill molasses clarifier. As shown in Fig 3., mineralization of sucrose at a flow rate of 2.5 ml/min was approximately 60-65%. Citric acid production accounted for 15-20% of the sucrose consumption. Biomass densities ranged from 5.0 - 8.5 million propagules /cm ^{3} carrier bed. An operating pH of 3.5 was consistently maintained in the system.

DISCUSSION

A biological treatment process employing immobilized microbial populations attached to inert surfaces with defined porosity has been field tested for many industrial effluent and ground water applications. These surfaces, consisting primarily of either diatomaceous earth or fused silica can provide the basic structural support for designing a modular bioreactor treatment system. With the ability to preselect the microbial consortia, the porous spaces within each extruded bead can provide an optimal microenvironment for multiple species maintenance within close proximity. The ability to sustain biomass density during upset events in wastewater treatment is a major contributing factor for implementation by industry.

More work is required to exploit immobilized cells fully in industrial/environmental remediation applications , particularly to gain an understanding of physiological and metabolic changes within the cells in response to substrate input. Polysaccharide materials may prove to be an inexpensive support surface for immobilizing whole cell adapted microbes for multiple toxicant biotransformation of polished effluents. Marine sources of cellulose and chitin provided similar efficiencies in both immobilization and biotransformation(Portier, 1986).

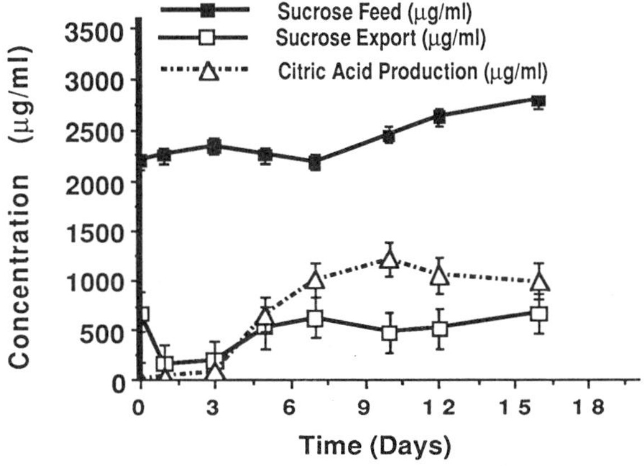

Fig. 3 Continuous treatment of molasses clarifier effluent with <u>Aspergillis niger</u>.

The economics of using adapted bacteria immobilized to inexpensive support surfaces, as compared to activated sludge systems, have been made(Portier et al, 1991). In comparing similar pilot scale tests, immobilized whole cell biocatalysts were more efficient carbon mineralization and/or complete biotransformation and equally efficient in final effluent (or clarifier) residual levels of targeted organics. Facility of immobilized cells in handling mixtures of organics associated with typical point source effluents has been demonstrated.

REFERENCES

Chakrabarty, A. 1982. Biodegradation and detoxification of environmental pollutants. CRC   Press, Inc., Boca Raton, Florida.

Chibata, I.1979. Immobilized microbial cells with polyacrylamine gel and carageenan, and their industrial applications. American Chemical Symposium Series 106, pp187-202

Edgehill, R. U. and R.K. Finn, 1983. Activated sludge treatment of synthetic wastewater containing pentachlorophenol. Bio-technology and Bioengineering, Vol. XXV, pp. 2165-2176.

Hood, M.A., 1973. Chitin degradation in the salt marsh environment. Ph.D. Dissertation Louisiana State University, Baton Rouge, 158 p.

Lewis, D.L., H. W. Holm and R. E. Hudson, 1984. Application of single and multiphasic Michaelis-Menten kinetics to predictive modeling for aquatic ecosystems. Environ. Toxicology and Chemistry, Vol. 3, pp. 563-574.

Mosbach, R., Koch-Smidt and K. Mosbach, 1976. Immobilization of enzymes to various acrylic copolymers. Enzymol. 44, 53-65.

Mathiasson, B., 1983. Immobilization methods. In Immobilized Cells and Organelles (B. Mattiasson, Ed.) CRC Press, Inc., Boca Raton, Fla., pp. 4-25.

McGhee , J. E. and J. Grant, 1982. Continuous and static fermentation of glucose to ethanol by immobilized Saccharomycetes cerevisiae cells of different ages. Appl. and Environ. Microbiol. 44(1): 19-22

Messing, R.A., Opperman, R.A. and Kolot, F.B., 1979. Pore dimensions for accumulating biomass in Immobilized Microbial Cells, ACS' Symp. Vol. 106, American Chemical Society, Washington.

Pfaender F.K. and G.W. Bartholomew, 1982. Measurement of aquatic degradation rates by determining heterotrophic uptake of radiolabelled pollutants. Appl. and Environ. Microbiol. 44(1) : 159-164

Portier, R.J., 1982. Correlative field and laboratory microcosm approaches in ascertaining xenobiotic fate and effect in diverse aquatic environments. Ph.D. Dissertation, Louisiana State Universtiy, Baton Rouge, 205 pages.

Portier, R.J., H.M. Chen and S.P. Meyers, 1983. Environmental effect and fate of selected phenols in aquatic ecosystems using microcosm approaches. Developments in Indust. Microbiol., Vol. 24, pp. 409-424.

Portier, R.J. and S.P. Meyers, 1984. Coupling of in situ and laboratory microcosm protocols for ascertaining fate and effect of xenobiotics. In. Toxicity Screening Procedures Using Bacterial Systems (D. Liu, B.J. Dutka, Eds.). Marcel Dekker, Inc., New York, pp. 345-379.

Portier, R.J., 1986. Chitin immobilization systems for hazardous waste detoxification and biodegradation.In Immobilization of Ions by Naturally Occurring Materials. (H. Eccles, Editor) Ellis Horwood Limited, Publishers, London. Chapter 6, 230-243.

Portier, R.J. , 1987. Enhanced biotransformation and biodegradation of polychlorinated biphenyls in the presence of aminopolysaccharides. American Society for Testing and Materials Special Technical Publication 971, Aquatic Toxicology, 10th volume). ,pp503-516

Portier, R.J. A. L. Zoeller and K. Fugisaki, 1990 Remediation of pesticide-contaminated   ground water using immobilized microbe biotreatment systems. Remediation ,.Vol 1, No 1, pp 41-60

Rosevear, A.1982. Improvements in or relating to composite materials. Eur Pat Appl 81304001.1

ORGANOPHOSPHORUS CHOLINESTERASE INHIBITORS:

DETOXIFICATION BY MICROBIAL ENZYMES

Joseph J. DeFrank

Biotechnology Division
U.S. Army Chemical Research, Development & Engineering Center
Aberdeen Proving Ground, Maryland 21010-5423

INTRODUCTION

Numerous organophosphorus compounds of importance in agriculture, medicine, military defense, and research have been shown to be potent inhibitors of cholinesterases and other enzymes with active serine residues in their active sites. Over the past 45 years a variety of enzymes that catalytically hydrolyze and detoxify these compounds have been described and characterized to varying degrees. Enzymes of this type have been found in both procaryotes and eucaryotes, but microbial sources have been of increased interest for a variety of reasons. These include their potential for production through large-scale fermentation, the relative ease of their physical and chemical manipulation, and their generally simpler genetic makeup which makes cloning of the enzyme genes potentially easier and more straightforward.

ORGANOPHOSPHORUS COMPOUNDS

The whole group of phosphorus-containing organic esters has been referred to by the generic term organophosphorus compounds. Most organophosphorus compounds are considered to be esters of alcohols with phosphoric acid or anhydrides of the phosphorus acid with some other acid. The structures of some of the most toxic organophosphorus compounds and the ones most commonly the target of enzymatic detoxification research are depicted in Figures 1 and 2. Figure 1 shows the G-type nerve agents such as soman, O-1,2,2-trimethylpropylmethylphosphonofluoridate; sarin, O-isopropyl methylphosphonofluoridate; and tabun, N,N-dimethylethyl phosphoroamidocyanidate. Also shown is VX, O-ethyl-S-(2-diisopropylaminoethyl) methylphosphonothioate, the most toxic of the nerve agents, and DFP, diisopropylfluorophosphate, which has been a commonly utilized model for the nerve agents and has been routinely used in research as a serine protease inhibitor. In recent years the compound NPEPP, p-nitrophenylethylphenylphosphinate, has also been used as a model for the organophosphorus nerve agents.

Figure 2 shows some of the organophosphorus pesticides that have received the most attention in the literature. They are parathion, O,O-diethyl-O-p-nitrophenylphosphorothioate, and the related paraoxon, O,O-diethyl-O-p-nitrophenylphosphate. Others that will be discussed are coumaphos, O,O-diethyl-O-(3-chloro-4-methyl-2-oxo-2H-1-benzopyran-7-yl)phosphorothioate; diazinon, O,O-diethyl(2-isopropyl-6-methyl-4-pyrimidyl) phosphorothionate; EPN, O-ethyl-O-(p-nitrophenyl)phenylphosphonothioate; fensulfothion, O,O-diethyl-O-(4-[methylsulfinyl]phenyl)phosphorothioate; methyl parathion, O,O-dimethyl-O-p-nitrophenylphosphorothioate; and mipafox, N,N'-diisopropyl phosphorodiamidofluoridate.

Applications of Enzyme Biotechnology, Edited by J.W. Kelly and
T.O. Baldwin, Plenum Press, New York, 1991

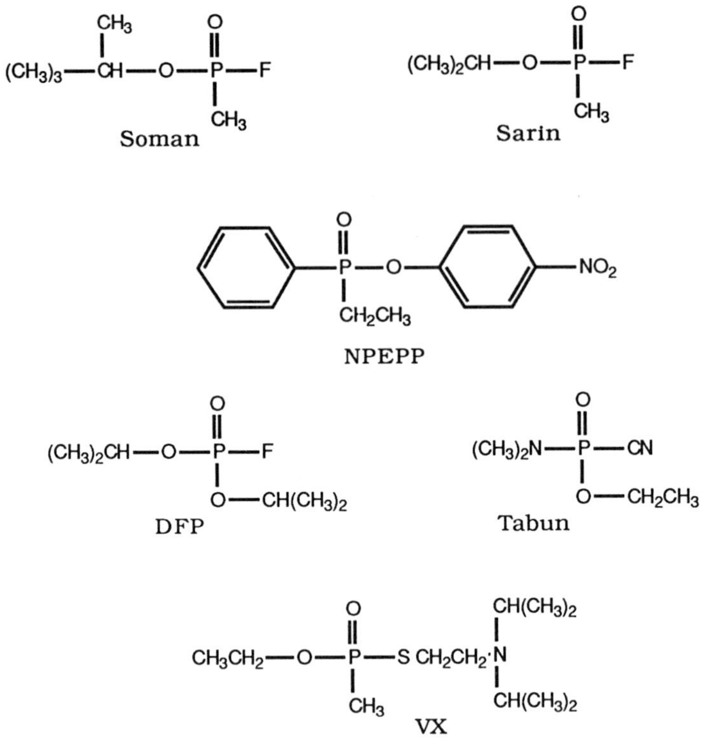

Soman

Sarin

NPEPP

DFP

Tabun

VX

Figure 1. Organophosphorus Nerve Agents and Related Compounds

Table 1. Toxicities of Organophosphorus Compounds and Toxins [58, 59]

Compound	Approx. LD_{50} (mg/kg, i.v.)
Diazinon	150-600
Coumaphos	90-110
EPN	35-45
Methyl parathion	14
Parathion	13
Fensulfothion	4.6-10.5
Paraoxon	0.5
DFP	0.3
Sarin	0.01
Soman	0.01
Tabun	0.01
VX	0.001
Palytoxin	0.00015
Maitotoxin	0.00013
Botulinum toxin	0.000001

166

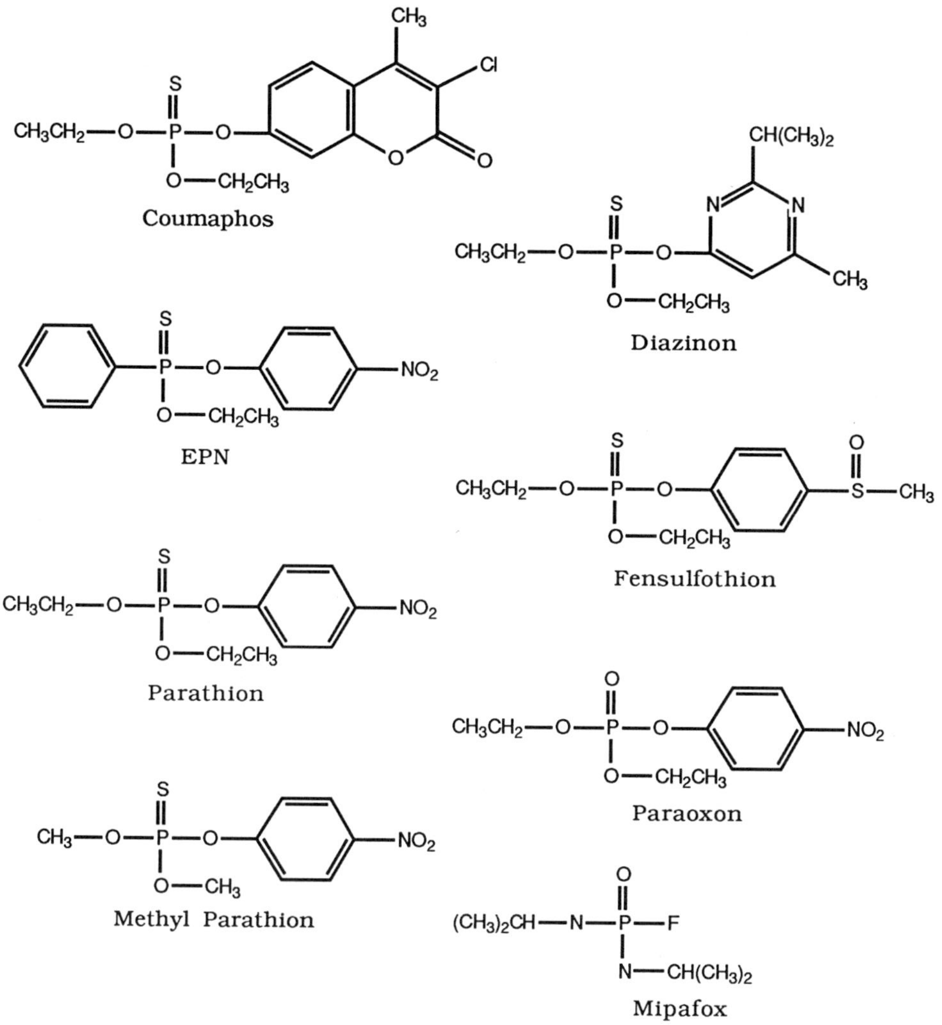

Figure 2. Organophosphorus Pesticide Substrates for Detoxifying Enzymes

Although all of these materials are considered toxic, their degree of toxicity varies considerably as is shown in Table 1.

As a point of contrast, some of the more toxic naturally occurring biomaterials have been included in Table 1. While extremely dangerous, the toxicity of even the nerve agents pale in comparison to maitotoxin, palytoxin and especially botulinum toxin. The toxicity of the latter being three orders of magnitude greater than for VX.

As yet VX has not been confirmed to be a substrate for the type of enzymes to be discussed here. Although less toxic than VX, the other nerve agents (soman, sarin and tabun) are at least an order of magnitude more toxic than DFP and paraoxon, and several orders of magnitude more toxic than most of the pesticides under study. However, the widespread use of many of these pesticides makes them potentially a more serious health and environmental problem than does the nerve agents. Because of this use and misuse of the pesticides worldwide, as well as the potential for demilitarization of chemical agents, organophosphorus degrading enzymes have been receiving a great deal more attention in recent years.

Although organophosphorus esters of the type with which this paper deals have been in existence for nearly a century, the modern era of interest can be thought to have begun in 1946 with the publishing of the synthesis of DFP [56]. The same year the earliest work dealing with enzymes capable of hydrolyzing these materials was reported by Mazur [55]. He described the hydrolysis of DFP, and thus its detoxification, to form hydrofluoric and diisopropylphophoric acids by enzymes in human and rabbit tissues. Using partially purified enzyme preparations from rabbit kidney, he was able to determine that the activity was not related to phosphatase, cholinesterase or esterase.

During the decade of the 1950's, the bulk of the work in this field was carried out by three investigators, Aldridge [5], Augustinsson [8-14] and Mounter [60, 62-66]. As had Mazur, Aldridge and Augustinsson concentrated on enzymes from mammalian sources Aldridge reported on what he designated an A-serum esterase that could hydrolyze para-oxon. This enzyme, more recently referred to as a phosphotriesterase or paraoxonase, was shown to differ from the phosphatases in that phosphatases hydrolyze monoesters of orthophosphoric acid. In addition, Aldridge showed that his A-esterases could be stereo-specific, hydrolyzing (+)-sarin but not (-)-sarin.

Augustinsson confirmed the findings of Aldridge and extended the work to include enzymes that would hydrolyze organophosphorus compounds such as tabun which has a P-N bond. He determined that this phosphorylphosphatase or tabunase cleaved the P-CN bond of tabun to release hydrocyanic acid   He also supported the idea that there were a number of phosphorylphosphatases. Using a combination of electrophoretic separation, substrate specificities, and sensitivity to inhibition, he concluded that there were three types of esterases in plasma: arylesterase (aromatic esterase, A-esterase), aliesterase (carboxylesterase, B-esterase, "lipase") and cholinesterase. He noted that there were species variation with respect of the properties of individual enzymes. He also confirmed the observation made by Aldridge that the phosphorylphosphatases showed stereo-selectivity (for hog kidney with tabun).

During this same time period, Mounter was attempting to further purify and characterize the enzyme from hog kidney originally reported by Mazur. He referred to this enzyme as dialkylfluorophosphatase (DFPase, fluorophosphatase). After partial purification by ethanol fractionation, it was determined that its activity was activated by Co^{2+} and Mn^{2+} ions. Reagents that reacted with metal ions or sulfhydryl and carbonyl groups were found to inhibit DFPase activity. Of particular importance for this paper, Mounter was the first to report on the DFPases in microorganisms. Of the bacteria tested, the greatest activity was observed with the gram-negative *Proteus vulgaris* and *Pseudomonas aeruginosa*, which were also Mn^{2+} stimulated. Based on the responses to metal ions and inhibitors, it was also demonstrated that there were a number of different DFPases. Studies conducted with preparations from *Escherichia coli*, *Pseudomonas fluorescens*, *Streptococcus faecales*, and *Propionibacterium pentosaceum*, showed that while differences were found in the relative rates of hydrolysis of a variety of organophosphorus compounds, they were comparable to those observed with the hog kidney enzyme. In addition to his own investigations, Mounter published an excellent review which covered the area of research from Mazur through the early 60's [61].

During the late 50's and early 60's, a number of additional groups became involved in the investigation of the hydrolysis of DFP, paraoxon, sarin and tabun [1-3, 21, 32, 45, 53, 54]. While considerable advances were made during this time by several of these groups in comprehending the diversity of enzymes, one of the most significant events was the beginning of the research efforts of Hoskin in this field. Of particular importance is his work, beginning in 1966, in the purification and characterization of the DFPase from squid [29, 30, 33-35, 40-44]. The significance of the squid enzyme lies in the fact that it has major differences from all the other DFPases. The differences were great enough that in 1984, Hoskin proposed that the DFPases could be grouped into two categories, the squid-type (for which there was one example) and all others which were referred to as Mazur-type DFPases [39]. A summary of the properties of these enzyme types is shown in Table 2. As will be seen later in this paper, these categories are no longer as applicable as

Table 2. Properties of Organophosphorus Acid (OPA) Anhydrases [36]

Squid-Type	Mazur-Type
Narrow distribution, squid nerve, saliva, hepatopancreas	Ubiquitous
Molecular weight, 30,000	Variable, 45-90,000
Soman/DFP $\approx$ 0.25	Soman/DFP, 5-50 and higher
Hydrolyzes all isomers of soman; some stereoselectivity in rates	Stereoselectivity variable: often quite stereospecific
Mn^{2+} indifferent or slightly inhibited	Mn^{2+} stimulated 2- to 20-fold and as high as 80-fold
Ca^{2+} requiring, not Ca^{2+} stimulated	May be Mg^{2+} requiring and stimulated
$(NH_4)_2SO_4$ indifferent	$(NH_4)_2SO_4$ labile
Mipafox indifferent	Mipafox inhibited

originally believed, due primarily to the results obtained with a number of microbial enzymes.

The interest in microbial enzymes for the degradation of organophosphorus compounds received a boost in the early 70's by the isolation of bacteria capable of growing on a variety of pesticides. The initial report was by Sethunathan and Yoshida who isolated a diazinon-degrading *Flavobacterium* sp. (ATCC 27551) from rice paddy soil [80, 81]. Cell-free extracts of this organism could also hydrolyze the insecticides chlorpyrifos [diethyl (3,5,6-trichloropyridyl) phosphorothionate] and parathion, the aromatic or heterocyclic products of which were not further metabolized. In 1973 a pseudomonad capable of hydrolyzing parathion and utilizing the *p*-nitrophenol product as a source of carbon and nitrogen was isolated by Siddaramappa et al. [84]. A *Pseudomonas stutzeri* capable of hydrolyzing parathion was isolated by Daughton and Hsieh from a chemostat culture [23]. Two *Pseudomonas* spp. were described by Rosenberg and Alexander [75] that were capable of hydrolyzing a variety of organophosphorus compounds and using the products as sole phosphorous source. Strains of *Bacillus* and *Arthrobacter* that can hydrolyze parathion were also described [71] as well as a *Pseudomonas* capable of utilizing isophenfos as sole energy and carbon source [73]. With the exception of the *Flavobacterium*, little if any additional information has been reported about these organisms.

In addition to the *Flavobacterium* mentioned above, the other major parathion-degrading bacteria described in the literature is *Pseudomonas diminuta* MG which was isolated in 1976 [69]. The parathion hydrolases from both the *Flavobacterium* sp. and *Pseudomonas diminuta* MG will be discussed in more detail in a following section.

At about this same time Zech et al. [89] reported that the organophosphorus decontaminating DFPase and paraoxonase in *E. coli* K_{12} could be separated from one another by gel filtration and isoelectric focusing. The enzymes showed no overlapping activity and appeared to have different pH optima (9.3 versus 8.3 respectively). Both enzymes were rather unstable, losing about 10 to 30% of their activity per day at room temperature.

More recently, the research efforts in this field have been divided into two major areas: the isolation and characterization of microorganisms (and their enzymes) capable of growth on variety of organophosphorus pesticides and the more random search for organisms that possess enzymes that will hydrolyze DFP and the related nerve agents. Examples of the former are the isolation of *Pseudomonas alcaligenes* C_1 which can hydrolyze and grow on fensulfothion [82]; the isolation of additional *Pseudomonas* sp. and other unidentified bacteria that hydrolyze and grow on parathion and/or methyl parathion [17, 67]; and the isolation of three distinct bacteria capable of metabolizing coumaphos [83].

In the search for nerve agent degrading enzymes, the most recent investigations have gone in a number of directions. Landis and co-workers have examined the ciliate protozoan *Tetrahymena thermophila* [48-50] and the clam *Rangia cuneata* [6]. Partial purification of extracts from *Tetrahymena* has revealed that this organism has at least five enzymes with DFPase activity and molecular weights ranging from 67-96,000. The ratios of rates for soman and DFP hydrolysis as well as the effect of Mn^{2+} on activity varied considerably from one enzyme to another. None of them appear to fit neatly into the squid-type or Mazur-type enzyme categories. The preliminary investigations on the clam have also resulted in the detection of several DFPases with differing characteristics of substrate specificity and metal stimulation. Of particular interest was the presence of and enzyme in the clam digestive gland that appears to have significant activity on the DFP analog mipafox. Most enzymes described to date are either indifferent to mipafox or subject to fairly strong competitive inhibition. In the latter case little or no hydrolysis of mipafox is generally observed, thus making the clam enzyme quite unique.

Little et al. [52] have reported on the characterization of an enzyme from rat liver that has a substrate preference of sarin > soman > tabun > DFP but with no activity on paraoxon. Preliminary investigations have shown the enzyme to have a molecular weight of 40,000 and to be stimulated by Mg^{2+} [15]. All stereoisomers of soman were hydrolyzed at equal rates [15]. Once again we have an enzyme that does not fit neatly into either of the squid-type or Mazur-type categories.

In the description of the types of organophosphorus compounds, it was mentioned that VX has not been determined to be a substrate for the organophosphorus-degrading enzymes. Preliminary studies have indicated that VX appears to be degraded in soil [46] as well as act as the carbon, sulfur, and phosphorus source for mixed bacterial cultures [4]. Therefore, it does appear likely that organisms and enzymes for the degradation of this highly toxic material will eventually be found.

The search for bacteria capable of degrading DFP and the nerve agents has yielded mixed results. A screen of 18 gram-negative bacterial isolates by Attaway et al. [7] resulted in the finding that of the organisms tested, while all showed at least some activity on DFP, only parathion hydrolase producing cultures gave significant levels. Only two bacterial enzymes that have activity against either DFP or the nerve agents, in addition to parathion hydrolase, have been purified to any significant extent and will be discussed in more detail below. These enzymes, which differ considerably in a number of parameters, were obtained from thermophilic and halophilic bacteria, and as will be seen, again do not necessarily conform to the squid-type or Mazur-type categories.

Much of the current work going on in this field deals with the cloning and sequencing of the genes coding for these enzymes. The sequence information, plus an examination of their activity with a variety of additional substrates may help in the attempt to gain an understanding of what the native function and substrate(s) for these enzymes might be. Another major area of investigation involves the development of decontamination formulations or immobilization systems to make practicaluse of these enzymes. There are two references that could be of assistance to anyone interested in getting a greater depth of information on more recent developments in this field [47, 74].

NOMENCLATURE

As illustrated in the discussion above, the nomenclature of these enzymes has been unsystematic and confusing. The names utilized have been representative of the particular substrate used by an individual investigator. Hence, the literature is filled with references to enzymes such as phosphorylphosphatase, fluorophosphatase, DFPase, paraoxonase, parathion hydrolase, phosphotriesterase, phosphofluorase, somanase, sarinase, and tabunase. At the First DFPase Workshop held 4-6 June 1987 at the Marine Biological Laboratory, Woods Hole, MA, it was agreed to use the name organophosphorus acid (OPA) anhydrase as a generic name for all of the enzymes that are capable of catalytically hydrolyzing the subject organophosphorus compounds. The name OPA anhydrase more adequately defined the function and substrate specificities of these enzymes and could be utilized until the natural substrates have been identified. For the remainder of this paper therefore, the term OPA anhydrase will be used except in the case of parathion hydrolase, which has been the name used most uniformly in the considerable body of literature dealing with its characterization.

COMPARISON OF THREE BACTERIAL OPA ANHYDRASES

<u>Parathion Hydrolase</u>

By far, the bacterial OPA anhydrase that has been most studied is parathion hydrolase, with much of the work conducted here at Texas A&M in the laboratories of Frank Raushel and Jim Wild. The majority of the work with this enzyme has been conducted with two bacterial strains, *Flavobacterium* sp. (ATCC 27551) and *Pseudomonas diminuta* MG. In both organisms the gene coding for this enzyme (designated as the *opd* gene) was found to reside on large plasmids [68, 78]. Subsequent cloning and sequencing of the *P. diminuta* MG [57, 77, 79] and *Flavobacterium* [31] *opd* gene have shown that while the plasmids in the organisms are highly dissimilar, the *opd* genes are essentially identical. In both organisms the native parathion hydrolase is a membrane bound protein with a molecular weight of 35,000. There are indications that in the cell membrane the enzyme forms dimeric complexes of ~65,000 molecular weight. It has been determined that most of the enzyme activity can be removed by washing whole cells with 1 M NaCl and 1% Triton X-100 [26]. Parathion hydrolase has a broad substrate range with the organophosphorus pesticides [70]. A sampling of some of these compounds and their relative rates of hydrolysis is given in Table 3.

This enzyme is the only bacterial OPA anhydrase for which the mechanism has been determined. In an elegant series of experiments utilizing ^{18}O-labelled water and paraoxon, as well as the chiral substrate EPN, it was demonstrated that the reaction was a single in-line displacement by an activated water molecule directly at the phosphorus [51]. Since this was the subject of a paper in the 8th Annual IUCCP Symposium (Caldwell

Table 3. Hydrolysis of Organophosphorus Pesticides by Parathion Hydrolase [20]

Compound	Hydrolysis (%)
Coumaphos	100.0
Fensulfothion	100.0
Paraoxon	100.0
Parathion	98.4
Diazinon	72.1
Chlorpyrifos	37.2
Methyl parathion	31.1
EPN	29.3
Malathion	16.4
Fenitrothion	7.9

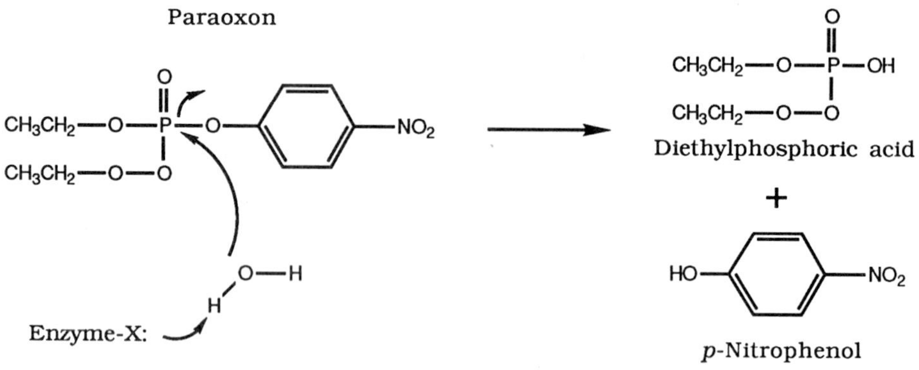

Figure 3. Detoxification of Paraoxon by Parathion Hydrolase [51]

and Raushel), the discussion of this feature of the enzyme will be limited to illustrating the mechanism of reaction with paraoxon (Figure 3).

Because of the potential difficulties in dealing with a membrane bound enzyme and in an effort to increase production of the enzyme, much of the more recent work in this area has focused on the cloning and expression of the *opd* gene into better host systems. While the earlier efforts involved cloning into *E. coli* to allow sequencing of the gene [57, 77, 79], more recently the *opd* gene has been placed into more novel hosts. The groups at Texas A&M have inserted the gene into a baculovirus expression vector which is used to infect insect tissue culture cells from the Fall Armyworm, *Spodoptera frugiperda* (sf9 cells) [26]. While good yields of the enzyme are reported [88], there is a question of whether or not the enzyme has had its signal sequence removed by posttranslational modification as occurs in the native *Flavobacterium* or *Pseudomonas*.

Work conducted at the University of Maryland in Baltimore has concentrated on cloning and expressing the parathion hydrolase from *Flavobacterium* in *Streptomyces lividans* [22, 72, 76, 85]. Using the plasmid pRYE1 with the *opd* gene, *S. lividans* was shown to produce the enzyme in a soluble form that is secreted into culture medium at a level of 10 to 15 U/ml. The high level of production of a soluble protein, in addition to its secretion, greatly simplifies the process of purification.

The considerable diversity of organophosphate pesticides that parathion hydrolase can utilize as substrates makes it very attractive for use in the detoxification of these materials as has been proposed by a number of groups [16, 22, 87]. However, the relatively low activity observed with DFP and the nerve agents sarin and soman, would indicate that parathion hydrolase is not as well suited for the detoxification of these compounds which are of military interest [27, 72]. The two bacterial OPA anhydrase to be discussed next may be of greater utility in this area.

<u>Thermophilic Bacterial OPA Anhydrase</u>

A gram-positive, aerobic, spore-forming, rod-shaped, obligate thermophile was isolated from the soil of Aberdeen Proving Ground and found to possess activity against DFP. This organism, designated as isolate JD-100, has tentatively been identified as a strain of *Bacillus stearothermophilus* [28]. In most respects JD-100 appears to possess most of the common characteristics of this species. Its temperature range for growth is from 40 to 70°C, with an optimum in the vicinity of 55°C. At this temperature the doubling time of a log phase culture is approximately one hour. The initial screening of the organism for OPA anhydrase activity was done with DFP and whole cell extracts, rather than just the soluble fraction, which turned out to be fortunate. Although whole cell crude extracts of JD-100 have low levels of DFP activity [24], the partially purified

172

Table 4. Enzymatic Degradation of Soman [19, 44]

OPA Anhydrase Source	Hydrolysis (%)[a]	Detoxification (%)[b]
Squid	73	52
Thermophile JD-100	49	39
Hog kidney	53	3
E. coli	51	0

[a] Assay by fluoride-selective electrode
[b] Assay by acetylcholinesterase (AChE) inhibition

OPA anhydrase showed no detectable hydrolysis of this compound [19, 37, 38]. It appears that the DFP activity was due to a enzyme associated with the cell membrane so that if the soluble fraction had been tested, no activity would have been observed and the organism would not have been examined further. This demonstrates the fact that some of the substrates that have been used as models for the nerve agents are in reality not nearly as good as was once believed. This result also shows that, like many organisms, JD-100 possesses more than one enzyme.

One of the characteristics of the JD-100 OPA anhydrase that is of considerable importance is its ability to apparently degrade both the toxic as well as non-toxic isomers of soman. This is illustrated in Table 4 in which the enzyme is compared to several OPA anhydrases from various sources. Although some stereoselectivity can be inferred from the values for the squid and JD-100, they are not nearly as dramatic as for the hog kidney and *E. coli.*

As with many of the Mazur-type OPA anhydrases, the JD-100 enzyme shows a considerable stimulation by the addition of Mn^{2+}. However, the observed 80-fold stimulation is much higher than generally observed with other enzymes [38]. Its molecular weight, 82-84,000, would also put in the Mazur-type category. However, this enzyme shows other properties that are more the classical squid-type in nature. It is stable to treatment with $(NH_4)_2SO_4$ to at least 50% saturation and the DFP analog mipafox does not inhibit the hydrolysis of soman.

The JD-100 enzyme has been immobilized on both cotton and agarose and found to retain its hydrolytic activity against soman and show to good stability [28].

<u>Halophilic Bacterial OPA Anhydrase</u>

Other than parathion hydrolase, the only bacterial OPA anhydrase that has been purified to homogeneity and characterized is an enzyme from an obligately halophilic bacterial isolate designated JD6.5 [25]. This organism was isolated from Grantsville Warm Springs, which is located approximately 48 km west of Salt Lake City, Utah, and just south of the Great Salt Lake. Isolate JD6.5 is a gram-negative, aerobic, short rod and requires at least 2% NaCl for growth. Fatty acid analysis (Microbial ID, Newark, DE) has tentatively identified JD6.5 as a species of *Alteromonas*, but further characterization is required. This isolate was found to possess very high levels of DFP-hydrolyzing OPA anhydrase activity.

As shown in Figure 4, the bulk of the DFP activity resides in a peak obtained by DEAE-Sephacel chromatography and designated as OPAA-2. This enzyme was purified 1,000-fold to homogeneity and was found to be a single polypeptide with a molecular weight of 60,000. This enzyme, with DFP as substrate, has a pH optimum of 8.5, a temperature optimum of 50°C, and has maximal activity with either manganese or cobalt.

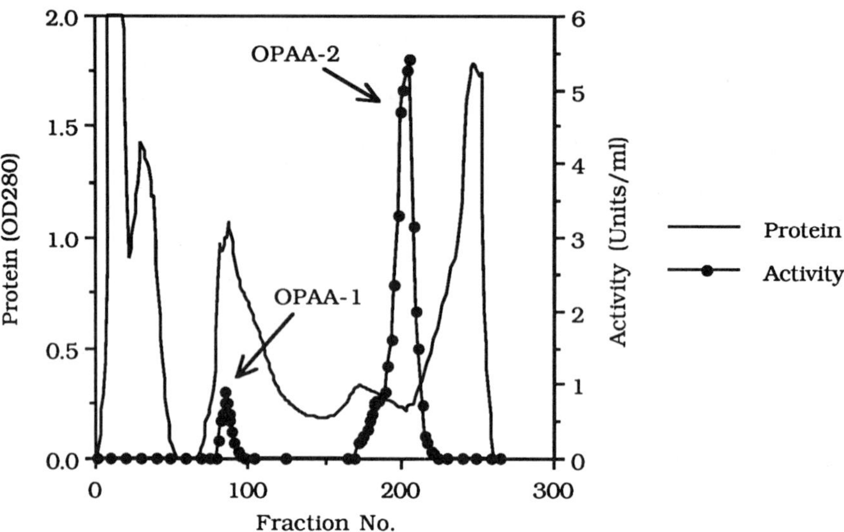

Figure 4. DEAE-Sephacel Chromatography of Halophile JD6.5 Cell Extract

The pH studies, stimulation by reducing agents such as DTT or β-mercaptoethanol, and inhibition by *p*-chloromercuribenzoate (PCMB), iodoacetic acid (IAA), and *N*-ethyl-maleimide (NEM), all suggest the presence of an active sulfhydryl group (cysteine) in the active site of OPAA-2 [25].

Through the use of monoclonal antibodies and polyclonal serum raised against purified OPAA-2, the enzyme preparations at different stages in the purification process were analyzed by Western blotting after SDS-PAGE (Figure 5). The two protein bands with molecular weights of 74,000 and 78,000 correspond to the OPAA-1 peak in Figure 4. The fact that both the monoclonal antibody and antiserum react more strongly to OPAA-1 than OPAA-2 suggests two possible explanations. OPAA-1 may be the precursor(s) for OPAA-2 and produced at high concentrations but with low specific activity against DFP. Alternatively, these proteins may all be descended from a common ancestral protein and still retain antigenic similarity while possibly differing significantly in specificity. Preliminary results in efforts to clone and sequence the gene for OPAA-2 appear to indicate that the first explanation, that OPAA-2 is derived from OPAA-1, is the correct one [18]. This would also be similar to the posttranslational modification that parathion hydrolase undergoes.

As with many of the OPA anhydrases, the natural substrate for this enzyme has yet to be determined. A variety of potential substrates for esterases, phosphatases, phosphodiesterases, phosphotriesterases, and phospholipases were examined, but all showed little or no activity with OPAA-2. Six substrates have been demonstrated to be hydrolyzed at significant levels. The highest activity is observed with soman, followed by DFP, NPMPP (the methyl analog of NPEPP), NPEPP, sarin and paraoxon. The specific activity observed with soman at pH 7.2, 25°C and without Mn^{2+} was 619 μmoles/min/mg (k_{cat} = 619 s⁻¹) Thiis the highest reported in the literature for any OPA anhydrase. Preliminary NMR experiments indicate that all stereoisomers of soman and sarin were hydrolyzed at equal rates by OPAA-2 [86].

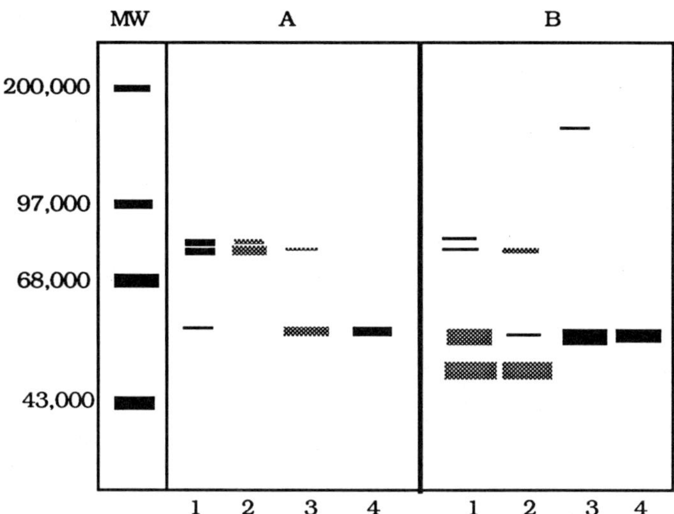

Figure 5. Western Blot of JD6.5 OPAA-2 Preparations. (A) Monoclonal antibody.
(B) Polyclonal serum. (1) Crude extract. (2) DEAE fractions (OPAA-1).
(3) DEAE fractions (OPAA-2). (4) Purified OPAA-2. (MW) Molecular weight
standards.

<u>SUMMARY</u>

A comparison of the substrate specificities and other properties of the three bacterial enzymes discussed is shown in Table 5. Because of differences in assay conditions (pH, reaction mixture, substrate concentration, etc.), the substrate which has given the highest activity for a particular enzyme has been arbitrarily set at 100%. Unlike most previously described bacterial enzymes, the three detailed here are all stable to treatment with $(NH_4)_2SO_4$.

As can be seen, none of these enzymes fit neatly into the squid-type or Mazur-type categorie of OPA anhydrases described previously. The parathion is squid-type in its molecular weight, soman/DFP ratio, and lack of effect by Mn^{2+}. However, inhibition by mipafox is a Mazur-type property. The thermophilic JD-100 enzyme is Mazur-type in regard to its molecular weight, preference for soman, and stimulation by Mn^{2+}. Its lack of inhibition by mipafox or $(NH_4)_2SO_4$ are squid-type traits however. The halophilic enzyme from JD6.5 is much more Mazur-type with its molecular weight, soman/DFP ratio, stimulation by Mn^{2+}, and competitive inhibition by mipafox. However, its stability to $(NH_4)_2SO_4$ has in the past been associated with the squid-type enzymes.

Two general conclusions can be drawn from the work that has been summarized above. First, the diversity of enzymes encountered, especially in the past decade, suggest that these OPA anhydrases are members of many different classes between which there may be only minimal similarity. That similarity may be restricted to the fact that many of them can react to some extent with the organophosphorus compounds described in this paper, but may have entirely different natural substrates and functions. Second, there is such a variety of enzymes being identified that it should be feasible to develop decontamination formulations or systems that can deal with most, if not all, of the various toxic organophosphorus compounds that are of importance in agriculture, industry, military defense and environmental restoration.

Table 5. Comparison of Three Bacterial OPA Anhydrases [19, 25, 27]

	Parathion Hydrolase	Thermophilic (JD-100)	Halophilic (JD6.5)
Molecular Weight	35,000	84,000	60,000
Relative Activity (%)			
DFP	2.7	0.0	41.7
NPEPP	32.4	~2.0	16.7
NPMPP	N.R.	~16.0	20.5
Paraoxon	100.0	0.0	1.6
Sarin	1.9	N.R.	8.4
Soman	0.2	100.0	100.0
Metal Involvement	Zinc	Manganese	Manganese
Mipafox Inhibition	Yes	No	Yes

N.R. = not reported

REFERENCES

1.   Adie, P. A. "The purification of sarinase from bovine plasma." Can. J. Biochem. Physiol. **34**: 1091-1094, 1956.
2.   Adie, P. A., F. C. G. Hoskin and G. S. Trick. "Kinetics of the enzymatic hydrolysis of sarin." Can. J. Biochem. Physiol. **34**: 80-82, 1956.
3.   Adie, P. A. and J. Tuba. "The intracellular localization of liver and kidney sarinase." Can. J. Biochem. Physiol. **36**: 21-24, 1958.
4.   Albizo, J. J. and W. E. White. Personal communication.
5.   Aldridge, W. N. "Serum esterases 1. Two types of esterase (A and B) hydrolyzing p-nitrophenyl acetate, propionate and butyrate and a method for their determination." Biochem. J. **53**: 110-124, 1953.
6.   Anderson, R. S., H. D. Durst and W. G. Landis. "Initial characterization of the organophosphate acid anhydrase activity in the clam, Rangia cuneata." Comp. Biochem. Phys. **91C**: 575-579, 1988.
7.   Attaway, H., J. O. Nelson, A. M. Baya, M. J. Voll, W. E. White, D. J. Grimes and R. R. Colwell. "Bacterial detoxification of diisopropyl fluorophosphate." Appl. Envir. Microbiol. **53**: 1685-1689, 1987.
8.   Augustinsson, K.-B. "The enzymatic hydrolysis of organophosphorus compounds." Biochem. Biophys. Acta. **13**: 303-304, 1954.
9.   Augustinsson, K.-B. and G. Heimburger. "Enzymatic hydrolysis of organophosphorus compounds I, Occurrence of enzymes hydrolyzing dimethylamido-ethoxy-phosphoryl cyanide (tabun)." Acta Chem. Scand. **8**(5): 753-761, 1954.
10.   Augustinsson, K.-B. and G. Heimburger. "Enzymatic hydrolysis of organophosphorus compounds II. Analysis of reaction products in experiments with tabun and some properties of blood plasma tabunase." Acta Chem. Scand. **8**(5): 762-767, 1954.
11.   Augustinsson, K.-B. and G. Heimburger. "Enzymatic hydrolysis of organophosphorus compounds IV. Specificity studies." Acta Chem. Scand. **8**(9): 1533-1541, 1954.
12.   Augustinsson, K.-B. and G. Heimburger. "Enzymatic hydrolysis of organophosphorus compounds V. Effect of phosphorylphosphatase on inactivation of cholinesterases by organophosphorus compounds in vitro." Acta Chem. Scand. **9**(2): 310-318, 1955.
13.   Augustinsson, K.-B. and G. Heimburger. "Enzymatic hydrolysis of organophosphorus compounds VI. Effect of metallic ions on the phosphorylphosphatases of human and swine kidney." Acta Chem. Scand. **9**(3): 383-392, 1955.

14. Augustinsson, K.-B. and G. Heimburger. "Enzymatic hydrolysis of organophosphorus compounds VII. The stereospecificity of phosphorylphosphatases." Acta Chem. Scand. **11**(8): 1371-1377, 1957.
15. Broomfield, C. A. Personal communication.
16. Caldwell, S. R. and F. M. Raushel. "Detoxification of organophosphate pesticides using an immobilized phosphotriesterase from *Pseudomonas diminuta*." Biotechnol. Bioeng. **37**: 103-109, 1991.
17. Chaudry, G. R., A. N. Ali and W. B. Wheeler. "Isolation of a methyl parathion-degrading *Pseudomonas* sp. that possesses DNA homologous to the *opd* gene from a *Flavobacterium* sp." Appl. Environ. Microbiol. **54**: 288-293, 1988.
18. Cheng, T.-c. "Unpublished data." :
19. Chettur, G., J. J. DeFrank, B. J. Gallo, F. C. G. Hoskin, S. Mainer, F. M. Robbins, K. E. Steinmann and J. E. Walker. "Soman-hydrolyzing and -detoxifying properties of an enzyme from a thermophilic bacterium." Fund. Appl. Toxicol. **11**: 373-380, 1988.
20. Chiang, T., M. C. Dean and C. S. McDaniel. "A fruit fly bioassay with phosphotriesterase for detection of certain organophosphorus insecticide residues." Bull. Environ. Contam. Toxicol. **34**: 809-814, 1985.
21. Cohen, J. A. and M. G. P. J. Warringa. "Purification and properties of dialkylfluorophosphatase." Biochem. Biophys. Acta. **26**: 29-39, 1957.
22. Coppella, S. J., N. DelaCruz, G. F. Payne, B. M. Pogel, M. K. Speedie, J. S. Karns, E. M. Sybert and M. A. Connor. "Genetic engineering approach to toxic waste management: Case study for organophosphate waste treatment." Biotechnol. Prog. **6**: 76-81, 1990.
23. Daughton, C. G. and D. P. H. Hsieh. "Parathion utilization by bacterial symbionts in a chemostat." Appl. Environ. Microbiol. **34**: 175-184, 1977.
24. DeFrank, J. J. "Unpublished data." :
25. DeFrank, J. J. and T.-c. Cheng. "Purification and properties of an organophosphorus acid anhydrase from a halophilic bacterial isolate." J. Bacteriol. **173**: 1991.
26. Dumas, D. P., S. R. Caldwell, J. R. Wild and F. M. Raushel. "Purification and properties of the phosphotriesterase from *Pseudomonas diminuta*." J. Biol. Chem. **264**: 19659-19665, 1989.
27. Dumas, D. P., H. D. Durst, W. G. Landis, F. M. Raushel and J. R. Wild. "Inactivation of organophosphorus nerve agents by the phosphotriesterase from *Pseudomonas diminuta*." Arch. Biochem. Biophys. **277**: 155-159, 1990.
28. Gallo, B. J. Personal communication.
29. Garden, J. M., S. K. Hause, F. C. G. Hoskin and A. H. Roush. "Comparison of DFP-hydrolyzing enzyme purified from head ganglia and hepatopancreas of squid (*Loligo pealei*) by means of isoelectric focusing." Comp. Biochem. Physiol. **52C**: 95-98, 1975.
30. Gay, D. D. and F. C. G. Hoskin. "Stereospecificity and active site requirements in a diisopropylphosphorofluoridate-hydrolyzing enzyme." Biochem. Pharmacol. **28**: 1259-1261, 1979.
31. Harper, L. L., C. S. McDaniel, C. E. Miller and J. R. Wild. "Dissimilar plasmids isolated from *Pseudomonas diminuta* MG and a *Flavobacterium* sp. (ATCC 27551) contain identical *opd* genes." Appl. Environ. Microbiol. **54**: 2586-2589, 1988.
32. Hoskin, F. C. G. "The enzymatic hydrolysis products of sarin." Can. J. Biochem. Physiol. **34**: 75-79, 1956.
33. Hoskin, F. C. G. "Possible significance of "DFPase" in squid nerve." Biol. Bull. **137**: 389-390, 1969.
34. Hoskin, F. C. G. "Diisopropylphosphorofluoridate and tabun: Enzymatic hydrolysis and nerve function." Science. **172**: 1243-1245, 1971.
35. Hoskin, F. C. G. "Distribution of diisopropylphosphorofluoridate-hydrolyzing enzyme between sheath and axoplasm of squid giant axon." J. Neurochem. **26**: 1043-1045, 1976.
36. Hoskin, F. C. G. "An organophosphorus detoxifying enzyme unique to squid." Squid as Experimental Animals. Gilbert and Adelman ed. in press Plenum. New York.
37. Hoskin, F. C. G., G. Chettur, S. Mainer, K. E. Steinmann, J. J. DeFrank, B. J. Gallo, F. M. Robbins and J. E. Walker. "Soman hydrolysis and detoxication by a thermo-

philic bacterial enzyme." Enzymes Hydrolysing Organophosphorus Compounds. Reiner, Aldridge and Hoskin ed. 1989 Ellis Horwood. England.

38. Hoskin, F. C. G., J. J. DeFrank, B. J. Gallo and J. E. Walker. Isomers of soman as research tools for the study of organophosphorus acid (OPA) anhydrases (formerly DFPases). Third International Symposium on Protection Against Chemical Warfare Agents. 187-194, 1989.

39. Hoskin, F. C. G., M. A. Kirkish and K. E. Steinmann. "Two enzymes for the detoxification of organophosphorus compounds - sources, similarities and significance." Tox. Appl. Pharmacol. **4**: S165-S172, 1984.

40. Hoskin, F. C. G. and R. J. Long. "Purification of a DFP-hydrolyzing enzyme from squid head ganglion." Arch. Biochem. Biophys. **150**: 548-555, 1972.

41. Hoskin, F. C. G. and R. D. Prusch. "Characterization of a DFP-hydrolyzing enzyme in squid posterior salivary gland by use of soman, DFP and manganous ion." Comp. Biochem. Physiol. **75C**(1): 17-20, 1983.

42. Hoskin, F. C. G., K. S. Rajan and K. E. Steinmann. "Organophosphorus acid (OPA) anhydrase from squid: a calcium-dependent P-F splitting enzyme." Biol. Bull. **175**: 305-306, 1988.

43. Hoskin, F. C. G., P. Rosenbery and M. Brzin. "Re-examination of the effect of DFP on electrical and cholinesterase activity of squid giant axon." Proc. Nat. Acad. Sci. USA. **55**: 1231-1235, 1966.

44. Hoskin, F. C. G. and A. H. Roush. "Hydrolysis of nerve gas by squid type diisopropylphosphorofluoridate hydrolyzing enzyme on agarose resin." Science. **215**: 1255-1257, 1982.

45. Hoskin, F. C. G. and G. S. Trick. "Stereospecificity in the enzymatic hydrolysis of tabun and acetyl-ß-methylcholine chloride." Can. J. Biochem. Physiol. **33**: 963-969, 1955.

46. Kaaijk, J. and C. Frijlink. "Degradation of S-2-diisopropylaminoethyl O-ethyl methylphosphonothiolate in soil. Sulphur-containing products." Pestic. Sci. **8**: 510-514, 1977.

47. Landis, W. G. and J. J. DeFrank. Enzymatic hydrolysis of toxic organofluorophosphate compounds. International Workshop on Biotechnology and Biodegradation. **4**: 183-201, 1989.

48. Landis, W. G., D. M. Haley, M. V. Haley, D. W. Johnson, H. D. Durst and R. E. Savage Jr. "Discovery of multiple organofluorophosphate hydrolyzing activities in the protozoan *Tetrahymena thermophila*." J. Appl. Toxicol. **7**(1): 35-41, 1987.

49. Landis, W. G., M. V. Haley and D. W. Johnson. "Kinetics of the DFPase activity in *Tetrahymena thermophila*." J. Protozool. **33**: 216-218, 1986.

50. Landis, W. G., R. E. Savage Jr. and F. C. G. Hoskin. "An organofluorophosphate-hydrolyzing activity in *Tetrahymena thermophila*." J. Protozool. **32**(3): 517-519, 1985.

51. Lewis, V. E., W. J. Donarski, J. R. Wild and F. M. Raushel. "Mechanism and stereochemical course at phosphorus of the reaction catalyzed by a bacterial phosphotriesterase." Biochemistry. **27**: 1591-1597, 1988.

52. Little, J. S., C. A. Broomfield, L. J. Boucher and M. K. Fox-Talbot. "Partial characterization of a rat liver enzyme that hydrolyzes sarin, soman, tabun and DFP." Fed. Proc. **45**(4): 791, 1986.

53. Main, A. R. "The differentiation of the A-type esterases in sheep serum." Biochem. J. **75**: 188-195, 1960.

54. Main, A. R. "The purification of the enzyme hydrolyzing diethyl *p*-nitrophenyl phosphate (paraoxon) in sheep serum." Biochem. J. **74**: 10-20, 1960.

55. Mazur, A. "An enzyme in animal tissue capable of hydrolyzing the phosphorus-fluorine bond of alkyl fluorophosphates." J. Biol. Chem. **164**: 271-289, 1946.

56. McCombie, H. and B. C. Saunders. "Alkyl fluorophosphonates: Preparation and physiological properties." Nature (London). **157**: 287-289, 1946.

57. McDaniel, C. S., L. L. Harper and J. R. Wild. "Cloning and sequencing of a plasmid-borne gene (*opd*) encoding a phosphotriesterase." J. Bacteriol. **170**: 2307-2311, 1988.

58. McEwen, F. L. and G. R. Stephenson. "The Use and Significance of Pesticides in the Environment." 1979 Wiley-Interscience. New York.

59. Middlebrook, J. L. and R. B. Dorland. "Bacterial toxins: Cellular mechanisms of action." Microbiol. Rev. **48**: 199-221, 1984.

60. Mounter, L. A. "The complex nature of dialkylfluorophosphatases of hog and rat liver and kidney." J. Biol. Chem. **215**: 705-709, 1955.

61. Mounter, L. A. Metabolism of organophosphorus anticholinesterase agents. Hanbuch de Experimentellen Pharmakologie: Cholinesterases and Anticholinesterase Agents. 486-504, 1963.

62. Mounter, L. A., R. F. Baxter and A. Chanutin. "Dialkylfluorophosphatases of microorganisms." J. Biol. Chem. **215**: 699-704, 1955.

63. Mounter, L. A. and L. T. H. Dien. "Dialkylfluorophosphatase of kidney V. The hydrolysis of organophosphorus compounds." J. Biol. Chem. **219**: 685-690, 1956.

64. Mounter, L. A., L. T. H. Dien and A. Chanutin. "The distribution of dialkylfluorophosphatases in the tissues of various species." J. Biol. Chem. **215**: 691-697, 1955.

65. Mounter, L. A., C. S. Floyd and A. Chanutin. "Dialkylfluorophosphatase of kidney I. Purification and properties." J. Biol. Chem. **204**: 221-232, 1953.

66. Mounter, L. A. and K. D. Tuck. "Dialkylfluorophosphatase of microorganisms II. Substrate specificity studies." J. Biol. Chem. **221**: 537-541, 1956.

67. Mulbry, W. W. and J. S. Karns. "Purification and characterization of three parathion hydrolases from Gram-negative bacterial strains." Appl. Environ. Microbiol. **55**: 289-293, 1989.

68. Mulbry, W. W., J. S. Karns, P. C. Kearney, J. D. Nelson, C. S. McDaniel and J. R. Wild. "Identification of a plasmid-borne parathion hydrolase gene from *Flavobacterium* sp. by Southern hybridization with *opd* from *Pseudomonas diminuta*." Appl. Environ. Microbiol. **51**: 926-930, 1986.

69. Munnecke, D. M. "Enzymic hydrolysis of organophosphate insecticides, a possible pesticide disposal method." Appl. Environ. Microbiol. **32**: 7-13, 1976.

70. Munnecke, D. M. "Enzymatic detoxification of waste organophosphate pesticides." Agric. Food Chem. **28**: 105-111, 1980.

71. Nelson, L. M. "Biologically-induced hydrolysis of parathion in soil: isolation of hydrolyzing bacteria." Soil Biol. Biochem. **14**: 219-222, 1982.

72. Pogell, B. M., S. S. Rowland, K. E. Steinmann, M. K. Speedie and F. C. G. Hoskin. "Genetic and biochemical evidence for the lack of significant hydrolysis of soman by a *Flavobacterium* parathion hydrolase." Appl. Environ. Microbiol. **57**: 610-611, 1991.

73. Racke, K. D. and J. R. Coats. "Enhanced degradation of isofenphos by soil microorganisms." J. Agric. Food Chem. **35**: 94-99, 1987.

74. Reiner, E., W. N. Aldridge and F. C. G. Hoskin. Enzymes Hydrolyzing Organophosphorus Compounds. 1989.

75. Rosenberg, A. and M. Alexander. "Microbial cleavage of organophosphorus insecticides." Appl. Environ. Microbiol. **37**: 886-891, 1979.

76. Rowland, S. S., M. S. Speedie and B. M. Pogell. "Purification and characterization of a secreted recombinant phosphotriesterase (parathion hydrolase) from *Streptomyces lividans*." Appl. Environ. Microbiol. **57**: 440-444, 1991.

77. Serdar, C. M. and D. T. Gibson. "Enzymatic hydrolysis of organophosphates: cloning and expression of a parathion hydrolase gene from *Pseudomonas diminuta*." Bio/Technology. **3**: 567-571, 1985.

78. Serdar, C. M., D. T. Gibson, D. M. Munnecke and J. H. Lancaster. "Plasmid involvement in parathion hydrolysis by *Pseudomonas diminuta*." Appl. Environ. Microbiol. **44**: 246-249, 1982.

79. Serdar, C. M., D. C. Murdock and M. F. Rohde. "Parathion hydrolase gene from *Pseudomonas diminuta* MG: Subcloning, complete nucleotide sequence, and expression of the mature portion of the enzyme in *Escherichia coli*." Bio/Technology. **7**: 1151-1155, 1989.

80. Sethunathan, N. and T. Yoshida. Degradation of parathion in flooded acid soils. Institute of Environmental Science 18th Technical Meeting. 255-257, 1972.

81. Sethunathan, N. and T. Yoshida. "A *Flavobacterium* that degrades diazinon and parathion." Can. J. Microbiol. **19**: 873-875, 1973.

82. Sheela, S. and S. B. Pai. "Metabolism of fensulfothion by a soil bacterium, *Pseudomonas alcaligenes* C_1." Appl. Environ. Microbiol. **46**: 475-479, 1983.

83. Shelton, D. R. and C. J. Somich. "Isolation and characterization of coumaphos-metabolizing bacteria from cattle dip." Appl. Environ. Microbiol. **54**: 2566-2571, 1988.

84. Siddaramappa, R., K. P. Rajaram and N. Sethunathan. "Degradation of parathion by bacteria isolated from flooded soil." Appl. Microbiol. **26**: 846-849, 1973.

85. Steiert, J. G., B. M. Pogell, M. K. Speedie and J. Laredo. "A gene coding for a membrane-bound hydrolase is expressed as a secreted, soluble enzyme in *Streptomyces lividans*." Bio/Technology. **7**: 65-68, 1989.

86. Szafraniec, L. L. and W. T. Beaudry. Personal communication.

87. Talbot, H. W., L. Johnson, S. Barik and D. Williams. "Properties of a *Pseudomonas* sp.-derived parathion hydrolase immobilized to porous glass and activated alumina." Biotechnol. Letters. **4**: 209-214, 1982.

88. Wild, J. R. Personal communication.

89. Zech, R. and K. D. Wigand. "Organophosphate-detoxicating enzymes in *E. coli.* Gel filtration and isoelectric focusing of DFPase, paraoxonase and unspecific phosphohydrolases." Experientia. **15**: 157-158, 1975.

APPLICATIONS OF MOLECULAR BIOLOGY TECHNIQUES TO

THE REMEDIATION OF HAZARDOUS WASTE

Burt D. Ensley

Envirogen, Inc.
4100 Quakerbridge Road
Lawrenceville, New Jersey 08648

INTRODUCTION

There are a growing number of applications of molecular biology tools to the manipulation of microorganisms involved in the degradation of hazardous compounds. These tools, including pathway cloning, polymerase chain reaction gene amplification, the use of hybrid promoter and regulatory sequences, broad host range and high copy number plasmids, DNA sequencing and synthesis methods, and enzyme recruitment can all be used to construct highly sophisticated biological catalysts that offer potentially superior performance in degradation.   The use of genetically engineered microorganisms for the degradation of hazardous molecules is not yet a widely accepted approach. At the same time, there are many advantages in using an organism containing a precisely engineered pathway that encodes enzymes capable of rapidly attacking highly recalcitrant or toxic molecules. Some of the potential benefits in the application of existing molecular biology techniques to the construction of hazardous waste degrading organisms are described in this chapter.

ALTERED REGULATION

Metabolic pathways are sometimes subject to complex regulation at both the transcriptional and translational level and can be sensitive to catabolite repression during growth on simple carbohydrates such as glucose.   Most pathways involved in the degradation of hazardous molecules can only be induced by growth in the presence of that molecule itself or a close analog. Molecular biology techniques can be used to change the regulation of a metabolic pathway by introducing new promoter and regulatory sequences that present improved flexibility and control of the desired metabolic pathway.

Induction

Promoter systems currently available can be activated by the addition of chemical inducers at low concentrations to the medium or even by simply changing the fermentation temperature by approximately 12°C.[1] These promoter systems are also extremely well regulated in that there is a large difference in the levels of the metabolic pathway enzymes in the cell before and after induction. Effective regulation enhances genetic stability of the desired metabolic function since the pathway can be induced at the end of the growth cycle, reducing selective pressure against the recombinant microorganism.

Resistance to Catabolite Repression

Promoter and regulator systems can be made resistant to catabolite repression, so that the promoters can be fully induced while the cells are growing rapidly on a simple carbohydrate such as glucose. Cells containing fully active degradative pathways can then be grown to extremely high cell densities. Recombinant *Escherichia coli* containing a fully active naphthalene metabolic pathway has been grown to densities in excess of 80 g/L dry weight of cells in 48 hours (unpublished data). It is highly unusual for a wild type pseudomonad growing with naphthalene or salicylate to reach densities in excess of 5 g/L dry weight of cells in the same period of time. While it is possible to generate mutants in a wild type metabolic pathway that are constitutive and insensitive to catabolite repression, these mutants are very often unstable and quickly revert to wild type. Unregulated expression of a superfluous metabolic pathway puts an organism at a distinct growth disadvantage compared to wild-type in medium containing carbohydrates or other rich nutrients. These constitutive and repression resistant mutants can only be maintained if they are grown in suboptimal conditions such as growth with nitrogen limitation[2] as patented by workers at Celanese.

Uncoupling Metabolism From Growth

Altering regulation permits expression of pathways metabolizing substrates that do not provide carbon and energy for growth. Microorganisms harboring these pathways can degrade a toxic molecule while growing on another carbon source, either indigenous to the environment or directly added for the purpose.

Another advantage offered by uncoupling metabolism and its regulation from growth on a particular molecule is that the molecule can then be degraded to extremely low concentrations. The mineralization of a number of organic compounds at low concentrations has been described by Martin Alexander and co-workers.[3,4] The degradation rates decrease at low ppb concentrations of compounds such as 2, 4-D. It was inferred that below these concentrations the degradative pathway was no longer induced, the enzymes responsible for metabolism were turned over, and the culture could no longer degrade the target substrate. Recombinant microorganisms have been used to degrade aromatics such as naphthalene and toluene and chlorocarbons such as trichloroethylene to parts per billion concentrations.[5] Because the regulation of enzyme activity in such cells is indifferent to the concentration of the substrate in the medium, a stable, metabolically active enzyme system can degrade a target molecule nearly to extinction given enough time (Fig. 1). The metabolism of trichlorethylene by wild-type *Pseudomonas mendocina* ceases if the organism is not incubated in the presence of a co-substrate such as toluene. At the same time, toluene is a competitive inhibitor of TCE degradation, and the presence of toluene in the reaction mixture limits the initial rates of TCE degradation and still does not provide sufficient energy for complete metabolism of this toxic molecule. The recombinant microorganism, on the other hand, can use glucose or other simple compounds as a co-substrate and displays greater than 99.9% metabolism of TCE under the same reaction conditions.

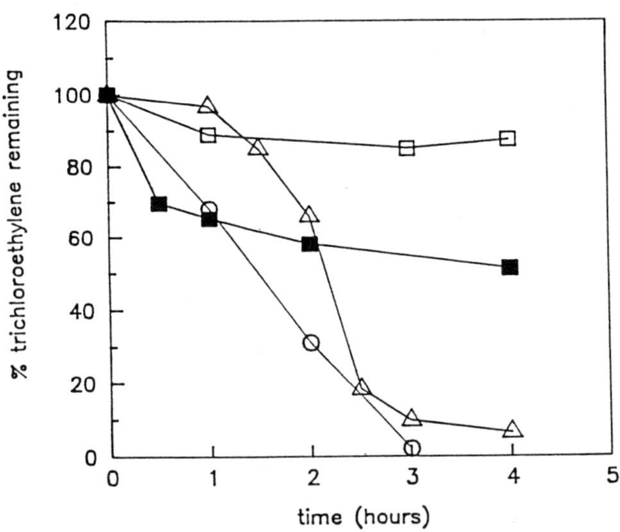

Figure 1.   Degradation of 2ppm TCE by *Pseudomonas dimunita* and Recombinant *Escherichia coli*

Cultures were incubated at a fixed cell density in sealed vials with 2 ppm TCE. Head space samples were withdrawn at various time points and analyzed by gas chromatography. □ Control experiments using no cells or cells lacking the degradative enzyme. ■ *P. dimunita* incubated with TCE alone. ▲ *P. dimunita* incubated with TCE plus toluene as an exogenous reducing source. ○ Recombinant *E. coli* with glucose as a carbon and energy source.

INCREASED EXPRESSION OF DEGRADATIVE PATHWAYS

Strong promoters and gene dose effects can be used to significantly amplify the expression of a metabolic pathway well above the levels of activity observed in a wild type cell. Examples exist in the literature of genetic manipulation being used to elevate the levels of enzyme activity in a cell from a few-fold to over 100-fold above wild type.[6,7]

Increased Metabolic Rates

Amplifying the expression of a degradative pathway in a cell can also increase the rate of degradation by whole cells and cell extracts.[8,9] We have observed as much as an order of magnitude increase in the specific rate of degradation over wild type by recombinant microorganisms (unpublished data). This elevated rate of activity can be maintained over time and bears directly on the efficiency and productivity of a culture. These highly active and inexpensively grown cells could reduce from days or weeks to hours the time required for degradation of a toxic compound in a given amount of water or soil.

Amplify Particular Steps

Operon engineering and strong promoters can be used to amplify the rate-limiting or least stable metabolic step in an entire pathway. This has the advantage of increasing the functional stability of the overall degradative pathway and allows metabolism to proceed efficiently by avoiding the accumulation of potentially toxic intermediates through amplification of the rate limiting step.

Access to Metabolic Pools

Higher concentrations of the appropriate metabolic enzymes in the cell have access to a larger proportion of the pools of metabolic energy in the cell. The oxidation of many hydrocarbons requires the presence of cofactors such as NADH. Studies have shown that both biosynthetic and biodegradative enzymes that require cofactors show rates of activity that increase with amplification of the proteins.[9,10] These findings suggest that a cell containing an amplified pathway for the degradation of a target molecule can direct more of the cellular pools of required cofactors toward the metabolism of that compound.

CHOICE OF HOST CELLS

Many of the sophisticated molecular biology tools have been developed for use in *Escherichia coli* or *Saccharomyces cerivisae.* This technology is maturing and the tools are being adapted for use in a wide range of host organisms. The ability to choose the appropriate host offers a number of advantages.

High Productivity

Host cells can be chosen that grow rapidly to high cell densities with simple carbon sources. *E. coli,* for example, can be grown to cell densities in excess of 80 g dry weight of cells per liter in less than 48 hours.[11] Other organisms amenable to genetic manipulation can grow at least as fast as *E. coli.* The advantage of rapid and efficient growth on inexpensive carbon sources is significant because large amounts of biomass with a high degradative capacity can be generated with efficient volumetric productivities and a low cost.

Defined Growth and Biochemical Characteristics

Organisms with known growth requirements and defined biochemical pathways offer a distinct advantage in use over indigenous degradative microorganisms because enhancing growth of the latter is often hit or miss, depends largely on guesswork, and can consume considerable time. Alternatively, the introduction of degradative genes into a microorganism that is known to thrive at a particular site increases the level of confidence that the organism will survive on that site long enough to be effective.

Environmental Impact

A host organism may be chosen because it is known to be indigenous to a particular site or because the introduction of this organism in large numbers into the environment can be shown to be innocuous. Some wild-type organisms may be able to degrade a toxic molecule but also have the capacity to cause disease in plants, insects or mammals or otherwise degrade environmental quality. This potential problem can be overcome using a recombinant microorganism that harbors the desired degradative activity but is benign in the environment.

<u>Resistance to Environmental Stress</u>

A new host organism for a degradative pathway may carry a natural resistance to stresses present in a particular environment. An organism that is naturally resistant to heavy metals such as lead would be an example. Some organisms are known to grow in the presence of high concentrations of toxic compounds such as 50% toluene and would be a logical choice in such an environment.

DIRECTED FLOW OF METABOLISM

Molecular biology techniques can be used to add and delete genes in a metabolic pathway and thereby control the flow of degradative intermediates. This type of fine metabolic control has already been shown to be important in the degradation of some chlorinated aromatics.

<u>Avoid Toxic Intermediates</u>

K.N. Timmis and co-workers have shown that certain metabolic pathways for chloroaromatics result in the formation of toxic intermediates that poison the degradative machinery. By using the techniques of enzyme recruitment, it has been possible to prevent the formation of toxic intermediates and thereby ensure the complete mineralization of the chloroaromatics.[12] This is an example of the applications of genetic engineering in controlling metabolic flow. These techniques can also be used to control the accumulation of degradative intermediates that are known to be toxic to man or other vertebrates. This approach permits the choice and use of the least noxious degradative pathway for a given compound.

<u>Control of Oxygen Consumption</u>

The complete aerobic degradation of a molecule such as toluene or benzene has a very large oxygen consumption requirement, and oxygen is a limiting factor in many biodegradation milieus. By deleting certain genes from a metabolic pathway, it may be possible to limit the aerobic degradation of a hydrocarbon such as benzene to an intermediate that can be easily degraded under anaerobic conditions. In this way, the most efficient use of available oxygen is assured, and aerobic metabolism only proceeds far enough to ensure detoxification of the molecule.

<u>Multiple Metabolic Pathways in One Host</u>

The classic approach of combining multiple metabolic pathways in a single organism as patented by Ananda Chakrabarty[13] is still a viable idea. The inclusion of several metabolic pathways can make a single organism multipotent against a range of hydrocarbon substrates. The combination of degradative pathways for benzene, xylene and toluene in a single organism is one example of such a combination that would be useful. Genetic engineering techniques are being used to combine multiple metabolic pathways in a single organism for specialty chemical synthesis.[14] These capabilities can be adapted in the short term toward combining degradative pathways in a single host.

<u>Ancillary Functions</u>

Molecular biology techniques can also be used to introduce into a host organism genes encoding functions, that while not directly involved in degradation, can stimulate the degradative capability of the recipient cell. An example would be the addition of genes that encode the synthesis of bioemulsifiers. Many hydrocarbons such as polynuclear aromatics are highly insoluble and this probably limits their rate of degradation. An organism that was able to secrete a bioemulsifier during growth in the environment could mobilize these insoluble molecules and increase availability for degradation.

PROTEIN ENGINEERING

The techniques of site-directed mutagenesis and gene synthesis have been used successfully and will increasingly be used to improve desirable properties of enzymes. These techniques offer a very powerful tool for effecting key parameters in enzyme activity for hazardous waste degradation.

Enzyme Stability

Protein engineering techniques have been used to significantly increase the stability of the industrial enzyme subtilisin.[15,16] In addition, site-directed mutagenesis has recently been used to increase the stability of a critical component in a hydrocarbon degradative pathway.[17] This may be an advantage in bioremediation applications because some molecules apparently have a deleterious effect on the enzymes degrading them[18] and the degree of degradation could be enhanced by a more stable enzyme.

Broaden Substrate Specificity

Protein engineering has been used to increase the substrate specificity of subtilisin by workers at Genentech.[19,20] This application of protein engineering is still in its infancy but holds significant promise in that a well characterized enzyme system could be recruited to attack additional recalcitrant toxic molecules. These proprietary, altered enzyme systems would permit rapid development on existing technology for degrading new molecules. Some enzymes already have a broad substrate specificity and expression of these genes in a recombinant microorganism can result in significant degradation of several toxic compounds by the same cultures (Table 1).

Increased Reaction Rates

In addition to amplifying the expression of a particular enzyme or enzyme system, one may also use the techniques of protein engineering to increase reaction rates against a particular substrate by manipulation of the active site. Site-directed mutagenesis has shown promise as a means to increase reaction rates in a biosynthetic pathway[17]. An added advantage is that increased expression of a more active protein has a cumulative effect on the overall rates and thus provides the opportunity for extremely rapid degradation.

PEPTIDE AND ENZYME SYNTHESIS

Most degradative pathways are complex, multi-enzyme systems and many require cofactors. However, some toxic molecules can be degraded or detoxified by the action of a single enzyme that requires very simple or no cofactors. Some examples could include ligninase, horseradish peroxidase and parathion hydrolase involved in the degradation of organophosphates.

Table 1.   Degradation of Chlorinated Ethenes by
Toluene Monooxygenase in
Recombinant *E. coli*

Chlorinated Ethene	Initial Concentration (ppm)	% Degradation after 20 Hours
vinyl chloride	5	97
<u>cis</u>-dichloroethylene	90	98
<u>trans</u>-dichloroethylene	90	62
trichloroethylene	8	95
tetrachloroethylene	1	18

A recombinant *E. coli* containing the cloned toluene monooxygenase enzyme system was incubated with the above chlorinated ethenes in teflon-sealed vials for 20 hours at 30°C. Pentane extracts of the incubation mixtures were analyzed by gas chromatography, and the percent degradation was compared to vials that did not contain bacteria.

High Level Synthesis

The over expression of single enzymes or peptides by recombinant microorganisms is among the most developed technologies in genetic engineering. Bacteria have been constructed that produce up to 40% of their total protein as a single molecule[11]. These methods can be used to synthesize important enzymes at extremely high levels that are only available in trace amounts from the natural isolate. A good example of enzyme overproduction is the synthesis in recombinant *E. coli* of the organophosphate-degrading enzyme parathion hydrolase. One version of this enzyme is synthesized in *Pseudomonas dimunita* at levels of approximately two units per milligram of crude cell extract protein. The gene encoding this enzyme has been cloned, sequenced, and the enzyme purified and its N-terminal sequence determined. This data was used to establish that the protein is processed after translation by the wild-type organism. A new N-terminal sequence of this gene was synthesized to encode expression of the mature, processed protein after cloning in *E. coli*. The use of a strong *E. coli* promoter and an *E. coli* ribosomal binding site, coupled with the synthetic gene[6] resulted in expression of parathion hydrolase activity in recombinant *E. coli* at levels of approximately 50 units per milligram of crude extract protein. Since *E. coli* can be grown to extremely high cell densities, this organism would produce 2,000-2,500 units per milliliter of fermentation volume. A highly efficient and useful source of parathion hydrolase for both scientific investigation and commercial application for organophosphate hydrolysis in the environment is provided by this recombinant *E. coli*.

Purification

High level synthesis of many enzymes in *E. coli* can result in the incorporation of these proteins into refractile bodies called inclusion bodies. Inclusion bodies are 70-90% pure recombinant protein, which makes the purification of these enzymes usually very simple. Other proteins are produced in a soluble form inside the cell, and the high level synthesis of these proteins makes batch type purification processes practical.

<u>Secretion</u>

Genetic engineering techniques for microorganisms that secrete proteins have been developed in several labs. Certain bacteria such as the *Bacilli* can secrete some proteins into the surrounding medium at 10-20 g/L concentrations. Secretion of a rare but important enzyme at close to those levels would result in a several orders of magnitude reduction in the cost of manufacture. This cost advantage would be very significant in large-scale bioremediation efforts.

DNA PROBE TECHNOLOGY

The techniques of gene synthesis have permitted the development of numerous precise and sophisticated DNA probes for the identification of particular DNA sequences present in very low concentrations. This technology could significantly benefit some applications of bioremediation.

<u>Rapid Identification of Hydrocarbon Degrading Microorganisms</u>

The presence and extent of hydrocarbon contamination in the environment could be estimated by enumerating rapidly the number of organisms present in the soil or water that contain genes encoding the degradation of that hydrocarbon. This could be developed into a rapid diagnostic test for the presence of toxic waste in particular sites.

<u>Evaluation of Biodegradative Potential</u>

DNA probes can be used to identify and enumerate the types of microorganisms present in contaminated soil. This information could be used to predict the biodegradative potential of indigenous microorganisms. If the genes that encode degradation of a certain key compound cannot be found among the microorganisms present in the site, it is unlikely that population will be able to degrade that compound. This technique would greatly simplify the evaluation of biotreatability at a given site.

<u>Monitoring of DNA</u>

DNA probe technology can be used, because of its high sensitivity, to monitor the spread and survival of microorganisms introduced into an environment. These organisms could be either wild-type cells that had grown in fermentors or recombinant microorganisms. These probes would also be a good way to measure the decline in a particular population of organisms after the targets waste has been degraded and are a direct way of measuring the fate of recombinant DNA in the environment.

REFERENCES

1.   W. N. Burnette, V. L. Marr and W. Cieplak, Direct expression of *Bordetalla pertussis* toxin subunits to high levels in *Escherichia coli*, <u>Bio/Technology</u>. 6:699-706 (1988).

2.   P. Maxwell, J. Hsieh, and J. Fieschko, Stabilization of a mutant microorganism population, European patent, EP0098750, (1984).

3. R. V. Subba-Rao, H. E. Rueben, and M. Alexander, Kinetics and extent of mineralization of organic chemicals at trace levels in fresh water and sewage, Appl. and Environ. Microbiol. 43:1139-1150 (1982).

4. M. Alexander, Ecologic constraints on genetic engineering, in: "Genetic Control of Environmental Pollutants," G.S. Ullman, and A. Hollaender, Plenum Press, New York (1984).

5. R. B. Winter, K. M. Yen and B. D. Ensley, Efficient degradation of trichloroethylene by a recombinant *Escherichia coli,* Bio/Technology. 7:282-285 (1989).

6. C. M. Serdar, B. C. Murdock, and M. F. Rohde, Parathion hydrolase gene from *Pseudomonas dimunita* mg: subcloning complete nucleotide sequence and expression of a mature portion of the enzyme in *Escherichia coli,* Bio/Technology, 7:1151-1155 (1989).

7. K. Nagahari, Deletion plasmids from transformants of *Pseudomonas aeuroginosa* trp cells RSF1010 trp hybrid plasmid and high levels of enzyme activity from the gene on the plasmid, J. Bacteriol., 136:312-318 (1978).

8. D. B. Janssen, F. Pries, J. Ploeg, B. Kavemier, P. Terpstra, and B. Witholt, Cloning of 1,2-dichloroethane degradation genes of *Xanthobacter autotrophicus* GJ10 and expression and sequencing of the dhlA gene, J. Bacteriol., 171:6791-6799 (1989).

9. B. D. Ensley, T. D. Osslund, M. Joyce, and M. J. Simon, Expression and complementation of naphthalene dioxygenase activity in *Escherichia coli,* in: "Microbial Metabolism and the Carbon Cycle," S.R. Hagedorn, R.S. Hanson, and D. A. Kunz, ed., Harwood Academic Publishers, New York (1988).

10. T. Isogai, M. Fukagawa, I. Aramori, M. Iwami, H. Kojo, T. Ono, Y. Ueda, M. Kohsaka, and Imanaka, Construction of a 7-aminocephalosporanic acid (7ACA) biosynthetic operon and direct production of 7ACA in *Acremonium chrysogenum,* Bio/Technology, 9:188-191 (1991).

11. J. Fieschko, and T. Ritch, Production of human alpha consensus interferon in recombinant *Escherichia coli,* Chem. Eng. Commun., 45:229-240 (1986).

12. K. N. Timmis, F. Rojo, and J. L. Ramos, Prospects for laboratory engineering of bacteria to degrade pollutants in: "Environmental Biotechnology; Reducing Risks from Environmental Chemicals Through Biotechnology," Ullman, G.S., ed. Plenum Press, New York (1988).

13. A. M. Chakrabarty, Microorganisms having multiple compatible degradative energy generating plasmids and preparation thereof, US patent 4,259,444 (1981).

14. J. F. Grindley, M. A. Payton, H. Pole and K. G. Harding, Conversion of Glucose to 2-keto-1-gulonate, an intermediate in 1-ascorbate synthesis by recombinant strain of *Erwinia citrus,* Appl. and Environ. Microbiol. 54:1770-1775 (1988).

15. L. O. Narhi, Y. Stabinski, M. Levitt, L. Miller, R. Sachdev, S. Finley, S. Park, C. Kolvenbach, T. Aurakawa, and M. Zukowski, Enhanced stability of

subtilisin by three point mutations, <u>Biotechnol. Appl. Biochem.</u> 13:12-24 (1991).

16.   M. Zukowski, Y. Stabinski, L. Narhi, J. Mauck, M. Stowers, and M. Fiske, An engineered subtilisin with improved stability: applications in human diagnostics, <u>in</u>: "Genetics and Biotechnology of Bacilli, Vol. 3," M.N. Zukowski, A. T. Ganesan, and J. A. Hoch, ed., Academic Press, New York (1990).

17.   D. Murdock, M. Thalen, C. Serdar, and B. Ensley, Manipulation of metabolic operons catalyzing the biochemical synthesis of indigo, <u>in press</u>.

18.   B. G. Fox, J. G. Borneman, L. P. Wackett, and J. D. Lipscomb, Haloalkene oxidation by the soluble methane monooxygenase from *Methylosinus trichosporium* OB3b: mechanistic and environmental implications, <u>Biochem.</u>, 29:6419-6427, (1990).

19.   J. A. Wells, B. C. Cuningham, T. P. Graycar and D. A. Estell, Recruitment of substrate specificity properties from one enzyme into a related one by protein engineering. <u>Proc. Natl. Acad. Sci.</u> 84:5157-5174 (1987).

20.   J. A. Wells, D. B. Bowers, R. R. Bott, T. P. Graycar, and D. A. Estelle, Designing substrate specificity by protein engineering of electrostatic interactions, <u>Proc. Natl. Acad. Sci.</u> 84:1219-1223 (1987).

DEHALOGENATION OF ORGANOHALIDE POLLUTANTS BY
BACTERIAL ENZYMES AND COENZYMES

Lawrence P. Wackett

Gray Freshwater Biological Institute
University of Minnesota
Navarre, MN 55392

1.   INTRODUCTION

Organohalides are widespread environmental pollutants which typically contain a carbon-halogen bond. The halogen substituent can be fluorine, chlorine, bromine, or iodine but chlorine is most common. The environmental fate of organohalides is dominated by the chemistry of the carbon-halogen bond of particular compounds. For example, fluorocarbons are particularly inert. This is due in large measure to the high bond dissociation energy of the carbon-fluorine bond which ranges from 106-115 Kcal/mol (Reinecke, 1984). Chlorinated compounds differ markedly in their environmental persistence. Generally, aryl and alkenyl chlorides decompose much more slowly than alkyl chlorides. The former compounds undergo hydrolytic and photolytic cleavage of the carbon-halogen bond much less readily. Environmental organohalides often derive from industrial sources, but many halogenated organic natural products are known as well. Commodity organic chemicals that contain chlorine include vinyl chloride, trichloroethylene, dichloromethane, 1,2-dichloroethane, and chlorobenzene. Each of these compounds are used by United States industries at levels exceeding ten million pounds annually (Hutzinger & Veerkamp, 1981). Several natural products such as methyl chloride (Wuosmaa & Hager, 1990) and tribromomethane (Gschwend, *et al.*, 1985) are released into the environment at comparable levels on a global scale by fungi and algae, respectively. As an illustration of the complexity of this group of compounds, over 700 halogenated natural products have been identified (Neidleman & Geigert, 1986).

Although organohalides have enjoyed widespread applications, several detrimental properties make them currently the largest group of Environmental Protection Agency priority pollutants (Leisinger, 1983). These compounds negatively impact mammals, including humans, because of three general properties. First, the uptake of organohalides by mammals can cause immediate toxicity or cancer. Second, organohalides are often environmentally persistent so chances for long-term human exposure are magnified. Third, many halogenated organic compounds bioaccumulate and increase in organisms feeding in high levels up the food chain (Atlas, 1989). For example, polychlorinated biphenyls (PCBs) partition readily into biomembranes and are very poorly metabolized. They accumulate in greater concentrations in higher members of a food web. Humans are at the top of several food chains, and this accentuates adverse effects from environmental PCB contamination.

Applications of Enzyme Biotechnology, Edited by J.W. Kelly and
T.O. Baldwin, Plenum Press, New York, 1991

Some bacteria and higher organisms have developed effective strategies for metabolizing organohalides. In general, mammals metabolize organohalides to detoxify the compounds whereas some bacteria can feed on selected halogenated molecules as a carbon and energy source. As an example, 1,2-dihaloethanes are processed in mammals (Anders & Pohl, 1985) via the scavenging tripeptide glutathione (GSH in Figure 1).

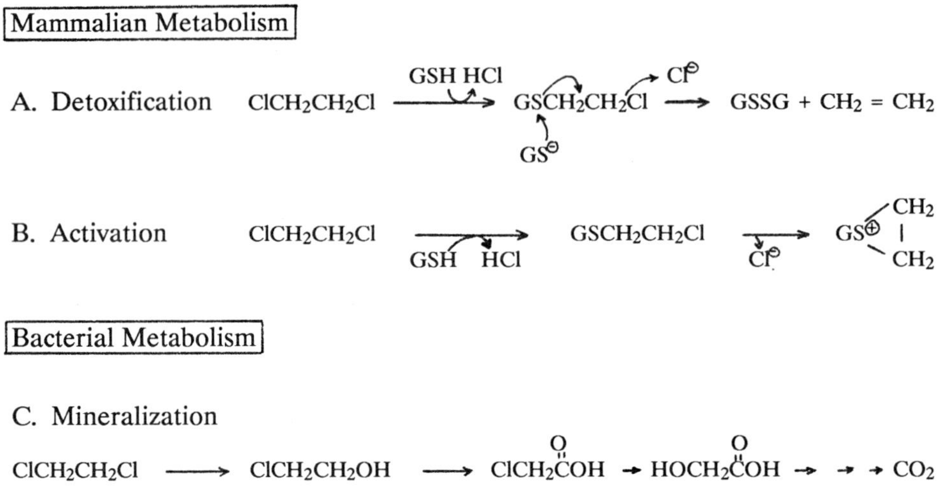

Mammalian Metabolism
A. Detoxification
GSH HCl
$ClCH_2CH_2Cl$
$GSCH_2CH_2Cl$
$Cl^{\ominus}$
$GSSG + CH_2 = CH_2$
$GS^{\ominus}$
B. Activation
$ClCH_2CH_2Cl$
GSH HCl
$GSCH_2CH_2Cl$
$Cl^{\ominus}$
$GS^{\oplus}$
CH_2
CH_2
Bacterial Metabolism
C. Mineralization
$ClCH_2CH_2Cl$
$ClCH_2CH_2OH$
$ClCH_2\overset{O}{\overset{\|}{C}}OH$
$HOCH_2\overset{O}{\overset{\|}{C}}OH$
CO_2

Figure 1. Bacterial and mammalian metabolism of 1,2-chloroethane.

This can result in either detoxification (Fig. 1A) or yield reactive intermediates such as S-(chloroethyl)glutathione or the episulfonium ion that can alkylate DNA (Fig. 1B). In contrast, a number of bacteria have been identified which are capable of growth on 1,2-dichloroethane as the sole source of carbon and energy (Janssen, *et al.*, 1990). The compound undergoes two distinct dehalogenation steps (Fig. 1C) enroute to complete oxidation to carbon dioxide. In this example, two different dehalogenases are biosynthesized in response to exposure to 1,2-dihaloethanes. In other cases, dehalogenation reactions of a non-specific nature also occur in nature. These will be discussed in the next section.

2. MECHANISMS OF CARBON-HALOGEN BOND CLEAVAGE

Given the widespread environmental distribution of natural product and synthetic organohalides, it is logical that multiple mechanisms of bacterial carbon-halogen bond cleavage have evolved. The known reactions can be classified into four main mechanistic types. These are shown in Figure 2 below. Substitutive dehalogenation reactions are largely hydrolytic (Goldman, *et al*, 1968; Kawasaki, 1981). A prominent example is shown in Figure 1C, in which two distinct hydrolytic dehalogenases function in the metabolism of 1,2-dichloroethane. Even carbon-fluorine bonds undergo enzymatic hydrolysis. For example, the stereochemical course of bacterial fluoroacetate hydrolysis was investigated and shown to proceed with net inversion of configuration (Au & Walsh). In another type of substitutive reaction, vicinal haloalcohols undergo enzyme-catalyzed intramolecular halide displacement to yield epoxides (van den Wijngaard, *et al.*, 1991).

Eliminative reactions figure most prominently in the dechlorination of pesticides such as 1,1-*bis*-(4-chlorophenyl)-2,2,2-trichloroethane (DDT). Oxidative dehalogenation reactions are involved in metabolic pathways that supply carbon for cell growth (Markus, *et al*, 1984). Another class of oxidative reactions occur as the result of non-specific oxidation of non-growth substrates by broad-specificity oxygenases (Wackett, *et al.*, 1989). Examples of this will be discussed later. Similarly, reductive dechlorination can occur as a result of specific metabolic reactions and by gratuitous side reactions. This will be discussed below.

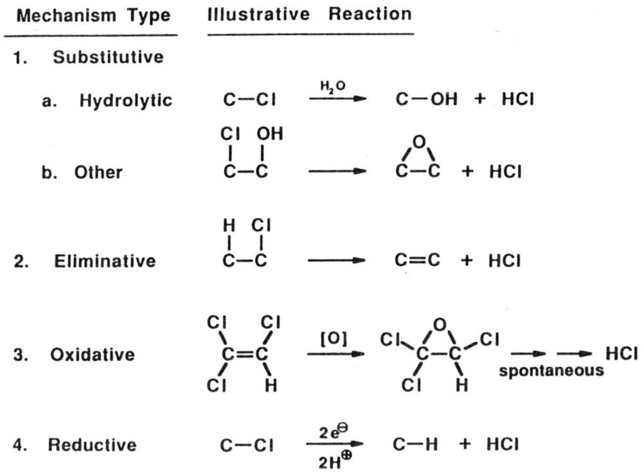

Figure 2. **Four general mechanisms of bacterial carbon-halogen bond cleavage.**

Hydrolytic Dehalogenation - Dichloromethane Dehalogenase.

As illustrated in Figures 1 and 2, typical hydrolytic halogen displacement reactions yield alcohols from alkyl halides. In other cases where the carbon atom undergoing hydrolysis is bound to a second halogen atom, a carbonyl functionality is formed (Berry, *et al.*, 1979; Stucki, *et al.*, 1981; Anders & Pohl, 1985). Following displacement of the first halogen by hydroxide anion, the second undergoes *gem* elimination of hydrogen halide:

$$H_2CCl_2 + HO^- \longrightarrow Cl^- + H_2CClOH \longrightarrow HCHO + HCl$$

The reaction sequence shown above is catalyzed by an enzyme found in some methylotrophic bacteria. This allows them to grow on dihalomethanes including the industrial solvent dichloromethane. Formaldehyde, the product of the dehalogenation reaction, is rapidly metabolized further. It can be assimilated into cell carbon or oxidized to formic acid and carbon dioxide to supply energy for cell sustenance.

The enzyme catalyzing the net hydrolysis of dihalomethane to formaldehyde has been given the name dichloromethane dehalogenase (Kohler-Staub & Leisinger, 1985). It is active with dichloro-, dibromo-, and diiodomethane. Two different forms of the dehalogenase have been purified and their properties compared (Scholtz, *et al.*, 1988). Both enzymes show structural and catalytic similarities although they are clearly different proteins. Most strikingly, the tripeptide glutathione (GSH) is an obligate participant in the dehalogenation reaction catalyzed by both enzymes. Steady-state kinetic studies indicate that glutathione is a saturatable enzyme-bound cosubstrate. However, it is not consumed as shown in the overall reaction stoichiometry:

$$\boxed{CH_2Cl_2 + H_2O} + GSH \longrightarrow \boxed{HCHO + 2HCl} + GSH$$

The stoichiometry suggests nucleophilic displacement of the first chloride atom by the glutathione thiolate (Kohler-Staub & Leisinger, 1985). The resultant chloromethylthioether would likely undergo rapid hydrolysis in an aqueous environment (Bohme, *et al.*, 1949). The hemiacetal thus formed is in equilibrium with glutathione and free formaldehyde:

$$CH_2Cl_2 + GSH \longrightarrow HCl + GSCH_2Cl \xrightarrow{H_2O} HCl + GSCH_2OH \rightleftharpoons GSH + HCHO$$

The putative reaction pathway above is rendered more plausible by the observation that rat liver glutathione transferases catalyze a similar reaction (Ahmed & Anders, 1978). Glutathione S-transferases typically catalyze reactions by enhancing the nucleophilicity of the glutathione thiolate for reaction with various electrophilic substrates (Keen, *et al.*, 1976). However, glutathione S-transferases convert dichloromethane to formaldehyde with a second order rate constant of only 10^{-6} that of dichloromethane dehalogenase as obtained by comparing relative V_{max}/K_m (Blocki & Wackett, unpublished data). Furthermore, dichloromethane is not very reactive with nucleophiles (Shaik, 1985). Dichloromethane hydrolyzes in potassium hydroxide at 79°C with a rate constant of 2.1×10^{-5} $Lmol^{-1}S^{-1}$ (Salomaa, 1966). The enzyme catalyzed V_{max}/K_m is 4.0×10^5 $Lmol^{-1}S^{-1}$, more than 10^{10} fold greater than the uncatalyzed rate (Scholtz, *et al.*, 1988). Thus, it is important to consider enzyme reaction pathways in addition to those involving direct chloride displacement by hydroxide or glutathione thiolate anions. An alternative role of glutathione might involve its facilitating the removal of formaldehyde from the active site to protect critical amino acid side chain groups.

Initially, it is important to differentiate between nucleophilic mechanisms and a potential deprotonation/halide elimination sequence leading to a carbene intermediate (Figure 3). Monochlorocarbene could then undergo hydrolysis to yield formaldehyde. These two general mechanisms differ with respect to the fate of the two hydrogen atoms on dichloromethane. In the carbene pathway, the first step involves cleavage of a carbon-hydrogen bond. Nucleophilic pathways preserve both original hydrogen atoms of dichloromethane in the product formaldehyde. The two pathways can be clearly distinguished by incubating dideutero-dichloromethane (CD_2Cl_2) with enzyme and analyzing the deuterium content of the product. Formaldehyde is readily trapped as a lutidine derivative (Nash, 1953) that can be examined for deuterium content by mass spectrometry and nuclear magnetic resonance spectroscopy. Dichloromethane dehalogenase from strain DM11 yielded

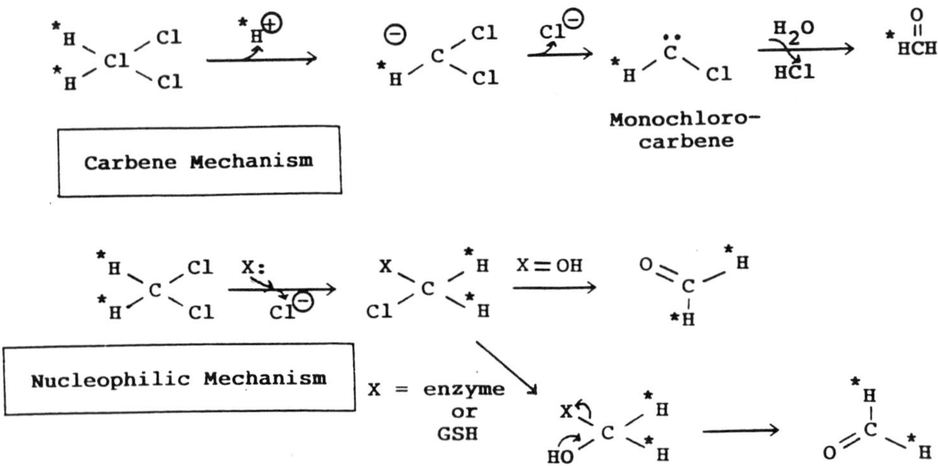

Figure 3.   **Potential reaction pathways of dichloromethane dehalogenase.**

dideuteroformaldehyde (DCDO) (Bao-li & Wackett, unpublished data). A similar experiment conducted with a different dichloromethane dehalogenase from strain DM2 also showed deuterium retention in the product (Gälli, *et al.*, 1982). These data argue for a direct halide displacement pathway.

Currently, investigations are focussed on determining the identity of a putative nucleophile displacing the initial chloride ion from dichloromethane. This could (1) be hydroxide ion, (2) the thiolate anion of glutathione, or (3) oxygen, nitrogen, sulfur nucleophiles on active site amino acid side chains (Figure 3). There is currently no evidence for the presence of metals or other prosthetic groups in dichloromethane dehalogenase.

It is difficult to distinguish among the different nucleophilic pathways because monohalo reaction intermediates from dihalomethanes would likely be short-lived. To overcome this, substrate analogs have been utilized with the hope of stabilizing intermediate states for identification of their structures. Potential chloromethyl substrates can be envisaged which stabilize hydrolysis, glutathione conjugate, and enzyme alkylation products, respectively (Figure 4). This strategy is somewhat more difficult to implement in practice due to the stringent substrate specificity of dichloromethane dehalogenase. For example, methylchloride and chloroform are not processed by the enzyme (Bao-li and Wackett, unpublished data).

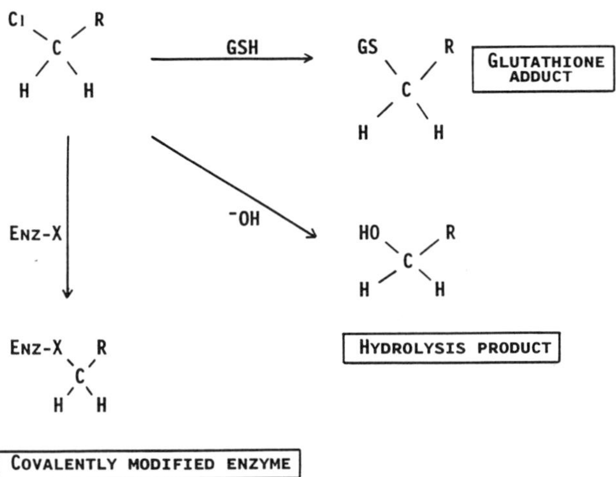

Figure 4.   **Substrate analogs to elucidate enzyme reaction intermediates.**

Against this backdrop, haloacetonitriles have emerged as a useful class of substrate analogs for beginning to unravel the enzyme reaction pathway. They have been observed to be potent inhibitors at concentration of ~100 nM. For example, a dilute enzyme solution shows a 50% loss of enzyme activity following incubation with chloro-, bromo-, or iodoacetonitrile at 70 nM, 96 nM, and 85 nM, respectively. Activity loss is essentially irreversible. Extended periods of dialysis fail to restore enzyme activity. It is particularly interesting that the inhibitory effects are observed in the absence of glutathione. Thus, the enzyme apparently processes haloacetonitriles with lethal effects. Further studies on the nature of enzyme inhibition might yield important information on the mechanism of catalysis.

<u>**Oxygenative Dehalogenation of Halogenated Ethylenes.**</u>

Bacterial oxygenases are important in both specific and non-specific dehalogenation of organohalides, usually via gem hydrogen halide elimination of oxygenated intermediates (Figure 5). In one example, 3,4-dioxygenation of 4-chlorophenylacetic acid yields 3,4-dihydroxyphenylacetic acid which can support growth of a soil bacterium (Markus, *et al.*, 1984). Other cases appear to arise from gratuitous oxidation of organohalides by non-specific catabolic oxygenases biosynthesized for other metabolic purposes (Wackett, *et al.*, 1989). This has been most extensively studied for the bacterial oxidation of trichloroethylene (TCE). TCE is a widespread industrial solvent and a major groundwater pollutant (Storck, 1987; Parsons, *et al.*, 1984). Efforts to use bacteria in TCE clean-up efforts are focussing on oxygenase-dependent mechanisms of TCE metabolism.

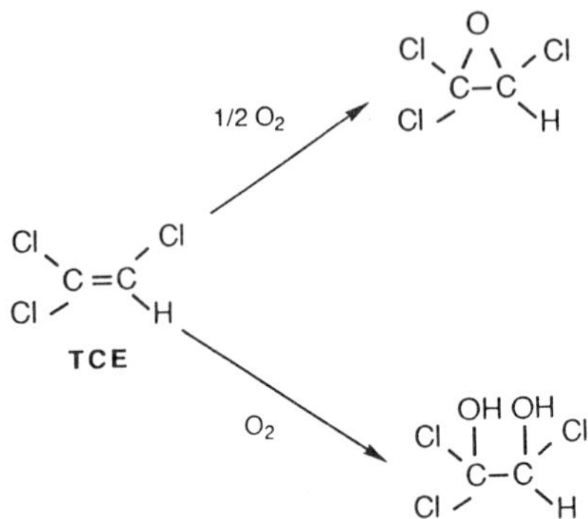

Figure 5.   **General routes of TCE oxidation by oxygenases.**

Two types of oxygenases, monooxygenases and dioxygenases, are known and both types have been implicated in TCE oxidation. Monooxygenases incorporate one oxygen atom and dioxygenases incorporate both oxygen atoms from a molecule of diatomic oxygen into various organic substrates. The oxidation of TCE by monooxygenases and dioxygenases yield unstable intermediates which decompose spontaneously with the liberation of chloride ions (Figure 5). Toluene dioxygenase is reported to oxidize TCE (Nelson, *et al*, 1988; Wackett & Gibson, 1988). Monooxygenases identified as being active *in vivo* with TCE as a substrate include soluble methane monooxygenase (Oldenhuis, *et al.*, 1989; Tsien, *et al.*, 1989), toluene 2-monooxygenase (Folsom, *et al.*, 1990), toluene 4-monooxygenase (Winter, *et al.*, 1989), ammonia monooxygenase (Arciero, *et al.*, 1989), propane monooxygenase (Wackett, *et al.*, 1989), and phenol hydroxylase (Harker & Kim, 1990). By far, the oxygenase with the highest specific activity directed against TCE is soluble methane monooxygenase (sMMO) from *Methylosinus trichosporium* OB3b (Fox, *et al.*, 1990). Purified enzyme components were used to rigorously determine the reaction products. As expected, sMMO oxidizes haloethylenes predominantly to epoxides (Fox, *et al.*, 199). However, minor products are observed and these likely derive from intramolecular halide or hydride migration during haloalkene oxidation. For example, TCE oxidation yields 94% TCE-epoxide and 6% 2,2,2-trichloroacetaldehyde. It is proposed that both products arise via a common catalytic pathway with steps as follows: (1) a resonance stabilized monoatomic oxygen species is generated; (2) an electron is abstracted from the haloalkene π-electron system and the oxygen and organic radicals recombine to form a cation; and (3) the cation intermediate rearranges to TCE-epoxide or to 2,2,2-trichloroacetaldehyde by chloride migration (Fox, *et al.*, 1990).

Reductive Dehalogenation Catalyzed by Transition Metal Coenzymes.

The environmental and human health significance of reductive dechlorination has prompted a large number of studies designed to understand these anaerobic microbial processes. Highly chlorinated molecules such as per-chloroethylene (PCE), hexachlorobenzene, and heavily chlorinated PCBs are environmentally persistent and degrade biologically only by reductive mechanisms. These processes are poorly understood at present. In one example, a specific membrane-bound protein, biosynthesized by *Desulfomonile tiedjei*, is proposed to mediate the reductive dechlorination of 3-chlorobenzoate and perhaps PCE (Dolfing, 1990; DeWeerd & Suflita, 1990).

Evidence is emerging that some biologically-mediated reductive dechlorination reactions may be catalyzed by transition metal cofactors which are often found in anaerobic bacteria. Krone, *et al.* (1989a,b) have presented evidence that the reduction of carbon tetrachloride to methane by methanogenic bacteria is catalyzed by cobalt-containing cobalamins, the nickel-containing coenzyme F_{430}, or both. While coenzyme F_{430} is unique to methanogens, participating in the major biological methane-yielding reaction (Wolfe, 1984; Walsh & Orme-Johnson, 1987), the cobalt-containing cobalamins (Hogenkamp, 1975) and the iron coenzyme hematin (Hambright, 1975) are found in a broad spectrum of anaerobic bacteria. Thus, there is potential for participation of these cofactors in non-specific reactions with hydrophobic chlorinated pollutants that could gain entry into bacterial cells by partitioning through membranes. Cobalamins, coenzyme F_{430}, and hematin undergo reversible redox transitions allowing multiple turnovers with chlorinated methanes in the presence of excess reductant (Krone, *et al.*, 1989a,b; Klecka & Gonsior, 1984). A more recent study reports the catalytic reductive dechlorination of chlorinated ethylenes and benzenes by the same transition metal coenzymes (Gantzer & Wackett, 1991). These data along with those of Krone, *et al.* (1989a,b) suggest that non-specific reactions with transition metal cofactors may be important in the reductive dehalogenation of many xenobiotic compounds.

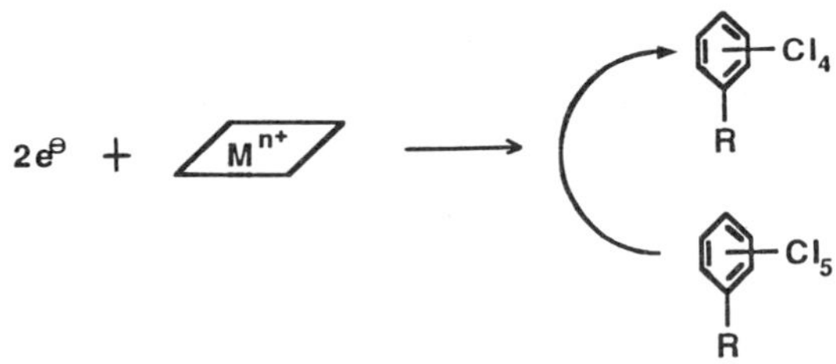

$$2\,e^- + \boxed{M^{n+}} \longrightarrow$$

Figure 6.   **Reductive dechlorination of chlorobenzenes by metallocofactors.**

A general mechanism for the reductive dechlorination of chlorobenzenes by reduced coenzymes is shown in Figure 6. Hexa- and pentachlorobenzene are dechlorinated and the former reacts more rapidly. Vitamin B_{12}, coenzyme F_{430}, and hematin all reduce hexachlorobenzene to pentachlorobenzene. This is of particular interest because cobalamins are reduced by one or two electrons ($Co^{3+} \rightarrow Co^{2+} \rightarrow Co^{1+}$) whereas coenzyme F_{430} ($Ni^{2+} \rightarrow Ni^{1+}$) and hematin ($Fe^{3+} \rightarrow Fe^{2+}$) typically undergo one electron cycling. Elucidating the details of electron transfer requires further study. There is evidence for the formation of carbon metal bonds during the reductive dechlorination of chlorinated ethylenes by cobalamins (Gantzer & Wackett, 1991; Schanke & Wackett, unpublished data). Stoichiometric reduction of cyanoaquacobinamide(III) to its cobalt(I) form is monitored spectrophotmetrically. Addition of excess perchloroethylene yields a new species with an absorption maximum at 470 nm. A similar intermediate is observed by reaction of reduced cobalamins with vinyl chloride (Figure 7). In this latter example, the organocobalt intermediate (I) shows identical absorption maxima and decay kinetics as for vinyl cobalamin prepared by reaction of cobalamin with acetylene (Hogenkamp, 1965). This suggests that vinyl chloride is dechlorinated by a net substitution process rather than by addition-elimination pathways (Figure 7). Additional experiments are required to understand the routes of dechlorination for more heavily chlorinated ethylenes. In previous studies with bromo and fluoroethylenes, both mechanisms are known.   The reaction of Co(I)-cobalamins with vinyl bromide and tetrafluoroethylene have been reported to yield vinyl-cobalamin (Johnson, *et al.*, 1963) and tetrafluoroethyl-cobalamin, respectively (Mays, *et al.*, 1964).

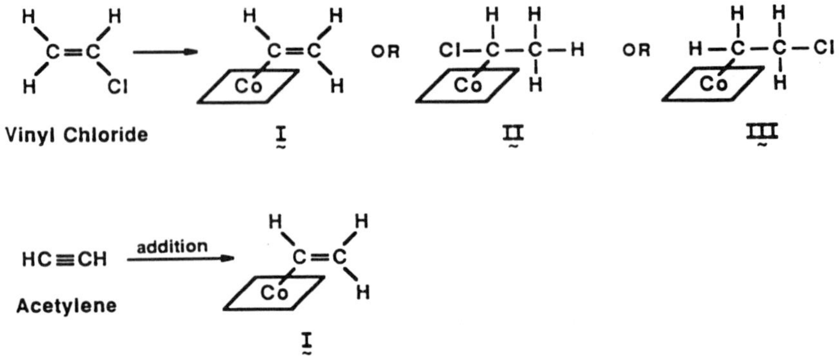

Figure 7.   **Organocobalt intermediates prepared from vinyl chloride and acetylene.**

Further studies on reductive dechlorination mechanisms should pay important dividends for understanding relevant bacterial reactions *in vivo*.   The rates of biological reductive dehalogenation are very low.   A greater knowledge of rate determining steps in these reactions could yield clues for rate enhancement of anaerobic bioremediation processes.

ACKNOWLEDGEMENTS

This work was supported by National Institutes of Health Grant GM41235. The assistance of Louise Mohn in the preparation of the manuscript is gratefully acknowledged.

REFERENCES

Ahmed, A. E. & Anders, M. W.   1978.   Metabolism of dihalomethanes to formaldehyde and inorganic halide II.   Studies on the mechanism of the reaction. *Biochem. Pharmacol.* **27**:2021.

Anders, M. W. & Pohl, L. R.   1985.   Halogenated alkanes.   In "Bioactivation of Foreign Compounds," ed. M. W. Anders.   Academic Press, New York.

Arciero, D., Vannelli, T., Logan, M. & Hooper, A. B.   1989.   Degradation of trichloroethylene by the ammonia-oxidizing bacterium *Nitrosomonas europaea. Biochem. Biophys. Res. Commun.* **159**:640-643.

Atlas, R. M. and Bartha, R.   1987.   "Microbial Ecology," Benjamin/Cumming Pub., Menlo Park, CA.

Au, K. & Walsh, C. T.   Stereochemical studies on a plasmid-encoded fluoracetate halidohydrolase. *Bioorg. Chem.* **12**:197-205.

Berry, E. K. M., Allison, N., Skinner, A. J. & Cooper, R. A.   1979.   Degradation of the selective herbicide 2,2-dichloropropionate (dalapon) by a soil bacterium. *J. Gen. Microbiol.* **110**:39-45.

Bohme, H., Fischer, H. & Frank, R.   1949. *Justus Liebig Annln. Chem.* **563**:54.

DeWeerd, K. A. & Suflita, J. M.   1990.   Anarobic aryl reductive dehalogenation of halobenzoates by cell extracts of *Desulfomonile tiedjei, Appl. Environ. Microbiol.* **56**:2999.

Dolfing, J. 1990. Reductive dechlorination of 3-chlorobenzoate is coupled to ATP production and growth in anaerobic bacterium, strain DCB-1. *Arch. Microbiol.* **153**:264-266.

Folsom, B. R., Chapman, P. J. & Pritchard, P. H. 1990. Phenol and trichloroethylene degradation by *Pseudomonas cepacia* G4: Kinetics and interaction between substrates. *Appl. Environ. Microbiol.* **56**:1279-1285.

Fox, B. G., Borneman, J. G., Wackett, L. P. & Lipscomb, J. D. 1990. Haloalkene oxidation by the soluble methane monooxygenase from *Methylosinus trichosporium* OB3b: Mechanistic and environmental implications. *Biochemistry* **29**:6419-6427.

Gälli, R., Stucki, G. & Leisinger, T. 1982. Mechanism of dehalogenation of dichloromethane by cell extracts of *Hyphomicrobium* DM2. *Experientia* **38**:1378.

Gantzer, G. J. & Wackett, L. P. 1991. Reductive dechlorination catalyzed by bacterial transition-metal coenzymes. *Environ. Sci. Tech.* (in press).

Goldman, P., Milne, G. W. A. & Keister, D. B.. 1968. Carbon-halogen bond cleavage: Studies on bacterial halidohydrolases. *J. Biol. Chem.* **243**:428-434.

Gschwend, P. M., MacFarlane, J. K. & Newman, K. A. 1985. Volatile halogenated organic compounds released to seawater from temparate marine microalgae. *Science* **227**:1033-1035.

Hambright, P. 1975. In "Porphyrins and Metalloporphyrins," Smith, K. M., ed., Elsevier Scientific, Amsterdam, p. 233.

Harker, A. R. & Kim, Y. 1990. Trichloroethylene degradation by two independent aromatic-degrading pathways in *Alcaligenes eutrophus* JMP134. *Appl. Environ. Microbiol.* **56**:1179-1181.

Hogenkamp, H. P. C. 1975. In "Cobalamin: Biochemistry and Pathophysiology," Babior, B. M., ed., John-Wiley & Sons, NY, p. 21.

Hutzinger, O., Veerkamp, W. 1981. In "Microbial Degradation of Xenobiotic and Recalcitrant Compounds," Leisinger, T., Cook, A., Hutter, R., Nuesch, J., eds. Academic Press, London, p. 3.

Janssen, D. B., Pries, F., Van der Ploeg, J., Kazemier, B., Terpstra, P., & Witholt, B. 1989. Cloning of 1,2-dichloroethane degradation genes of *Xanthobacter autotrophicus* and expression and sequencing of the *dhl A.* gene. *J. Bacteriol.* **171**:6791.

Kawasaki, H., Miyoshi, K. & Tonomura, K. 1981. Purification, crystallization and properties of haloacetate halidohydrolase from *Pseudomonas* species. *Agric. Biol. Chem.* **45**:543-544.

Keen, J. H., Habig, W. H. & Jakoby, W. B. 1976. Mechanism for the several activities of the glutathione S-transferases. *J. Biol. Chem.* **251**:6183-6188.

Klecka, G. M. & Gonsior, S. J. 1984. Reductive dechlorination of chlorinated methanes and ethanes by reduced iron(II) porphyrins. *Chemosphere* **13**:391.

Kohler-Staub, D. & Leisinger, T. 1985. Dichloromethane dehalogenase of *Hyphomicrobium* sp. strain DM2. *J. Bacteriol.* **162**:676-681.

Krone, U. E., Laufer, K., Thauer, R. K. & Hogenkamp, H. P. C. 1989a. Coenzyme F_{430} as a possible catalyst for the reductive dehalogenation of chlorinated C_1 hydrocarbons in methanogenic bacteria. *Biochemistry* **28**:10061-10065.

Krone, U. E., Thauer, R. K. & Hogenkamp, H. P. C. 1989b. Reductive dechlorination of chlorinated C_1-hydrocarbons mediated by corrinoids. *Biochemistry* **28**:4908-4914.

Leisinger, T. 1983. Microorganisms and xenobiotic compounds. *Experientia* **39**:1183-1191.

Markus, A., Klages, V., Krauss, S., & Lingens, F. 1984. Oxidation and dehalogenation of 4-chlorophenylacetate by a two-component enzyme system from *Pseudomonas* sp. strain CBS3. *J. Bacteriol.* **160**:618.

Nash, T. 1953. The colorimetric estimation of formaldehyde by means of the Hantzsch reaction. *Biochem. J.* **55**:416-421.

Neidleman, S. L. & J. Geigert. 1986. "Biohalogenation: Principles, Basic Roles and Applications," John Wiley, New York.

Nelson, M. J., Montgomery, S. O. & Pritchard, P. H. 1988. Trichloroethylene metabolism by microorganisms that degrade aromatic compounds. *Appl. Environ. Microbiol.* **54**:604-606.

Oldenhuis, R., Vink, R. L., Vink, J. M., Janssen, D. B. & Witholt, B. 1989. Degradation of chlorinated aliphatic hydrocarbons by Methylosinus trichosporium OB3b expressing soluble methane monooxygenase. *Appl. Environ. Microbiol.* **55**:2819-2826.

Parsons, F., Wood, P. R. & DeMarco, J. 1984. Transformation of tetrachloroethene and trichloroethene in microcosms and groundwater. *J. Am. Water Works Assoc.* **76**:56-59.

Reineke, W. 1984. Microbial degradation of halogenated aromatic compounds. In "Microbial Degradation of Organic Compounds," Gibson, D. T., ed., Marcel Dekker, New York, pp. 319-360.

Salomaa, P. 1966. Formation of carbonyl groups in hydrolytic reactions. In "The Chemistry of the Carbonyl Group," S. Patai, ed. Wiley Interscience, New York, pp. 177-210.

Scholtz, R., Wackett, L. P., Egli, C., Cook, A. M. & Leisinger, T. 1988. Dichloromethane dehalogenase with improved catalytic activity isolated from a fast-growing dichloromethane-utilizing bacterium. *J. Bacteriol.* **170**:5698-5704.

Shaik, S. S. 1985. The collage of S_N2 reactivity patterns: A state correlation diagram model. In "Progress in Physical Organic Chemistry," R. W. Taft, ed., Wiley & Sons, NY.

Storck, W. 1987. Chlorinated solvent use hurt by federal rules. *Chem. Eng. News* **65**:11.

Stucki, G., Gälli, R., Ebersold, H-R. & Leisinger, T. 1981. Dehalogenation of dichloromethane by cell extracts of *Hyphomicrobium* DM2. *Arch. Microbiol.* **130**:366-371.

Tsien, H.-C., Brusseau, G. A., Hanson, R. S. & Wackett, L. P. 1989. Biodegradation of trichloroethylene by Methylosinus trichosporium OB3b. Appl. Environ. Microbiol. **55**:3155-3161.

Van den Wijngaard, A. J., Reuvekamp, P. T. & Janssen, D. B. 1991 Purification and characterization of haloalcohol dehalogenase from *Arthrobacter* sp. strain AD2. *J. Bacteriol.* **173**:124.

Wackett, L. P. & Gibson, D. T. 1988. Degradation of trichloroethylene by toluene dioxygenase in whole cell studies with Pseudomonas putida F1. *Appl. Environ. Microbiol.* **54**:1703-1708.

Wackett, L. P., Brusseau, G. A. & Hanson, R. S. 1989. Survey of microbial oxygenases: Trichloroethylene degradation by propane-oxidizing bacteria. *Appl. Environ. Microbiol.* **55**:2960-2964.

Walsh, C. T. & Orme-Johnson, W. H. 1987. Nickel enzymes. *Biochemistry* **26**:4901.

Winter, R. B., Yen, K.-M. & Ensley, B. D. 1989. Efficient degradation of trichloroethylene by a recombinant *Escherichia coli*. *Biotechnology* **7**:282-285.

Wolfe, R. S. 1985. Unusual coenzymes of methanogenesis. *Trends. Biochem. Sci.* **10**:396.

Wuosmaa, A. M. & Hager, L. P. 1990. Methyl chloride transferase: A carbocation route for biosynthesis of halometabolites. *Science* **249**:160-162.

IMMOBILIZED ARTIFICIAL MEMBRANE CHROMATOGRAPHY:
SURFACE CHEMISTRY AND APPLICATIONS

Charles Pidgeon,[1,*] Craig Marcus,[2] and Francisco Alvarez[3]

[1]Department of Medicinal Chemistry and Pharmacognosy
Purdue University
West Lafayette, Indiana 47907
[2]Department of Pharmacology and Toxicology
Purdue University
West Lafayette, Indiana 47907
[3]Schering Plough Corporation
200 Galloping Hill Road
Kenilworth, New Jersey 07033

Abstract

Immobilized Artificial Membranes are solid surfaces containing phospholipids immobilized on silica particles at surface densities similar to the ligand density of reversed phase chromatographic surfaces. Chromatographic and non-chromatographic applications of Immobilized Artificial Membrane surfaces are reviewed and compared to the chromatographic and non-chromatographic applications of reversed phase columns. The methodology for synthesizing Immobilized Artificial Membranes and the stability of Immobilized Artificial Membranes are also described. Several examples are presented regarding the ability of Immobilized Artificial Membrane surfaces to model biological processes. Examples include predicting the transport of solutes across human skin, predicting the transport of amino acids across the blood brain barrier, and predicting the binding of solutes to liposome membranes. In addition, the purification of several membrane proteins, including cytochrome P450 from rat adrenals and rat livers, NADH oxidase, and rabbit intestinal phospholipid binding protein, are discussed.

Introduction

Immobilized Artificial Membrane (IAM) surfaces have only recently been invented (1-11). The fundamental idea associated with IAM technology is that a mechanically stable surface, that emulates artificial membranes (i.e., liposomes), can be synthesized by covalently binding membrane-forming lipids to solid surfaces. IAM surfaces are thus confluent monolayers of immobilized membrane lipids, wherein each lipid molecule is covalently linked to the surface. Membrane lipids contain both a polar headgroup and two non-polar alkyl chains, and lipid immobilization requires synthesizing lipids with ω-carboxyl functional groups on at least one of the lipid alkyl chains. The carboxyl groups are used to covalently link each lipid molecule to the silica surface. Because the immobilized lipid

---

*Corresponding author

Applications of Enzyme Biotechnology, Edited by J.W. Kelly and
T.O. Baldwin, Plenum Press, New York, 1991

headgroups protrude away from this new chromatographic surface, they are the first contact site between biomolecules in solution and the IAM surface.

An important disadvantage associated with IAM technology is that lipid molecular dynamics (e.g., lateral diffusion, flip-flop, and axial displacement) are absent because each lipid molecule is covalently linked to the silica surface. Thus the molecular dynamics of lipids in fluid artificial membranes or cells are not modeled on surfaces prepared for IAM technology. One lab has noncovalently adsorbed phosphatidylcholine molecules to silica-based chromatographic surfaces in order to measure octanol/water partition coefficients by chromatographic methods (12). In other reports, the immobilization of intact liposome particles, with and without membrane proteins embedded in the liposome membrane, have been used for the purpose of (i) studying membrane-protein mediated solute transport across membranes, and (ii) purifying biomolecules (13-16). The immobilization of intact liposomes is quite clever and has unique applications as described in a recent review of the topic (17). Methods that immobilize liposomes accurately mimic the lipid dynamics of cell membranes and are well suited for studying solute-membrane interactions. However, in IAM technology, the 'inability' to emulate the lipid dynamics associated with fluid artificial membrane surfaces is traded for the increased stability gained by covalently immobilizing lipids to solid surfaces. Most applications of IAM surfaces 'require' covalent lipid immobilization as described below and therefore IAM surfaces have unique applications compared to immobilized liposomes. In particular, organic solvents and detergents can be included in the mobile phase during IAM chromatographic studies, but not during chromatographic studies using immobilized liposomes (14-18) because the immobilized liposomes would be destroyed. Regarding phosphatidylcholine adsorbed to silica (12), this system most likely does not emulate the lipid dynamics of cell or artificial membranes because the adsorbed-lipids are not organized similar to artificial or natural membranes.

Synthesis of IAM Surfaces

The prototype IAM surface contains a diacylphosphatidylcholine (PC) covalently bonded to silica propylamine through an ω-carboxyl group on the C2 fatty acid chain (Scheme 1). The synthetic lipid immobilized was 1-myristoyl-2-[(13-carboxyl)tridecanoyl]-sn-3-glycerophosphocholine denoted as Lecithin COOH (7), and the bonding densities are approximately 60 mg-PC/g-silica. The surface is referred to as IAM.PC and to date, all IAM.PC surfaces utilized silica propylamine containing 300 Å pores as the mechanically stable matrix for lipid immobilization. The molecular area of membrane lipids is greater than the molecular area of propylamine groups initially tethered to the silica surface, and therefore after coupling large PC molecules to silica propylamine, residual free propylamine groups still exist on the silica surface (Scheme 1). Approximately 2 moles of residual amines per mole of immobilized PC reside at a depth of approximately 15 Å below the immobilized PC headgroup and thus the IAM.PC sub-surface is chemically basic.

Residual amines on IAM.PC surfaces decreased the chemical stability of the bonded phase (11), irreversibly bound mobile phase citric acid (9), increased the retention of acidic compounds (9,11), and decreased the retention of basic compounds (11). Consequently a detailed study was performed to identify methods to eliminate the residual amines by end-capping (11). The primary purpose of the end-capping studies was to prepare IAM.PC surfaces that exhibit little or no phospholipid leaching when challenged with different solvents. Significant destruction of the immobilized lipids occurred when silyation reagents were used to end-cap the residual amines and therefore these conventional end-capping reagents can not be used. However, symmetric fatty acid anhydrides were found to be excellent end-capping reagents. Symmetric fatty acid anhydrides convert amines into amide bonds, i.e., chemically neutral functional groups, and were found to be the best end-capping reagents. End-capping with symmetric anhydrides: (i) resulted in very little phospholipid leaching during the end-capping reaction, and (ii) prevented most of the phospholipid leaching that occurs from the IAM.PC surface when exposed to different solvents.

Quantitatively end-capping the residual amines on IAM.PC packing material by alkyl anhydrides required two end-capping reactions. The primary end-capping reaction utilized decanoic or dodecanoic anhydride which converted ~85 % of the residual amines into amides, and also significantly increased the hydrocarbon density of the the IAM.PC

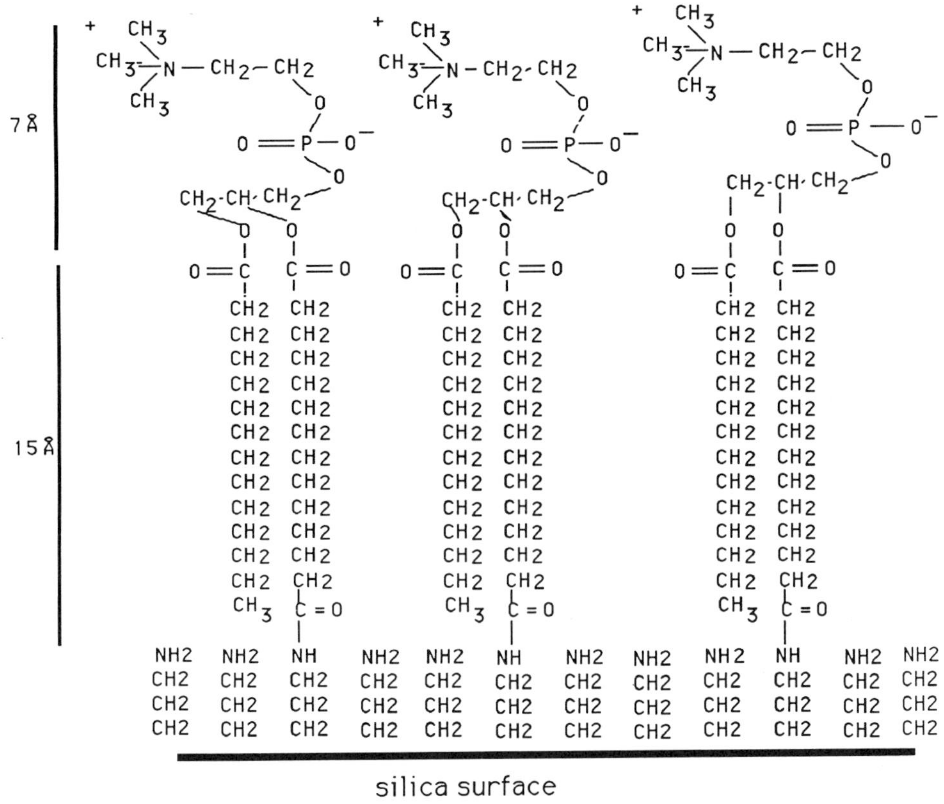

7 Å

15 Å

silica surface

Scheme 1. Immobilized Artificial Membranes are chromatographic packing material containing high surface densities of immobilized lipids. The bonded phase shown above contains a diacylated phosphatidylcholine immobilized on silica propylamine; after coupling residual propylamine groups exist near the silica subsurface as shown above. Some Immobilized Artifical Membranes have been end-capped to remove these residual amines (see Table 1).

interphase. This was followed by secondary end-capping with propionic anhydride which converts the residual amines into amides. IAM.PC surfaces end-capped with anhydrides are denoted by a superscript corresponding to the anhydride alkyl chain length used for end-capping. For example, IAM.PC^{C4} denotes that the IAM.PC surface was end-capped with butyric anhydride (Table 1). In our lab, end-capping reactions are always performed twice and usually the resultant end-capped surface is ninhydrin-negative or only slightly ninhydrin-positive, which indicates that virtually all residual amines have been eliminated from the IAM surface.

Three major limitations exist for the non-end-capped IAM.PC surface: (1) a diacylated phosphatidylcholine was immobilized, (2) the fatty acids are linked to the glycerobackbone by ester groups, and (3) lipid molecules are linked to the silica surface through the C2 fatty acid chain. Problem (1) is based on diacylated phospholipids immobilized on gold surfaces by only one alkyl chain (18); the surface density of diacylated phospholipids in this system is not maximal after bonding because the conformationally free fatty acid chain restricts the packing of lipids during immobilization. For problem (2) esters are intrinsically unstable relative to ethers and the fatty acid alkyl chains should be linked to the glycerobackbone by ether bonds to improve the stability. For problem (3), we speculate that anchoring the PC molecule to the silica surface by the C2 chain pulls the phospholipid molecule towards the silica surface and distorts the glycerobackbone into a conformation different than the glycerobackbone conformation found in cell membranes. Mobile

203

Table 1. Immobilized Artificial Membranes Surfaces.

Packing material	Commercially available (yes/no)	Description of surface
IAM.PC	YES	1-myristoyl-2-[(13-carboxyl)-tridecoyl]-sn-3-glycerophosphocholine bonded to silica-propylamine
IAM.PC.Glycidol	NO	IAM.PC end capped with glycidol
IAM.PC.MG	YES	IAM.PC end capped with methyglycolate
IAM.PCC2	NO	IAM.PC end capped with acetic anhydride
IAM.PCC3	NO	IAM.PC end capped with propionic anhydride
IAM.PCC4	NO	IAM.PC end capped with butyric anhydride
IAM.PCISOCC4	NO	IAM.PC end capped with isobutyric anhydride
IAM.PCC6	NO	IAM.PC end capped with hexanoic anhydride
IAM.PCC10	NO	IAM.PC end capped with decanoic anhydride
IAM.PCC12	NO	IAM.PC end capped with dodecanoic anhydride
IAM.PCC12/C3	NO	IAM.PCC12 end capped with proprionic anhydride
IAM.PCETHER	NO	1-O-[(11-carboxyl)-undecoyl]- 2-O-Methyl sn-3-glycerophosphocholine bonded to silica propylamine.
IAM.PCETHER/C10	NO	IAM.PCETHER end capped with decanoic anhydride

phospholipids in membranes typically have the sn-2 ester a few angstroms closer to the bulk aqueous media than the sn-1 ester, and this is due to the conformation of the glycerobackbone. In other words, the phospholipid alkyl chain in the sn-2 position of the glycrobackbone is pulled toward the membrane interface approximately 2 methylene groups relative to the alkyl chain in the sn-1 position (19). Thus for both cell and artificial membranes, the sn-1 and sn-2 ester groups of typical phospholipids are chemically nonequivalent. Although the effect of linking the phospholipid to the surface by the C2 chain may be a problem regarding the conformation of the glycerobackbone, the effect on the surface-recognition properties of these bonded phases is unknown. Ongoing infrared and NMR studies in our lab are aimed at evaluating the interfacial region of IAM surfaces.

To address problems 1, 2, and 3, we have recently synthesized a single-chain phospholipid analog which contains: an ether-linked fatty acid in the glycero sn-1 position, an ω-carboxyl group on the fatty acid chain, and a methyl ether in the glycero sn-2 position. This single chain analog is expected to have a higher surface bonding density than the diacylated analog described above. In addition, this analog links the phospholipid to the silica surface through the C1 alkyl chain, which is expected to not distort the glycerobackbone. Furthermore, the ether-linkage eliminates the chemical instability of IAM surfaces containing ester-linked alkyl chains. Table 1 describes all of the bonded phases synthesized in our lab to date, and a future publication will describe the synthesis of the ether-lipid used for immobilization and the chromatographic properties of this extremely stable IAM surface.

We note that the commercial manufacturer (Regis Chemical Co. Morton Grove IL) of IAM.PC has end-capped IAM.PC surfaces with glycidol and methylglycolate (MG) (12). End-capping with glycidol converts the residual silica amines into secondary or tertiary amines and therefore the silica subsurface remains chemically basic after end-capping. In addition, end-capping with glycidol causes new hydroxyl groups to exist approximately 10-

12 Å below the PC headgroup. IAM.PC end-capped with glycidol is not commercially available. End-capping IAM.PC surfaces with MG converts the residual amines into chemically <u>neutral</u> amides (highly desirable considering the hydrocarbon environment of artificial bilayers) but one mole of new hydroxyl groups are formed during end-capping. The 'new' immobilized hydroxyl groups from methylglycolate end-capping are ~10-12 Å below the immobilized PC headgroup. This bonded phase is commercially available as IAM.PC.MG. Artificial and biological membranes do not contain hydroxyl groups near the center of the membrane and for this reason we have not pursued the chromatographic applications of IAM.PC surfaces end-capped with glycidol or methylglycolate. However, because the residual amines participate in the selectivity (i.e., surface recognition properties) of the IAM.PC surface, these surfaces may have unique applications.

Stability of IAM.PC Surfaces

Diligent utilization of guard columns containing the same packing material as the IAM HPLC column significantly improves the column lifetime by reducing contamination by particulates and other insoluble material commonly found in biological samples. Guard columns also protect the main column from irreversible protein binding which can permanently alter the retention characteristics of the column [6].

Our studies show that approximately 2 % of the immobilized phospholipid was leached from the commercially available IAM.PC columns when they were challenged with 2000 milliliters of mobile phase containing 35 mM citric acid, 35 mM $NH_4H_2PO_4$, and 35 mM NH_4Cl [9,11]. However this lipid leaching depends on the solvent used for the perfusion [8]. Perfusing IAM.PC columns with either acetone or acetonitrile cause virtually no lipid leaching during prolonged exposure, but non end-capped columns stored in acetone require several column volumes to equilibrate to new aqueous mobile phase compositions. In contrast, IAM.PC^{C10} columns (Table 1) equilibrate rapidly (~2-3 column volumes) to new aqueous mobile phases.

Perfusion of nonendcapped IAM.PC columns with either 0.1 % TFA/H_2O or 0.1 %TFA/acetonitrile caused virtually the same amount of lipid leaching (i.e., ~ 12 % of the initially immobilized phospholipid leached from the column), and alcohols cause about the same amount of lipid leaching as 0.1%TFA/H_2O [8]. These mobile phases should be avoided if possible when using the non-end-capped IAM.PC bonded phase. In general, neutral or slightly acidic buffers (pH > 4) cause very little lipid leaching from non-end-capped IAM.PC surfaces after perfusing 2000 milliliters of mobile phase through the column. Small amounts of lipid leaching apparently do not affect the chromatographic properties of the IAM.PC for membrane proteins as evidenced by the observation that non-end-capped IAM.PC columns stored in totally aqueous detergent buffers pH ~ 7.25 exhibited similar chromatographic resolution of cytochrome P450 and other membrane proteins [6]. For membrane protein purification, IAM.PC columns are thus apparently stable for 1-2 years [6]. End-capping the IAM.PC surface eliminates virtually all lipid leaching [11]; thus very stable IAM surfaces can be prepared, but the chromatographic properties are expected to change if the analyte uses residual amines as a binding site (which is true for many analytes). We note that we have not evaluated the participation of residual amines in the purification of membrane proteins. Although lipid leaching from non-end-capped surfaces is significantly greater than end-capped surfaces, quantitative comparisons have not been made. The effect of temperature on the stability of the IAM columns has also not been evaluated.

Aside from the chemical stability of IAM surfaces described above, the stability of IAM surfaces in the presence of phospholipases has been evaluated. The stability of IAM surfaces (shown in Scheme 1) to phospholipase activity is critical to the use of IAM chromatographic surfaces for purifying membrane proteins or other biological samples that are expected to contain different phospholipases. We have tried several times to hydrolyze the immobilized phospholipid molecules comprising the IAM.PC surface with phospholipase A2 (snake venom). Phospholipase A2 (PLA2) will not hydrolyze even small amounts of immobilized phospholipids, at least with regard to IAM surfaces. This is not surprising based on the proposed mechanism of action of PLA2 recently suggested from the X-ray structure of three different PLA2 enzyme-substrate analogs [20-22]. Apparently, PLA2 adsorbed to a cell membrane or lipid aggregate requires the phospholipid substrate to be axially displaced from the lipid-surface into a highly conserved 14 Å hydrophobic pocket.

The 14 Å hydrophobic channel of PLA2 is sufficiently deep to accept the phospholipid headgroup and the first 8 methylene groups in the alkyl chains; this lipid molecular distance defines the approximate axial displacement a membrane lipid must incur to become a PLA2 substrate. Since IAM.PC contains immobilized phospholipids, they are not substrates for PLA2 under any solution conditions because the lipid molecules can not be axially displaced. Preliminary data in our lab have shown that phospholipase D can accept immobilized phospholipids as substrates, but the enzymatic formation of different products was extremely difficult to detected because the products remained immobilized on the IAM surface; however, phospholipase D will accept the immobilized lipids as substrates but we do not know to the efficiency or product distribution of the reaction. Currently we are trying to establish assays to quantitate the amount immobilized phosphatidylcholine converted to different phospholipids. We note that the stability of immobilized monolayers of membrane lipids on IAM surfaces to enzymes would be expected to differ if compared to the enzymatic stability of immobilized phospholipid molecules that are isolated. Isolated phospholipid molecules have been immobilized on soft gels for the purification of PLA2 by affinity chromatography, but the immobilized lipids contained ether linked alkyl chains (23, 24).

For reversed phase chromatography surfaces, it is well established that the chemical stability of packing material increases as the ligand bonding density increases and/or steric protection of the silica subsurface is provided by the immobilized ligands (25-27). Based on our recent work, this is also true for IAM surfaces (11). For reversed phase columns there is also an optimum ligand density for efficient partitioning of small molecules into the interphase (28) and there is also an optimum alkyl chain density for minimizing the time needed for column equilibration during gradient elution (29). These observations are explained by the concept that when the maximum alkyl chain density is grafted to silica, partitioning of solutes is eliminated because the configurational energy of the bonded phase ligands limits the ability of the solute to partition (28). In other words, solute partitioning requires that the grafted ligands elicit conformational changes to create a solute cavity inside the bonded phase interphase. The higher the grafted alkyl chain density on the surface, the higher the configurational energy associated with this process. We also expect that the ligand density on IAM surfaces will affect the partitioning of solutes. Thus another trade-off with IAM technology is that very high surface densities of ligands is desirable for improved chemical stability but this stability may be gained at the expense of minimizing the ability of solutes to partition into the membrane. However, for large molecules we do not expect the ligand density to significantly influence retention because in reversed phase chromatography of proteins, similar results are obtained on columns containing both short and long alkyl chains immobilized to silica. Large molecules predominantly adsorb to chromatographic surfaces; whereas small molecules can either adsorb or partition, or both. We note that it is extremely difficult to distinguish whether a particular solute, large or small, exhibits adsorption or partitioning as the major binding mechanism on reversed phase columns (30). However, with IAM.PC surfaces solute partitioning into the hydrocarbon interphase is demonstrably not ambiguous because residual amines participate in the chromatography; participation of the residual amines in retaining solutes (described below) indicates that solutes are partitioning into the IAM bonded phase as they migrate through the IAM column.

Chromatographic and Physical-Chemical Properties of IAM.PC Surfaces

All IAM surfaces synthesized to date mimic only 1/2 of a true bilayer in hydrocarbon thickness, and contain the phosphatidylcholine headgroup. The phosphatidylcholine headgroup is itself electrically neutral, but contains both a positive charge from a quaternary amine and a negative charge from phosphate. The charge separation is approximately 4.5 Å (19). Titration of the non-end-capped IAM.PC surface showed that the surface is neutral with a pKa of approximately 7 (9).

Retention mechanisms on IAM.PC include (i) partitioning into the hydrocarbon region, (ii) solute adsorption to the polar headgroups which contain both anion- and cation-exchange sites, and (iii) ionic bonding (or salt bonds) to the uncapped primary amines. The retention mechanism of solutes on IAM.PC surfaces depends on both the analyte's size and charge, and hence mixed-mode chromatography is expected on all IAM surfaces because the surface itself contains ~ 7 Å thick hydrophilic interfacial region covalently piggy-backed on top of ~ 14 Å of hydrocarbon (Scheme 1). Thus, typical analytes containing both hydrophilic and hydrophobic moieties are expected to interact with the IAM surface by a mixed mode chromatographic retention mechanism.

All nonendcapped IAM.PC surfaces, and IAM.PC surfaces end-capped with either glycidol or methylglycolate immediately wet when suspended in aqueous solvents indicating that these surfaces are hydrophilic. However, end-capping the IAM.PC surface with fatty acid anhydrides eliminates the ability of the IAM.PC surface to be wet by aqueous solvents. For instance, IAM.PC surfaces end-capped with either propionic anhydride or dodecanoic anhydride would not wet when suspended in aqueous media. Thus the alkyl chain length of the anhydride does not determine the aqueous wetability of the surface, but rather the elimination of the residual amines seems to be the dominant factor affecting the wetability of the IAM.PC surface. We note that immobilization of single chain phospholipids, followed by end-capping with alkyl anhydrides, is expected to generate a surface that wets with aqueous mobile phases. This is because the bonding density of the single-chain phospholipid is higher than the bonding density of the double-chain analog.

Small basic solutes (amphetamine analogs) were retained on non-end-capped analytical size IAM.PC columns approximately 2 to 25 minutes depending on the analyte. However, the retention of basic compounds on IAM.PC surfaces increases approximately 2-fold if the surface is end-capped (Table 2, amphetamine analogs). The retention of deoxynucleotides (acidic compounds) on IAM.PC surfaces depends on the number of phosphate groups. In general, the order of elution depends on the number of phosphate groups; triphosphate analogs elute after diphosphate analogs which elute after monophosphate analogs. This indicates that the IAM.PC surface functions in part as an anion exchange surface (Table 2, Deoxynucleotides). End-capping the IAM.PC surface virtually eliminates the retention of deoxynucleotides; this indicates that the residual amines on the IAM surface are the primary binding site for deoxynucleotides (Table 2, Deoxynucleotides). Similar results were found for oligonucleotides (unpublished). The non-end-capped IAM.PC column appears to be very efficient regarding the retention of deoxynucleotides as evidenced by the observation that 30,000 to 40,000 plates/meter were observed for the mobile phase compositions shown in Table 2. These high plate numbers demonstrate that minimal band dispersion occurs during analyte migration through the IAM.PC column.

Preliminary studies with peptides and proteins have also been performed and increased retention times are found for peptides that contain cysteine. For unknown reasons, peptides containing cysteine are retained approximately 2 fold longer than non-cysteine containing peptides (Table 2, Peptides). We note that these results were obtained with aqueous mobile phases containing trifluroacetic acid (TFA) which causes small amounts of lipid to leach from the IAM surface (8). At the time these data were obtained, this stability problem was unknown. Thus although the IAM column can discriminate cysteine groups on peptides, the data needs to be reacquired. Unlike peptides which can be eluted in totally aqueous mobile phases from IAM surfaces, soluble proteins require organic solvents for elution. However, the amount of organic solvent required for elution from the IAM surface is significantly less than that required for elution from typical C8 or C18 reversed phase columns; this is because the IAM.PC surface has the hydrophobicity of a reversed phase column containing immobilized propyl groups, i.e., 3 carbons (7). Thus mobile phase elution conditions on IAM.PC surfaces require significantly less amounts of denaturing components in the mobile phase.

IAM.PC columns can also be used for the analysis of fatty acids. IAM.PC columns give good separations of oleic acid from elaidic acid despite the fact that these analytes differ only in their cis/trans stereochemistry (Table 2, Fatty acids). In addition, linoleic acid and linolenic acid are also well resolved on IAM.PC columns; these fatty acids differ by only the number of double bonds. For analysis of fatty acids by IAM.PC chromatography, 50 % acetonitrile was mixed with isotonic phosphate buffered saline (PBS) and this mixture was used as mobile phase.

Table 2 IAM Solute Binding Studies

	Sample	Experimental objective	Mobile phase[1]	Comment (reference)
Amphetamine-analogs[2]	17 analogs	Effect of end-capping on the retention of basic compounds	isocratic (1.0 ml/min) 0.1 M $(NH_4)H_2PO_4$ pH 7.2	End-capping IAM.PC with methylglycolate (to make IAM.PC.MG) increased ~ 2-fold the retention of all basic compounds tested (ref. 11).
Deoxy-nucleotides[2]	d-AMP, d-ADP, d-ATP, d-CMP, d-CDP, d-CTP, d-TMP, d-TTP, d-GMP, d-GDP, d-GTP	characterization of deoxnucleotide binding to IAM surface and the Effect of end capping on deoxynucleotide binding to IAM surface	isocratic (1.0 ml/min) 35 mM Citric acid, 35 mM NH_4H_2PO4, 35 mM NH_4Cl, pH4.8	8 of 11 deoxynucleotides were resolved. Mobile phase citric acid irreversibly bound to the IAM.PC surface which caused changing retention times (refs. 9, 11). Decreasing the basicity of the IAM.PC surface by end-capping caused deoxynucleotide retention times to decrease. End-capping with methylglycolate eliminated the binding of all deoxynucleotides tested (ref. 9, 11).
Peptides[2]	NH_2MPSWTGGCOOH + NH_2MPSWTGGCCOOH	characterization of peptide binding to IAM.PC	isocratic (1 ml/min) 0.1 % $TFA/H_2$0	Peptides containing Cys have approximately double the retention of non-Cys containing peptides (ref. 7).
Peptides[2]	NH_2WGSTWPGCOOH + NH_2WGSTWPGCCOOH	characterization of peptide binding to IAM.PC	isocratic (1 ml/min) isotonic PBS pH 7.4	Peptides containing Cys have approximately double the retention of non-Cys containing peptides (ref. 7).
Soluble Proteins[2]	Trypsin inhibitor Albumin Ovalabumin	characterization of protein binding to IAM.PC	gradient (1 ml/min) A) 0.1 % $TFA/H_2$0 B) 0.1 %TFA/CH_3CN 0 % B to 50 % B in 15 min	Protein elution from IAM.PC columns required organic solvents (ref. 7).
Fatty acids[2]	oleic acid, trans-oleic acid (elaidic), linolenic acid, linoleic acid	Chemical analysis of fatty acids	Isocratic (1 ml/min) CH_3CN:PBS 50:50	The IAM.PC column was used to resolve non-derivatized fatty acids differing in cis/trans sterochemistry (oleic vs elaidic acids) and by number of double bonds (linolenic and linoleic acids).

Alcohols[2]	7 alcohol analogs C2 to C8 carbons	Correlation of alcohol binding to IAM.PC with alcohol transport through human skin	isocratic (1.0 ml/min) MeOH/CH$_3$CN/H$_2$O 1:1:8	Alcohol retention on IAM.PC was log linear with transport of alcohols through human skin r^2=0.996) Reverse phase columns gave a poor correlation r^2= 0.850.
Amino acids[2,3]	L, I, M, R, V, K, T, C, S, A, D, P, G, GABA, cycloleucine, ornithine, Citrulline	correlation of amino acid binding to IAM.PC.MG with brain uptake of amino acids in rats	isocratic (1 ml/min) isotonic PBS pH 7.4	Binding of amino acids to both reverse phase and IAM.PC.MG columns correlated with amino acid transport across the blood-brain barrier (in vitro) but the correlation on IAM.PC.MG (r^2 = 0.734 was significantly better than C18 columns (r^2 =0.593).
Steroids[2]	progesterone, cortexone testosterone cortexalone corticosterone hydrocortisone cortisone	Correlation of steroid binding to IAM.PC with steroid transport through skin	isocratric (0.7 ml/min) MeOH/CH$_3$CN/H$_2$O 1:1:8	Binding of steroids to both reverse phase and IAM.PC columns correlated with steroid transport transport across human skin but IAM.PC (r^2 = 0.906 was better than C18 columns (r^2 = 0.843).
Amphiphilic Drugs[2]	Amidone, Chlorpromazine, promethazine, Imipramine, Propranolol	Correlation of amphilic drugs binding to IAM.PC with to DPPC liposomes	isocratic (2 ml/min) CH$_3$CN/MeOH:PBS 15:15:70	Amphiphilic drug retention on IAM.PC was log linear with binding of drugs to DPPC liposomes r^2= 0.906.
Imidazolines[2]	Tetryzoline, Naphazol-ine, Oxymethazoline, Xylometazoline	Correlation of imidazoline ad-renoreceptor ag-onist binding to IAM.PC.MG with partitioning of imidazoline ana-logs into liposomes	Isocratic (1 ml/min) CH$_3$CN/PBS 20:80	Binding of imidazolines to both reverse phase and IAM.PC.MG columns correlated with imidazoline partitioning into liposomes but the correlation with IAM.PC.MG (r^2= 0.987) was significantly better than the correlation with octanol/water partition coefficients (r^2 = 0.820).
Benazepine analogs	~20 analogs	Correlation of drug partitioning into liposomes with drug binding to IAM.PC surface	Isocratic (1.5 ml/mm) CH$_3$CN/PBS 20:80 pH 5, 6, 7.4	High correlation of liposome partition factors (or coefficients) determined by NMR with binding to IAM.PC columns, r^2=0.881. Low correlation with drug binding and reverse phase columns r^2=0.234 (personal communication, Dave Floyd).

1   TFA, trifluroacetic acid; PBS, isotonic phosphate buffered saline pH~7.2; CH$_3$CN, acelonitrile, MeOH, Methanol.

2   All analytes were pure compounds prepared either by synthesis or were purchased from standard suppliers (e.g., Sigma Chemical Company, Aldrich etc.).

3   Single letter amino acid code was used. L, leucine; I, isoleucine; M, methionine; r, arginine; V, valine; K, lysine; T, threonine; C, cysteine; S, serine; A, alanine; D, aspartic acid; P, proline;G, glycine. GABA is γ amino butyric acid.

In-Vitro Modeling of Biological Processes

Historically, reversed phase chromatography columns have been used extensively to obtain a hydrophobic parameter that correlates with drug activity during studies pertaining to quantitative structure-activity-relationships (QSAR) (31-36). Typically, the hydrophobic parameter obtained from chromatographic methods is considered to be correlated with the classical octanol/water partition coefficient measured by a shake-flask method. Grafting only alkyl chains to silica to form reversed phase chromatography columns generates a column that can very accurately measure the octanol/water partition coefficient of many compounds. Unfortunately, most biological processes are not rate-limited by their ability to exhibit only hydrophobic-partitioning and consequently, correlation of solute activities in biological processes with binding to reversed phase columns tends to be only a first approximation. Reversed phase columns do not provide the polar headgroup interactions associated with membrane lipids and therefore IAM surfaces are expected to provide a better correlation than reversed phase columns when biological processes are being monitored. As described below (and shown in Table 2), IAM.PC surfaces always generate better correlations than reversed phase columns for all studies performed to-date. However, for very hydrophobic solutes, IAM surfaces are only slightly better than reversed phase columns in predicting biological processes. The main reason that IAM surfaces and reversed phase surfaces are similar for very hydrophobic solutes is that hydrophobic solutes have only minimal interactions with the many polar components of cell membranes and other cellular compartments.

An extremely useful application of IAM columns thus resides in their ability to predict the transport of solutes across human skin and other biological barriers. In particular IAM columns are useful for predicting the skin transport of straight chain alcohols, containing from 2 to 8 carbons, which are impossible to accurately model using reversed phase columns (Table 2, alcohols). For hydrophobic compounds (e.g., steroids, Table 2), both reversed phase columns and IAM.PC columns can be used for predicting the solute transport rates across human skin but the correlation on IAM.PC columns is significantly better. Several other classes of compounds were used for these skin studies and a detailed description of the topic will be submitted for publication very soon (F. Alvarez, personal communication). Another interesting application of IAM columns is their ability to predict the transport of amino acids across the blood brain barrier (Table 2, amino acids). The blood-brain barrier is difficult to study because of the complexity of the system; consequently, *in-vitro* methods that can predict solute transport across the blood brain barrier would be of significant value during drug development. If the correlation of solute transport across the blood brain barrier can be extended to biologically active compounds, this will significantly aid the design of drugs that can penetrate the blood brain barrier.

Several studies have been performed correlating the adsorption/partitioning of solutes to liposome membranes with the retention of solutes on IAM.PC and IAM.PC.MG columns (Table 2; Amphiphilic drugs, Imidazolines, Benazepine analogs). In all cases, IAM.PC columns were better than reversed phase columns at predicting the partitioning of drugs into liposomes. Measuring the partitioning of drugs into liposomes can be experimentally time consuming and/or difficult. For instance, the partitioning of benazepine analogs into dipalmitoylphosphatidylcholine (DPPC) liposomes were measured by NMR (Dave Floyd, personal communication). For liposomes to remain intact during the NMR measurement, totally aqueous solvents (buffers) were necessary. However, several of the benazepine analogs have limited aqueous solubility which made the NMR measurement difficult or impossible. Furthermore, the benazepine analogs are basic compounds and the pH-dependent binding to liposomes was feasible on IAM.PC columns but not by NMR because many of the analogs had limited solubility at different solution pH values. However, evaluating the binding of benazepine analogs to IAM.PC columns did not suffer from solubility problems because an organic phase modifier was included in the mobile phase. Cells contain membrane proteins that transport calcium, and benazepine analogs are an important class of calcium channel blockers (31); consequently partitioning of benazepine analogs into membranes is important for their activity. The ability to substitute IAM.PC columns for liposomes for the measurement of drug-membrane-binding interactions was a significant experimental advantage. Liposomes are routinely used to study the membrane binding of different solutes and IAM.PC columns are proving to be both superior and much more convenient than this traditional method.

Purification of Membrane Proteins by IAM Chromatography

One of the most important applications of IAM technology to date is its ability to function as one of the chromatographic steps for the purification of functional membrane proteins. Membrane proteins are very difficult to purify because detergents must be present during each chromatographic step for maintaining both functional activity and solubility of the desired membrane protein; maintaining protein solubility is necessary to prevent protein aggregation which is a common problem during the purification of membrane proteins. Currently only 4 reports exist that demonstrate that IAM surfaces can purify membrane proteins (3-6). High pressure liquid chromatography (HPLC) utilizing IAM columns has been used as either the initial or final step in the purification of membrane proteins. As an example of IAM chromatography used as the first purification step, cytochrome P450 has been purified from rat liver, rat kidney (6), and rat adrenals (5) and several other sources as well (unpublished). For IAM HPLC chromatography used as the final purification step, a phospholipid binding protein was purified from rabbit intestinal mucosal cells. Aside from conventional IAM HPLC chromatography, IAM particles have also been used in batch mode to purify the glucose transporter from human red cells in one step (4). However, the batch purification of the red cell glucose transporter resulted in irreversible binding of virtually all other red cell membrane proteins in the sample to the IAM packing material; protein-protein interactions were attributed to the irreversible binding. Thus the IAM.PC packing material was destroyed and until methods can be developed that allow regenerating the IAM particles and removing strongly adsorbed red cell membrane proteins, the batch binding purification of the glucose transporter will require that fresh IAM packing material be used for each purification. No other membrane protein samples tested to date appear to suffer from such irreversible binding to the packing material and a reason for the irreversible binding has been proposed (4). Table 3 lists the necessary experimental details to reproduce the purification of membrane proteins on IAM surfaces that have been reported to date.

The most important factors to consider during the purification of membrane proteins are: (i) sample preparation, (ii) sample loading on the column, (iii) total protein recovery from the column, (iv) recovery of protein functional activity from the column, and (v) most importantly, the selectivity of the column for the desired protein. Each of these factors are considered below.

Sample preparation

Normal precautions should be observed for preparing membrane proteins for IAM HPLC chromatography. The sample must be completely solubilized and free of particulates and protein aggregates. Membrane proteins can aggregate under many conditions including: short or long term storage, when concentrated, or when mixed in different solutions. Protein samples suspected of containing aggregated proteins should not be injected into the IAM column. It is recommended that solubilized samples for IAM.PC chromatography be centrifuged at 105,000g for 60 minutes at 4°C to remove all particulate material from the sample. If the samples have been stored for any length of time, the samples should also be either filtered (0.2 to 0.4 μ filter) or centrifuged again. Guard columns significantly increase the column lifetime and are highly recommended. A single injection of a sample containing aggregated protein can increase the column back pressure sufficiently to cause protein functional activity to decrease during IAM chromatography (6). Destruction of protein functional activity most likely depends on the sensitivity of the protein to column pressure difference and/or 'pressure-induced turbulent fluid flow' near the chromatographic surface rather than the ability of the column packing material or elution conditions to directly denature the protein. For example, a 4.6 mm x 15 cm IAM column exhibiting a back pressure of 900 psi (i.e., typical conditions used for the purification of cytochrome P450 on IAM.PC HPLC columns packed with 7 μ size chromatographic particles) has a pressure drop across the column of 60 psi/cm. Increasing the pressure to 1400 psi causes the column pressure drop to increase to ~ 100 psi/cm. This large pressure difference causes turbulent fluid flow near the surface of the packing material as the mobile phase flows over the particles. As the protein migrates through the column and undergoes reversible adsorption-desorption on the surface, the turbulent fluid generates shear forces that can denature the protein. We have found that cytochrome P450 is significantly destroyed at column pressures greater than about 900 psi but other proteins may not suffer from this limitation. Typically changing the guard

Table 3 Purification of membrane proteins by IAM chromatrography

Sample	Experimental objective	Mobile phase[1]	Comment[2] (reference)
rat liver microsomes	purification of cytochrome P450	gradient (0.5 ml/min) 100% A 50 min; 0%B to 40% B 15 min;40% B 20 min; 40% B to 65%B 40 min; 65 %B 40 min, 65%B to 0%B 20 min.	Purification of cytochrome P450 in a single step on IAM.PC; ~70 % of the protein content was P450 and exhibited increased specific content (ref. 5,6).
rat liver microsomes	effect of end capping on cytochrome P450 purification	gradient (0.1 ml/min or 0.5 ml/min) 100% A 50 min; 0%B to 40%B 15 min;40% B 20 min; 40%B to 65%B 40 min; 65 %B 40 min, 65%B to 0%B 20 min.	High column back pressure on IAM.PC^{C10} columns perfused at 0.5 ml/min (1400 psi). This caused loss of protein-functional activity eluting from the column. At 0.1 ml/min, maximum protein functional activity was recovered. End capping reduced column specificity for P450, i.e., several other proteins copurified.
rat kidney microsomes	purification of cytochrome P450	gradient (0.5 ml/min) 100% A 50 min; 0%B to 40% B 15 min;40% B 20 min; 40% to 65%B 40 min; 65 %B 40 min, 65%B to 0%B 20 min.	Rat kidney cytochrome P450 is more labile than liver forms and consequently during pruification on IAM.PC, the recovery of functional activity was only 40% (ref. 6).
rat adrenal microsomes	Purification of cytochrome P450	gradient (0.5 ml/min) 100% A 50 min; 0%B to 40% B 15 min;40% B 20 min; 40% B to 65%B 40 min; 65 %B 40 min, 65%B to 0%B 20 min.	Purification of adrenal cytochrome P450 in a single step; ~70 % of the protein content was P450 and exhibited high specific activity (refs. 5, 6).
crude microsomes from rat-liver cells	purification of NADH oxido-reductase	gradient (0.5 ml/min) 100% A 15 min; 0%B to 40 % B 5 min; 40% B 10 min; 40% B to 65%B 15 min; 65 %B 40 min, 65%B to 0%B 10 min.	NADH oxidase and NADH:ferricyanide oxidoreductase were partially purified by DE52 chromatography prior to IAM.PC chromatography. IAM.PC was used as a scond step and a 6-fold increase in specific activity was found (ref. 6).
human red cell membrane proteins	correlation of batch-binding to protein elution from IAM.PC	isocratic (0.75 ml/min) 1.5 % LysoPC/ 0.06 M tris pH 7.	Protein-protein interactions caused irreversible binding of red cell membrane proteins in both batch binding studies and column elution studies. The glucose transporter was purified to near homogeneity in one step but the column was destroyed because of irreversible protein adsorption. Lysolecithin was used as an affinity detergent for eluting the glucose transporter (ref. 4).

| pre-purified rabbit intestinal microsomal cholesterol-binding protein | purification of membrane protein | gradient (0.7 ml/min)

A) 10 mMNa$_2$PO$_4$ pH 7.3, 0.14 M NaCl, 2.5 mM EDTA, 0.02% NaN$_3$.0.125% CHAPS
B) 0.5 % CHAPS + mobile phase A.

100 %A 10 min, 0%B to 70 % B 5 min, 70%B to 90%B 60 min, 90 %B to 100%B 5 min, 100%B to 0%B 5 min. | IAM.PC column was used as the final purification step. The Cholesterol binding protein eluting from the IAM column was a single band on SDS gels (ref. 3). |

---

1   Unless indicated the two mobile phases used for the gradient elution of membrane proteins were: Mobile phase "A" 0.1 M KH$_2$PO$_4$, pH 7.25/20 % glycerol/1mM EDTA/ 0.6% sodium cholate; mobile phase "B" was mobile phase "A" plus 2 % Lubrol PX.

2.   All separations were performed on analytical size IAM.PC (15 cm x 4.6 mm) columns, and purification of rat-liver and rat-adrenal P450 was also performed on a semi-preparative column (10 cm x 21.1 mm).

column regularly eliminates the increased column pressure generated by injecting dirty samples on the IAM column. The back pressure can be reduced by lowering the mobile phase flow rate but this increases the analysis time. End-capped columns (i.e., IAM.PC^{C10}) were used to evaluate the effect of pressure on the recovery of functional protein from the IAM column because these columns exhibit a higher back pressure than non-end-capped columns. At the same flow rate (0.5 ml/min) IAM.PC^{C10} columns have ~500 psi higher pressure than IAM.PC. We note that IAM.PC.MG columns, which are IAM.PC columns end-capped with methylglycolate, do not exhibit an increase in pressure when compared to IAM.PC columns of the same dimensions.

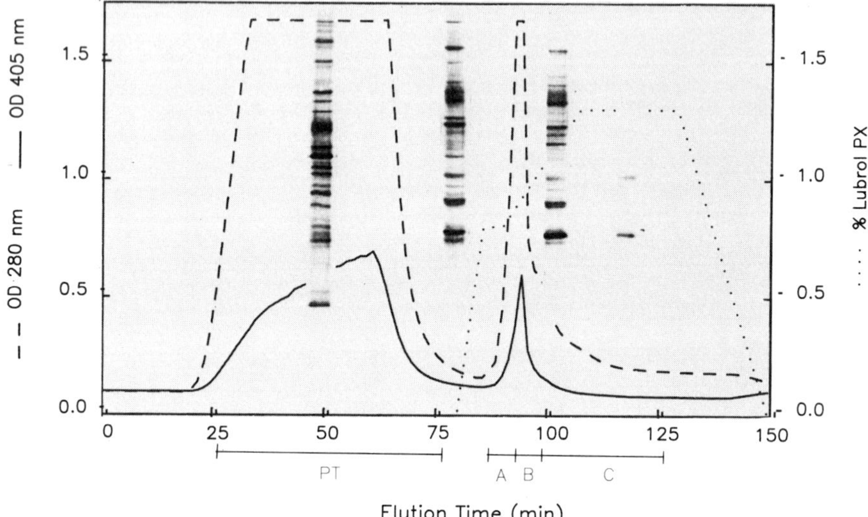

Figure 1. Purification of cytochrome P450 from rat adrenal microsomes by IAM.PC preparative chromatography profiles of OD$_{280}$ for total protein and OD$_{405}$ for heme protein are shown. An IAM.PC column 2.1 cm x 10 cm was loaded with 269 mg of solubilized microsomal proteins. Elution gradients are given in Table 3 and elution gradient used to purify rat-adrenal cytochrome P450 is shown above. Four regions of the chromatogram are labeled: **PT** for pass-through peak of proteins; **A** for the leading edge of the main heme peak; **B** for the main heme peak; and **C** for the trailing edge of the heme peak. Silver stain 7.5 % page gels of individual fractions in each region of the chromatogram are presented as inserts. Region **C** of the chromatogram corresponds to almost completely pure rat adrenal cytochrome P450. This figure is reprinted with permission from the publisher. [Anal. Biochem. 194, 163-173 (1991)]

Sample Loading on IAM Columns

Sample loading on IAM columns is unexpectedly high. We routinely inject 20-40 mg of total protein onto analytical IAM.PC columns and successfully purify cytochrome P450 to near homogeneity in a single step (6). At least 200 mg of total protein can be loaded onto semipreparative IAM.PC columns (2.5 cm x 10 cm) and perhaps as much as 0.5 g - 1 g can be loaded. Both analytical and semi-preparative IAM chromatograms obtained from injecting solubilized microsomes onto IAM.PC columns are characterized by a large pass-through protein peak that occurs during sample loading, and most of the non-heme proteins, present in the pass-through peak, are separated from the proteins of interest. Figure 1 shows a typical chromatogram obtained during the purification of rat-adrenal microsomal cytochrome P450 on a semipreparative scale. The large pass-through peak of proteins (Fraction PT in Figure 1) contains virtually all of the injected proteins, yet fraction C contains an almost pure adrenal cytochrome P450. Samples are loaded using mobiles phases that show low eluotropic strengths for the protein of interest. After sample loading, the eluotropic strength of the mobile phase is increased by an increasing detergent gradient to elute the protein of interest. The key concept is that the protein of interest should remain adsorbed to the column during sample loading because the immobilized PC molecules have high affinity for this class of proteins.

Although IAM columns, prepared by the immobilization of PC, provide a surface with good selectivity for cytochrome P450 and other proteins, the affinity of individual proteins is expected to depend on the particular membrane lipid immobilized. Thus we are currently synthesizing IAM bonded phases containing other lipid headgroups. In the case of P450's and IAM.PC, the target protein reversibly adsorbs at the top of the column while many of the contaminating proteins elute during loading; consequently high loadings are feasible. The loading step partially enriches the target protein concentration on the IAM column prior to eluting and separating the target protein with a detergent gradient. A large pass-through peak containing many proteins is common for many types of column chromatography and the development of additional IAM surfaces is expected to provide a wide selection of immobilized membrane surfaces to optimize the surface that may be selected for different target proteins. The purification of membrane proteins using high performance liquid chromatography has recently been reviewed (38).

Total Protein Recovered from IAM Columns

The recovery of total protein from IAM.PC columns is usually greater than 90% of the injected proteins for any given chromatographic experiment which really means that protein recovery is quantitative and within experimental error of the measurement. During the purification of the cholesterol-transfer protein from rabbit intestinal mucosal cells, 90-95 % of the applied protein was recovered from the IAM.PC column (3). During the purification of cytochrome P450 from rat livers, rat adrenals, rat kidneys, and other sources, the total protein recovered was virtually quantitative. Injecting red cell membrane proteins onto IAM.PC columns resulted in irreversible adsorption of virtually all of the injected proteins; this was attributed to strong protein-protein interactions as briefly discussed (4). It is very likely that irreversible binding of membrane proteins to IAM surfaces is a complex process and depends on several factors including the detergent used for solubilizing the biological sample, the detergent used for elution, ionic strength, protein-protein interactions, protein lipid interactions, and other variables.

Recovery of Functional Protein Activity from the IAM Column

In tests to date, the functional activity of proteins recovered from IAM columns generally distributes into two main peaks which elute at different times or regions of the chromatogram. Typically some functional activity elutes in the column pass though peak, but the majority of functional activity elutes afterwards and the elution time for the protein containing the functional activity depends on the mobile phase gradient. Since functional activity in the pass through peak is usually contaminated with all the proteins in the original sample, we do not include this in our calculations of 'recovered' functional protein. Thus % recovered functional protein refers to functional protein that is eluted in a peak after sample loading and well beyond the column void volume.

During the purification of the cholesterol-transfer protein (located in the intestinal brush border membranes of rabbits), the recovery of functional activity from the IAM.PC column was approximately 30 %. During the purification of several cytochrome P450 species, the recovery of functional activity varied from 60-95%. Typically, >70% recovery can be expected unless the column back pressure increases above ~900 psi (Table 3). When the back pressure of the column was 1400 psi, approximately 20 % of the injected P450 activity was recovered from the column (6). Overloading the analytical column (~100 mg of total protein injected) caused ~ 17 % decrease in the recovery of functional activity in the heme-peak eluting from the IAM.PC column.

Efficient purification of cytochrome P450 utilizes a Lubrol-PX/sodium cholate buffer system which allows the high recovery of functional activity from the IAM.PC column (Table 3). Substituting CHAPS (an analog of cholic acid) for cholate in the mobile phase caused a noticeable reduction in the purification of cytochrome P450 by the IAM.PC column. When CHAPS is substituted for cholic acid, most of the cytochrome P450 functional activity elutes in the pass through peak (6). This is most likely the result of CHAPS being able to prevent adsorption of this protein to the IAM.PC surface during injection (i.e., sample loading). The recovery of NADH oxidase activity from IAM.PC columns was 75% (6). The recovery of NADH reductase activity from IAM.PC columns was 50 % (6).

Mobile phase flow rate can also affect the recovery of functional activity. Flow rates between 0.25 ml/min to 1.0 ml/min were evaluated on analytical size IAM.PC columns and ~0.5 ml/min flow gave the optimum retention of P450 on the IAM.PC column, i.e., little P450 eluted in the column pass through peak. Flow rates greater than 0.5 ml/min increased the amount of P450 eluted in the pass through peak. Thus protein adsorption during sample loading depends on flow rate and therefore the recovery of functional activity from the IAM column also depends on flow rate. The effect of flow rate on the recovery of proteins from semipreparative IAM columns has not been evaluated. In general we use flow rates of 0.5 ml/min for analytical size IAM.PC columns and flow rates of 5.0 ml/min for semipreparative IAM.PC columns (Figure 1 shows results for semipreparative columns).

In summary, the recovery of functional proteins from IAM columns is usually very high but depends on the detergent conditions used during sample loading, mobile phase flow rate, and undoubtedly other factors still unknown.

Column Selectivity of IAM Surfaces

The column selectivity α is an important determinant of resolution. The concept of column selectivity is well suited for small molecule chromatography but not for the purification of membrane proteins because the desired protein is rarely free of contaminants after a single chromatographic step. Although it is routine for several proteins to co-elute from any given column in a single 'peak', many columns used for the purification of membrane proteins can quantitatively remove several important contaminating proteins; however, the desired protein is rarely removed from its 'nearest' contaminant. Thus for the purification of membrane proteins from complex mixtures, selectivity should be defined as the fraction of contaminating proteins removed from the sample during a given chromatographic step and we suggest that this be denoted as α_m. Its main purpose is to compare different chromatographic systems. Estimating α_m can be performed by gel electrophoresis of the sample before and after column chromatography. For example if approximately 300 proteins are present in the initial protein mixture injected into a column and 12 proteins co-elute with the peak of interest, then

$$\alpha_m = \frac{300 - 12}{300} = 0.96$$

An α_m value of 0.96 indicates that 96 % of the contaminating proteins were removed from the desired protein. Although it may be difficult to accurately calculate all of the protein bands in a given lane of an SDS polyacrylamine gel, estimates can be made because the error will be only relative when comparing different chromatographic systems. In other words, α_m should be calculated after injecting the <u>same sample</u> on several different columns (e.g., reversed phase, HIC, IAM, anion exchange, etc.). Comparing α_m for the same sample on different columns makes any error in calculating the number of proteins in the original sample cancel. Thus it is recognized that estimating the number of protein bands in samples by gel electrophoresis can be difficult or impossible for heavily overloaded gels, but these estimates will be underestimates of the total number of proteins and should only be a relative error in the calculation of α_m, particularly if the same sample is used for different chromatography columns.

Usually specific activity (i.e., total-activity/mg-total-protein) or specific content (nmoles of desired protein/mg-total protein) is used to estimate the increased purity of samples after column chromatography but neither of these calculations estimate the number of different proteins contaminating the desired protein during each chromatographic step. However, it is the number of contaminating proteins in a given sample that is frequently used to make a decision on the next chromatographic step. Thus calculating α_m allows direct comparison of individual purification processes regarding the removal of contaminating proteins. Perhaps reporting both α_m and specific activity would be the best description of the sample when protein samples have not been purified to homogeneity. For example, on both preparative and analytical sized IAM.PC columns, α_m ranges from 0.73 to 0.90 for microsomal P450's samples obtained from a variety of sources.. This is based on the polyacrylamide gel electrophoresis of microsomes containing cytochrome P450 and contaminating proteins, whereby the gel can resolve proteins in the 30-95 kD range.

Comparison of Reversed Phase Chromatography, IAM Chromatography, and Hydrophobic Interaction Chromatography

Virtually all purification strategies for membrane proteins require multiple chromatographic steps and typically include hydrophobic interaction (HIC) chromatography and/or reversed phase chromatography (RPC) as one of the chromatographic steps. Our studies to date on the purification of membrane proteins demonstrate that IAM surfaces possess chromatographic properties intermediate between the very hydrophobic RPC surfaces and the weakly hydrophobic surfaces characteristic of HIC bonded phases. The key difference between RPC and HIC columns is that protein loadings are usually greater on RPC columns compared to HIC columns, but stronger elution conditions are needed to remove the adsorbed protein from RPC columns. Another key difference is that HIC chromatography tends to be less denaturing to proteins than RPC chromatography. Thus the major advantage of RPC is high loading capacity and the major advantage of HIC columns is mild elution conditions. In this respect, IAM chromatography columns have the advantages of both HIC and RPC columns without the concomitant disadvantage of each system, i.e., high loading capacity and mild elution conditions are characteristic of the IAM surface. In other words IAM surfaces apparently do not denature proteins during chromatography because elution conditions used for purifying membrane proteins by IAM chromatography are similar to the conditions used to purify membrane proteins on HIC surfaces (as discussed (6)).

Batch Binding Studies of Proteins with IAM Particles

A batch binding assay was developed to study the interactions of membrane proteins with IAM surfaces (4). The purpose of these batch binding studies was to quickly determine if biological samples contained components that would irreversibly bind to the IAM.PC surface in the presence or absence of different detergents. The initial batch binding protocol was described in detail (4) and was initially performed in a microfilterfuge tube. Briefly, the IAM packing material is mixed with the protein sample followed by washes with buffers containing various detergents. The microfilterfuge tube is briefly centrifuged to **'filter'** the suspension. Unfortunately, even during the brief centrifugation, the IAM packing material may become dry if great care is not taken to prevent this. We have evaluated protein recovery from the IAM surface under several experimental conditions and found that if the packing material becomes dry during the centrifugation step of the batch binding assay, then most of the protein irreversibly binds to the IAM surface, regardless of protein affinity for the hydrated IAM surface. Consequently it is experimentally more convenient to perform the batch binding study in eppendorf tubes. If the IAM packing material is mixed in eppendorf tubes with protein mixtures, detergents, or non-detergent buffers, then centrifuging the eppendorf tube forms a **'supernatant'** which contains proteins not adsorbed to the IAM surface. The supernatant can be removed from the packing material that has pelleted to the bottom of the eppendorf tube, yet the packing material remains wet <u>during the entire experiment,</u> which requires washing the packing material several times with different detergents. Thus it is recommended that batch binding studies be performed in eppendorf tubes instead of microfuge tubes. The ability of a particular detergent to elute a protein from the IAM surface can be identified by the appearance of proteins in the supernatant (determined by gel electrophoresis). Only one publication is available describing the results from a batch binding experiment (4). However, this work demonstrated that (i) batch binding studies can be used to predict the irreversible binding of proteins to IAM surfaces, and (ii) the red cell glucose transporter can be purified in one step from using a batch extraction process and IAM.PC surfaces (Table 3).

Immobilization of Trypsin and α-Chymotrypsin on IAM Packing Material and Columns

Trypsin was immobilized on either IAM.PC or IAM.PC.MG loose packing material and on columns packed with these particles and the enzyme retained functional activity while immobilized (Dr. I. W. Wainer, personal communication). The general protocol for loose packing material is as follows. Trypsin was immobilized on IAM.PC.MG packing material by gently stirring a solution of trypsin with the IAM particles. The UV absorbance of trypsin in solution, not bound to the IAM particles, was determined by pelleting the particles before measuring the supernatant UV absorbance. The difference in UV absorbance before and after stirring trypsin with IAM.PC.MG particles was used to quantitate the amount of trypsin immobilized. The hydrolytic activity of 'immobilized' trypsin was determined by monitoring the amount of p-nitroaniline formed from the catalytic hydrolysis of Nα-benzoyl-DL-arginine-p-nitroanilide (BAPNA). It was found that 21.5% of the trypsin was immobilized on the IAM.PC.MG packing material which corresponds to 1.96 mg-trypsin/50 mg-IAM particles. The biological activity of the immobilized trypsin was compared to the activity of an equivalent amount of free trypsin. The 'immobilized' trypsin exhibited 41.5 % activity as compared to an equivalent amount of free enzyme.

The general protocol for immobilizing α-chymotrypsin on IAM HPLC columns is as follows. Either IAM.PC or IAM.PC.MG columns (12 μ particles, 15 cm x 4.6 mm) was initially equilibrated with phosphate buffer and a known concentration of α-chymotrypsin (ACHT) was perfused through the column until breakthrough enzyme activity was observed in the mobile phase eluting from the column. The difference in the UV absorbance in the mobile phase before and after perfusion through the column was used to quantitate the amount of ACHT immobilized on the column. The activity of the immobilized enzyme was evaluated by a bolus injection of the substrate DL-tryptophan ethyl ester. Approximately 62.9 % of ACHT was immobilized on the IAM.PC column (which corresponds to 26.4 mg-trypsin) and the immobilized ACHT retained its hydrolytic activity which was found to be stereoselective. The L-tryptophan ethyl ester was converted to L-tryptophan while the D-tryptophan ethyl ester remained unhydrolyzed. When L-tryptophan ethyl ester was injected onto the IAM.PC.MG (or IAM.PC) column containing immobilized ACHT, complete hydrolysis to L-tryptophan was observed.

Summary and Conclusions

The purpose of this review was to describe the strengths and limitations of IAM surfaces regarding their ability to mimic fluid artificial membranes, to obtain physico-chemical information, and to purify membrane proteins. IAM technology has only recently been introduced and several additional types of IAM surfaces are expected to soon be available. The IAM surfaces synthesized to date have proved very effective at purifying membrane proteins and obtaining solute binding data and solute transport data for small molecules. IAM surfaces have also been useful for reconstituting enzymes either in batch mode or packed in HPLC columns. The ability to both recover functional protein and reconstitute membrane proteins ion IAM surfaces indicates that IAM surfaces can be considered as 'soft' surfaces for labile proteins. Much additional work is needed to define the many applications of IAM technology and many of these studies are now in progress.

Acknowledgements

We are grateful to John Perry for pointing out that the pressure difference across the IAM column can cause the loss of functional activity of proteins during chromatography. We are also very grateful to Dr. Dave Floyd of Squibb Research Institute for allowing us to include the benazepine data in this review before he reports the data as a complete paper and Dr. Robert Markovich for critically reading the manuscript. We are also very grateful to Dr.

I. W. Wainer for allowing us to briefly describe the reconstitution of α-chymotrypsin and trypsin on IAM surfaces which will be published in the very near future. C. Pidgeon and C. Marcus are supported in part by NSF CTS 8908450 and C. Pidgeon is very grateful to the support provided by Eli Lilly and Company. The authors thank Judi Chadwell for her assistance in preparing this manuscript.

References

1. Pidgeon, C. "Immobilized Artificial Membrane" U.S. Patent 4,931,498 1990.

2. Pidgeon, C. "Method for Solid Membrane Mimetics" U.S. Patent 4,927,879 1990.

3. Thurnhofer, H., Schnabel, J., Betz, M., Lipka, G., Pidgeon, C, and Hauser, H. Biochim. et Biophys. Acta 1991, 1064, 275-286.

4. Chae, W.G., Luo, C., Rhee, D.M., Lombardo, C.R., Low, P., and Pidgeon, C. *Modern Phytochemical Methods* , in Recent Advances in Phytochemistry, 25:149-174. Plenum Press, N.Y. 1991. eds. N.H. Fischer, M.B. Isman, and H.A. Stafford.

5. Otto, S., Marcus, C., Pidgeon, C., and Jefcoate, C. "Purification of a Novel Adrenocorticotropin Inducible, Polycyclic Aromatic Hydrocarbon Metabolizing Cytochrome P450 from rat Adrenal Microsomes". Endocrinology in press.

6. Pidgeon, C., Stevens, J., Otto, S., Jefcoate, C., and Marcus, C. Anal. Biochem. 1991, 194, 163-173.

7. Pidgeon, C., and Venkatarum, U.V. Anal. Biochem. 1989, 176, 36-47.

8. Markovich, R. J., Stevens, J.M., and Pidgeon, C. Anal. Biochem. 1989, 182, 237-244.

9. Stevens, J.M., Markovich, R.J., and Pidgeon, C. Biochromatography 1989, 4, 192-205.

10. Pidgeon, C. "Solid Phase Membrane Mimetics". Enz. Microb. Technol. 1990, 12, 149-150.

11. Markovich, R.J., Qiu, X.-X, Invergo, B., Nichols, D.E., Alvarez, F.A., and Pidgeon, C. End-Capping Immobilized Artificial Membrane Surfaces: Silica Subsurface Amines Affect Both the Chemical Stability and Chromatographic Properties of IAM surfaces. Anal. Chem. in press.

12. Miyake, K., Kitaura, F., Mizuno, N., and Terada, H. J. Chrom. 1987, 389, 47-56.

13. Sandberg, M., Lundahl, P., Greijer, E., and Belew, M. Biochim.et Biophys. Acta 1987, 924, 185-192.

14. Yang, Q., Wallsten, M., and Lundahl, P. Biochim. et Biophys. Acta. 1988, 938, 243-256.

15. Wallsten, M., Yang, Q., and Lundhal, P. Biochem. et Biophys. Acta 1989, 982,47-52.

16. Yang, Q., Wallsten, M., and Lundhal, P. J. Chrom. 1990, 506, 379-389.

17. Lundhal, P., and Yang, Q., J. Chrom. 1991, 544, 283-304.

18. Diem, T., Czajka, B., Weber, B., and Regen, S.L. J. Am.Chem. Soc. 1986, 108, 6094-6095.

19.   Hauser, H., Pascher, I., Pearson, R.H., and Sundell, S. Biochem. et Biophys. Acta 1981, 650, 21-51.

20.   Scott, D.L., Otwinowski, Z., Gelb, M.H., and Sigler, P.B. Science 1990, 250 1541-1546.

21.   White,S.P., Scott, S.L., Otwinowski, Z., Gelb, M.H., and Sigler, P.B. Science 1990, 250, 1560-1563.

22.   Scott, D.L., Otwinowski, Z., Gelb, M.H., and Sigler, P.B. Science 1990, 250, 1563-1566.

23.   Rock, C.O., and Snyder, F. J. Biol. Chem. 1975, 250, 6564-6566.

24.   Kramer, R. M., Wuthrich, C., Bollier, C., Allegrini, P.R., and Zahler Biochim. et Biophysi. Acta 1978, 507, 381-394.

25.   Kirkland, J.J., Glach, J.L., and Farlee, R.D., Anal. Chem. 1989, 61, 2-11.

26.   Kohler, J. Chase, D.B., Farlee, R.D., Vega, A.J., and Kirkland, J.J. J. Chrom. 1986, 352, 275-305.

27.   Sagliano, N., G. Floyd, T.R., Hartwick, R.A., Dibussolo, J.M., and Miller, N.T. J. Chrom. 1988, 443, 155-172.

28.   Dill, K.A., J. Phys. Chem. 1987, 91, 1980-1988.

29.   Cole, L.A., and Dorsey, J.G. Anal. Chem. 1990, 62, 16-21.

30.   Colin, H., and Guiochon, G. J. Chrom. 1978, 158, 183-205.

31.   Klein, W., Koerdel, W., Weiss, M., and Poremski, H.J. Chemosphere 1988, 17, 361-366.

32.   Bodor, N., Gabanyi, Z., and Wong, C.K., J. Amer. Chem. Soc. 1988,111, 3783-3786.

33.   Valko, K. J. Liq. Chrom. 1987, 10, 1663-1686.

34.   Minick, D.J., Sabatka, J.J., and Brent, D.A. J. Liq. Chrom. 1987, 10, 2565-2589.

35.   Minick, D.J., Frenz, J.H., Pastrick, M.A., and Brent, D. J. Med. Chem. 1988, 31, 1923-1933.

36.   Kaliszan, R., Petrusewicz, J., Blain, R.W., and Hartwick, R.A., J. Chrom. 1988, 458, 395-404.

37.   Floyd, D.M., Moquin, R.V., Atwal, K.S., Ahmed, S.Z., Spergel, S.H., Gougoutas, J.Z., and Malley, M.F. J. Org. Chem. 1990, 55, 5572-5579.

38.   Rassi, Z.E. Biochrom. 1988, 3,188-200.

PERFUSION CHROMATOGRAPHY:

RECENT DEVELOPMENTS AND APPLICATIONS

Noubar B. Afeyan, Scott P. Fulton and Fred E. Regnier[+]

PerSeptive Biosystems, Cambridge, MA 02139
[+]Chemistry Dept., Purdue University, West Lafayette, IN 47907

Introduction

Liquid chromatography (LC) is, without question, the most important and widely used technique for the recovery and isolation of proteins and peptides. LC has proven to be not only highly selective, but also extremely flexible and very gentle. This unique combination allows good yield of both mass and biological activity together with extremely high purity.

However, LC as currently practiced has a number of critical limitations. Cycle times for chromatography are relatively long compared to other separations methods. This has tended to restrict the use of LC in time-critical applications such as on-line monitoring. In addition, the time required per run restricts the amount of work done in method development and optimization. Due to the complexity of chromatographic chemistry and scaleup engineering, many current applications, even large-scale processes, are run under sub-optimal, marginally economic conditions. Finally, the restricted throughput of chromatography columns has to date limited the use of LC for selective adsorption of material from dilute feed solutions.

Chromatographic Mass Transport

Most of the problems inherent in LC stem from restricted mass transport within packing materials.[1-3] In order to maximize the amount of available surface area (and therefore the binding capacity), highly porous particles with a high surface-to-volume ratio have generally been employed. A primary challenge in chromatography is to find a means of rapidly and efficiently exposing the mobile phase and sample molecules to the entire available surface area within the column, including the area inside the pores.

Applications of Enzyme Biotechnology, Edited by J.W. Kelly and
T.O. Baldwin, Plenum Press, New York, 1991

In conventional "diffusion chromatography" media, transport through the column to the outer surface of the packing particles is by rapid convective flow. Once the molecules reach the outer surface of the particles, however, mass transport to the inner surface area (where most of the binding capacity lies) is entirely by molecular diffusion, which is very slow relative to convection, especially for proteins and peptides.

Two serious problems can occur if the time required to diffuse into and out of the pores is much greater than the time required to flow past the particle. The first problem is a dramatic increase in bandspreading (loss in resolution), since the molecules still in the convective stream or binding near the outside of the particles will elute long before those that have diffused to the centers of the particles. The second problem is a loss in binding capacity, caused by the great range in access times to the active surface within the particles. It is possible for some of the sample to pass fully through the column well before all of the binding sites deep within the particles are occupied. Because of these problems, conventional chromatography has always involved a tradeoff between speed, capacity and resolution, primarily caused by the potential mismatch between mass transport rates inside (diffusion) and outside (convection) the particles.

Principle of Perfusion Chromatography^{TM}

In Perfusion Chromatography flow-through particle media, transport into the particles occurs through a combination of convection and diffusion in a way which eliminates the need for this compromise (Figure 1). Transport to the surface of the particles is by convective flow through a packed bed, just as in a conventional column. The particles themselves, however, contain two distinct classes of pores--**throughpores**, large enough to allow some convective flow through the particles and smaller **diffusive pores** lining the throughpores, which provide high adsorptive surface area.

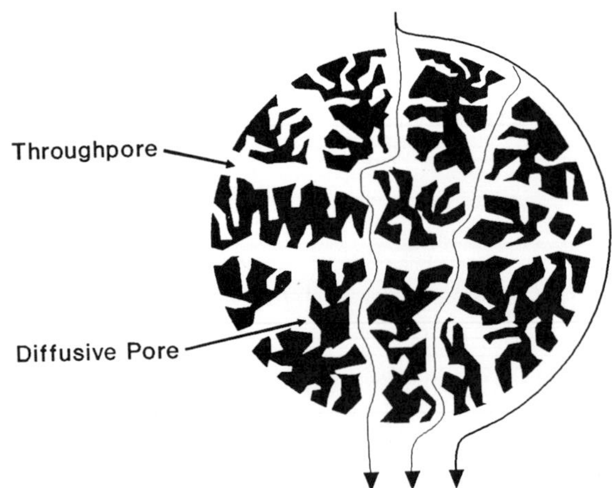

Figure 1: Schematic diagram of Perfusion Chromatography flow-through packing particle, showing throughpores for fast convective interparticle mass transport and diffusive pores for high surface area and binding capacity.

Above a critical flow rate, convection dominates over diffusion in the throughpores, allowing efficient access to the high surface area diffusive pores throughout the entire particle. The rapid "perfusive" transport within the throughpores and the ultra-short diffusion path lengths (< 1μm) of the diffusive pores combine to make both resolution and capacity essentially independent of flow rate. Detailed theory, experimental verification and comparisons of this technology with conventional chromatography are published elsewhere.[4,5]

Conventional Approaches

Perfusion Chromatography flow-through media contrast sharply with alternative ways used to solve the mass transport problems of chromatography. With conventional diffusion-limited media (Figure 2), the standard approach is to reduce the packing particle diameter, thereby reducing the diffusion path and thus the time required for molecules to diffuse in and out of the particles. This is the general principle of HPLC.[6,7] The constraint on this approach is caused by the fact that reducing the particle size dramatically increases the pressure drop needed to obtain a given flow rate. Current HPLC technology with 3-5 μm particles is at or near practical operating limits. These materials quite useful for analysis but are not practical for large-scale, preparative applications.

Two more recent approaches to solving the problem of slow diffusive mass transport have been the development of non-porous media and so-called membrane chromatography. In non-porous media (Figure 3A) the packing contains no pores, and all the binding occurs on the outer surface, entirely accessible by rapid convective transport.[8-10] In order to partially overcome the loss in surface area, very small particles (1-3 μm) are used. This allows much higher flow rates than those used even with small-particle conventional packings, but creates a serious pressure drop problem, as well as limited capacity. These factors have restricted the non-porous particle approach to analytical applications.

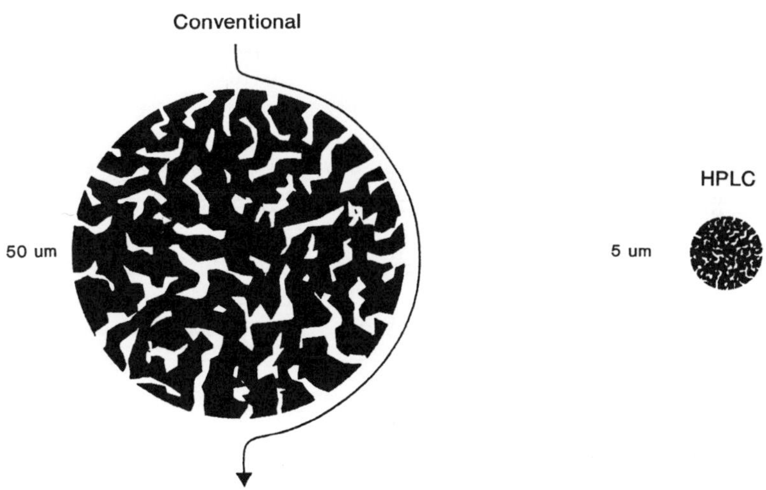

Figure 2: Schematic diagram of conventional and HPLC "diffusion" chromatography packing particles

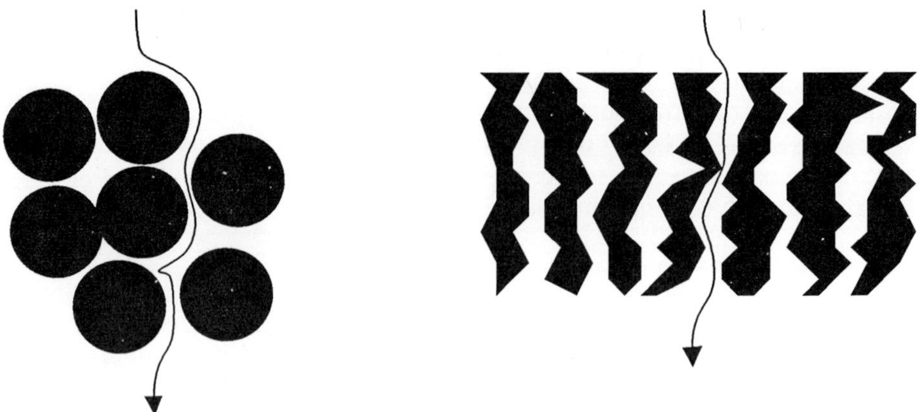

Figure 3: Schematic diagrams of "convection" chromatography media. A--non-porous particles, B--membrane.

In membrane chromatography (Figure 3B), microporous or ultrafiltration membranes are surface-derivatized to give selective binding. As with non-porous particles, the surface area is entirely accessible via convective flow through the membrane, allowing for rapid binding and elution, but also suffering from low capacity per unit volume. The physical form of membranes, however, allows devices to be fabricated with large "bed volume" and very short "bed length" (with reasonable pressure drop), thus making them practical for process applications. However, the very short "bed length" makes complete capture of binding molecules a problem. Membrane systems also usually have an extremely high ratio of extra-bed volume to bed volume, giving large dilution factors and high bandspreading, which severely limit application in high resolution separations. Flow rate declines due to membrane fouling and concentration polarization effects are also obstacles in this approach.

Implications of Perfusion on Chromatographic Performance

In standard diffusion-based media, as flow rates increase, the bandspreading (or theoretical plate height) increases linearly (i.e. chromatographic efficiency decreases linearly). With Perfusion Chromatography media, this relationship between flow rate and bandspreading no longer applies. Once the rate of perfusive transport exceeds that of diffusion (typically below 500 cm/hr linear velocity), bandspreading becomes virtually independent of flow rate, up to at least 5000 cm/hr linear velocity.[4,5] This contrasts sharply with the significant loss of resolution that occurs with conventional media.

In preparative chromatography, throughput (amount of product produced per unit time) is as important as resolution. One important variable affecting throughput is the dynamic loading capacity, normally measured by applying a sample feed continuously to the column at a given flow rate until a significant concentration is detected in the eluent ("breakthrough"). This technique is called frontal chromatography.

It has been shown that with "diffusion-limited" media, increasing the loading flow rate causes substantial loss in dynamic capacity as feed molecules begin emerging prematurely from the column due to slow mass transport.[11,12] Perfusion media have about 75-80% of the dynamic capacity of conventional supports at very low flow rates and show extremely sharp breakthrough curves with no change in dynamic capacity over a very broad flow rate range[4,5] (see Figure 9).

The table below summarizes the relative merits of each of the available approaches to chromatographic mass transport.

	Speed	Capacity	Resolution	Scaleability
Low Pressure LC	-	+	-	+
Conventional HPLC	+/-	+	+	+/-
Non-Porous LC	+	-	+	-
Membranes	+	-	-	+
Perfusion Chromatography	+	+	+	+

Applications

High Speed Analysis and On-Line Monitoring

Virtually any analytical method benefits from higher speed, since answers can be obtained more quickly and the number of samples processed per unit time can be increased. Perfusion Chromatography enables the reduction of chromatographic analysis times from the current 30-60 minute range down to a few minutes or less without loss in resolution (Figure 4) In the biotechnology field, this capability opens the possibility of using chromatography as an on-line monitoring technique for fermentation, primary recovery or even preparative chromatographic separations.

Rapid Method Development

The capability to complete runs in a few minutes also significantly affects the development of chromatographic methods. Entire studies of variables such as mobile phase composition, sample loading, etc. can be completed in the time ordinarily taken for a single run. Figure 5A shows a rapid gradient run used to measure the binding strength of two test proteins. When many of these runs are done in sequence at different pH's, the binding strength can be plotted (Figure 5B) as a function of pH for the protein of interest (in this case a monoclonal IgM antibody) and for the major contaminants (in this case transferring and insulin). These "pH maps" are extremely useful in the development of a purification or

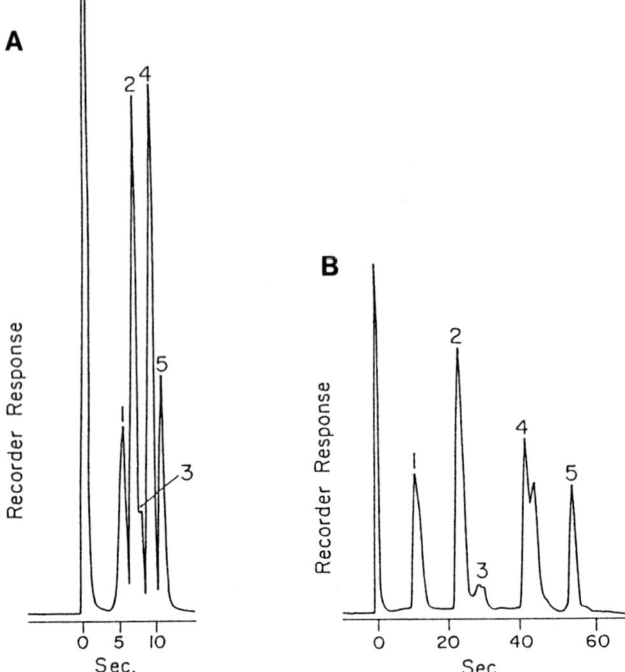

Figure 4: Analytical reversed-phase separation of standard test proteins on POROS perfusion packing.The two runs (A and B) differ only in gradient volume. Note the extremely rapid run (12 seconds) in A and the markedly improved resolution with only a 1 minute run time in by increasing the gradient volume in B.

Column:	POROS R/M 2.1 mmD/30 mmL
Sample:	1 μm 5 mg/ml each of
	1. Ribonuclease 3. Lysozyme
	2. Cytochrome C 4. ß-Lactoglobulin
	5. Ovalbumin
Eluent:	A-0.1% TFA/water
	B-0.1% TFA/acetonitrile
Gradients:	20-60 %B
	A-0.2 minutes (10 column volumes)
	B-1.2 minutes (60 column volumes
Flow Rate:	5.0 ml/min (9000 cm/hr)
Detection:	OD 220 nm

analytical process. However, the time required to produce them by conventional methods is quite prohibitive (see table).

Thus Perfusion Chromatography not only allows method development to be completed more quickly, but also it makes practical the exploration of process variables that could not previously be studied in any depth within normal time constraints. This is particularly important with complex, multi-stage separations, where even the sequence of steps can be critical, and changes in one step can dramatically affect steps downstream. The combined scaleability and very high speed of Perfusion Chromatography allow comprehensive chromatographic process optimization to be a practical option.

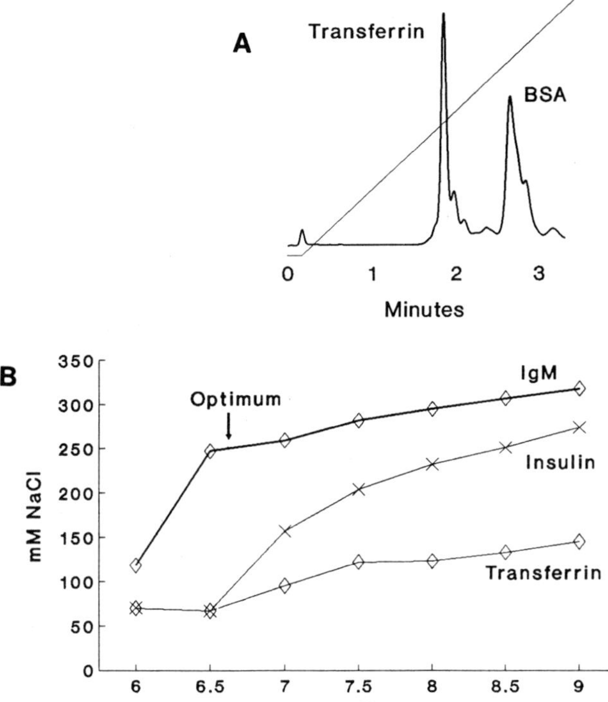

Figure 5: Development of a pH map for an IgM antibody purification on POROS Q/M strong anion exchanger. 5A shows a typical analytical-scale gradient run at one pH run with insulin and transferrin, the two major contaminants. Similar runs were done at 0.5 pH intervals from pH 6.0 to 9.0, and with a purified antibody standard. Figure 5B shows the pH map constructed by plotting the NaCl concentration at elution as a function of pH for each of three tested proteins. The optimal pH was the one with the greatest difference in binding strength (NaCl concentration for elution) between the antibody and the contaminants.

		POROS	HPLC	LC
Stage	**# Runs**	**hours**	**hours**	**hours**
pH Map	14	2.3	11.7	35.0
Optimize Elution	5	0.4	2.1	6.3
Loading Study	5	0.8	4.2	12.5
Prep Simulation	1	0.2	0.9	2.5
TOTAL	**25**	**3.7**	**18.9**	**56.3**

High Resolution Scaleup

The ability to do rapid method development is of limited use in preparative chromatography if the packing material does not successfully scale. Figure 6 shows a 600X scaleup of an antibody purification using POROS® perfusion columns. In the 20 µm particle size, these media are easily packed in large columns and operate at reasonable pressure drops, even at the high flow rates used in perfusion mode.

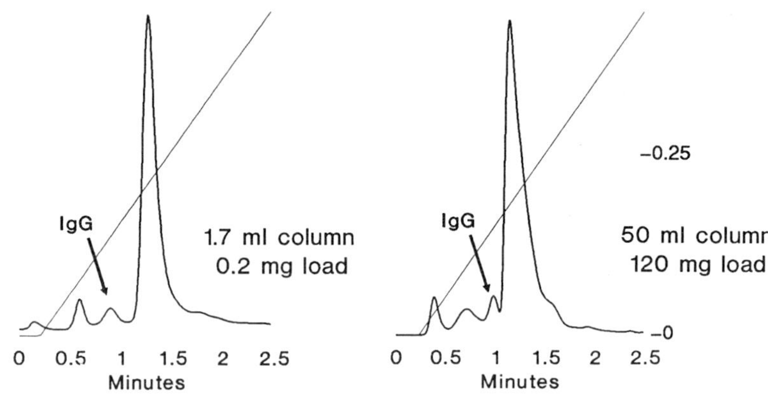

Figure 6: 600x scaleup from analytical to preparative of high speed, high resolution anion exchange purification of IgG from hybridoma cell culture supernatant on POROS perfusion columns. Note that sample load per ml column volume was increased 20-fold and the column volume increased 30-fold.

Columns:	A--4.6 mmD/100 mmL (1.7 ml) POROS Q/M
	B--25.4 mmD/100 mmL (50.7 ml) POROS Q/M
Sample:	4 mg/ml total protein
	hybridoma cell culture supernatant
Loading:	A--50 µl (0.2 mg total protein)
	B--30 ml (120 mg total protein)
Eluent:	20 mM Tris/HCl pH 8.0
Gradient:	0-0.5 M NaCl in 25 ml (15 column volumes)
Flow Rate:	A--10 ml/min (3600 cm/hr)
	B--340 ml/min (4000 cm/hr)
Detection:	OD 280 nm

In scaleup applications, the extraordinary throughput of Perfusion Chromatography columns allow an unusual degree of process design flexibility. One possiblility is to use a very small, highly economical column in rapid cycles to achieve a given overall production rate (Figure 7). Another possibility is to utilize the rapid cycle time simply to process the entire batch very quickly, which can in some cases greatly improve product yield and even eliminate the need for cold room operation.

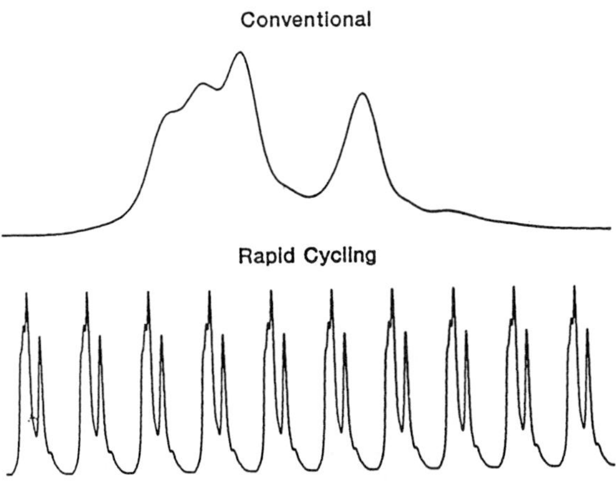

Figure 7: Comparison of a conventional chromatographic process with a rapid cycling process. In rapid cycling chromatography a smaller, very high speed column is used repeatedly in a given time period to produce the same amount of product as a larger, low speed column can produce in the same period. In this example, a rapid cycling column 10 times faster can be 1/10 the size and still have the same throughput.

Note in Figure 6 that peak resolution (i.e. analytical information) is nearly unchanged upon scaling up. In most preparative applications, resolution is sacrificed by a dramatic increase of the sample load to obtain a high enough throughput. The particle size is also typically increased to reduce the pressure drop in order to accomodate the very large columns needed. One interesting application of Perfusion Chromatography is to utilize "near-analytical" sample loads with a relatively small, high resolution column, run in the rapid cycling mode. In some cases this strategy may eliminate the need for collection of many fractions within unresolved peaks, each of which must be separately analyzed for determination of the final "cut". With perfusion in rapid cycling mode, the UV detection signal itself may serve as a sufficient fraction analysis.

Figure 8 compares the performance of alternative chromatographic techniques with a computer model of column productivity (unit product per unit column volume per unit time, calculated for a particular case). The results are based upon published and experimentally-derived mass transport coefficients for each of the three types of media shown. The improvement in productivity due to the more rapid mass transport of Perfusion Chromatography (with or without rapid cycling) is quite evident.

Dilute Feed Adsorption (Capture)

The limited throughput of conventional chromatography, especially with soft gels, has caused a need for upstream preconcentration steps (usually using ultrafiltration or precipitation) for large volume, dilute feed solutions such as cell culture supernatants. These steps add both processing time and product loss, while adding little to product purity. With Perfusion Chromatography, the combination of high flow rates with very high capture efficiency allows concentration and preliminary purification to be achieved in a single, rapid step.

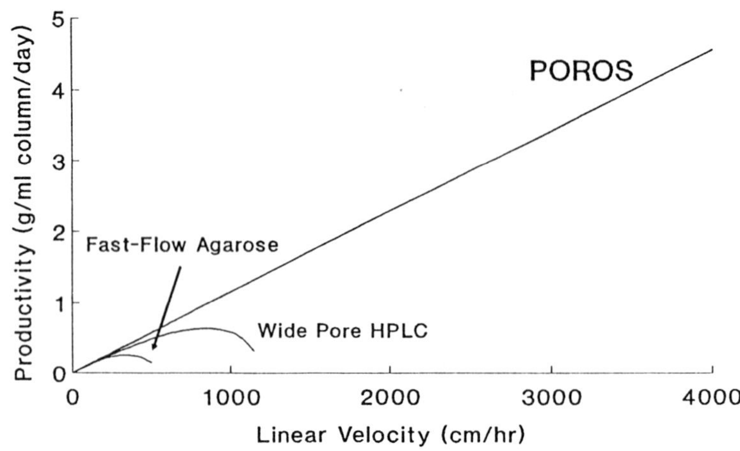

Figure 8: Calculated column productivity vs. flow rate (linear velocity) for a large molecule, gradient ion exchange separation. Curves for POROS perfusion media, wide pore HPLC and fast-flow agarose (soft gel) are compared. Note the reduced productivity in the two "diffusion" chromatography media at high flow rates due to the loss in binding capacity caused by slow mass transport rates. The perfusion packing has constant capacity over this flow range, and thus a linear increase in productivity.

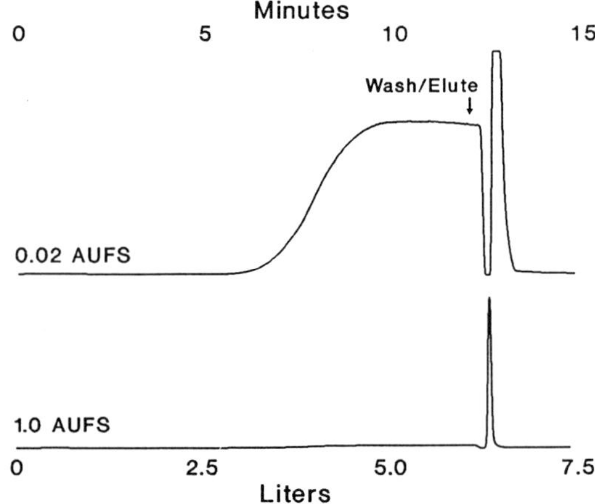

Figure 9: Frontal uptake curve of dilute protein solution on POROS anion exchange column.

Column:	50 mmD/10 mmL (20 ml) POROS Q/M
Sample Feed:	0.1 mg/ml bovine serum albumin in 20 mM Tris/HCl pH 8.0
Eluent:	0.5 M NaCl in buffer
Flow Rate:	500 ml/min (1500 cm/hr)
Pressure:	6.7 bar (100 psi)
Detection:	OD 280 nm

Figure 9 shows an example with a model protein. In this case a 20 ml column was able to capture all the protein (0.3 g) from 3 liters of dilute feed prior to breakthrough. This was accomplished in less than 15 minutes, operating at less than 7 bar (100 psi) pressure. This same column could also perform the same application in less than 30 minutes using a peristaltic pump.

Conclusions

Perfusion chromatography represents a fundamentally new approach to solving the "mass transport problem" of liquid chromatography. Perfusion allows the use of 10- to 100-fold higher flow rates (and lower cycle times) without significant loss in either resolution or capacity compared to conventional soft gel and HPLC columns. This technique can be applied to all conventional adsorption modes of chromatography for proteins and peptides, including ion exchange, reversed phase and affinity. Perfusion Chromatography has significant implications for analysis, on-line monitoring, method development, scaleup engineering and processing of dilute feed streams.

*POROS and Perfusion Chromatography are trademarks of PerSeptive Biosystems. U.S. and foreign patents pending.

References

1. J.C. Giddings, Dynamics of Chromatography, Part I, Principles and Theory. Marcel Dekker, New York (1965).

2. Cs. Horvath and H.-J. Lin, *J. Chromatogr. 149*:43 (1978).

3. J. Huber, *Ber. Bunsenges. Phys. Chem. 77*: 179 (1973).

4. N. Afeyan *et al., Bio/Technology, 8*:203 (1990).

5. N. Afeyan *et al., J. Chromatogr. 519*:1 (1990).

6. R. Majors, *Anal. Chem. 44*:1722 (1972).

7. K. Unger *et al., J. Chromatogr. 99*:435 (1974).

8. K. Unger *et al., J. Chromatogr. 359*:61 (1986).

9. K. Kalghatgi and C. Horvath, *J. Chromatogr. 398*:335 (1987).

10. K. Kalghatgi and C. Horvath, *J. Chromatogr. 443*:343 (1988).

11. H. Chase, *J. Chromatogr. 297*:179 (1984).

12. W. Kopaciewicz *et al., J. Chromatogr. 409*:111 (1988).

HIGH PERFORMANCE CAPILLARY ELECTROPHORESIS OF PROTEINS

AND PEPTIDES: A MINIREVIEW

Robert S. Rush
Barnett Institute
Northeastern University
Boston, MA 02115

INTRODUCTION

The purpose of this minireview is to summarize recent advances in peptide and protein high performance capillary electrophoresis (HPCE) for the protein chemist. The review is intended to be general and assumes that the reader is planning to enter the field to solve protein related problems. Therefore, I will focus on applications directed to protein and peptide chemistry, and attempt to provide the requisite HPCE background. HPCE is an instrumental approach to electrophoresis, introduced by Mikkers *et al.* (1) and Jorgenson and Lukas (2), that is finding increasing application in the area of protein and peptide chemistry. As such, it offers the scientist several advantages: (A) sensitive online detection without the normal postrun manipulations of staining, destaining or blotting to detect the proteins; (B) rapid analysis times; (C) automation with computer controlled HPCE systems, thus, increased quantitation and sample throughput and (D) requires small sample volumes. Diverse separation methodologies are being developed that produce greatly increased resolution compared to routine protein electrophoresis methodologies.

The development of the analytical instrumentation and methods has been reviewed extensively by others (2 - 7). Basically there are four major separation methods in capillary electrophoresis: (A) open tube or zone electrophoresis; (B) coated capillaries for protein separations and isoelectric focusing; (C) polyacrylamide gel filled capillary electrophoresis and (D) micellar electrokinetic chromatography. The first three modes of operation represent the main battery of methodologies that are directly applicable to the protein chemist. Specific examples will be presented after brief descriptions of operational theory, sample injection, detection and preparation of the capillary.

Operational Theory: Capillary electrophoresis can be distinguished from standard electrophoresis by operation at high electric fields within narrow bore capillaries, typically 5 to 150 μm inside diameter. The vast majority of applications presented to date have employed either 50 or 75 μm tubing. The

Applications of Enzyme Biotechnology, Edited by J.W. Kelly and
T.O. Baldwin, Plenum Press, New York, 1991

velocity (v) of migration is directly proportional to the apparent electrophoretic mobility (μ_{app}) and the electric field (E)

$$v = \mu_{app} E = \mu_{app} V/L \qquad [1]$$

where V is the applied voltage and L is the total length of the capillary. The apparent electrophoretic mobility is defined as

$$\mu_{app} = l\,L\,/\,V\,t \qquad [2]$$

where l is the effective length [i.e., length to the detector], L is the total length, and t is the migration time. In the open tube mode, the apparent mobility is modulated by the electroosmosis or the bulk flow of water caused by the negatively charged silica inner wall. Electroosmotic flow has been discussed by many investigators (2-8). A fundamental expression for μ_o is given by equation 3

$$\mu_o = \epsilon\,\zeta\,/\,4\,\pi\,\eta \qquad [3]$$

where ϵ is the dielectric constant, ζ the zeta potential at the slipping plane and η the solvent viscosity (9). Obviously, μ_o is linearily related to the dielectric constant which will vary with the buffer conditions employed and is inversely related to the viscosity, which is temperature dependent. Temperature and viscosity effects will be briefly discussed below. The polarity of the bulk flow is from positive to negative, thus if the species is positively charged it migrates faster than a neutral compound and negatively charged species migrate more slowly. The electrophoretic mobility (μ_e) is calculated from the apparent mobility by measuring the electroosmotic flow (μ_o) with a neutral compound as in equation 2, then calculating μ_e by equation 4.

$$\mu_{app} = \mu_e - \mu_o \qquad [4]$$

Electroosmotic flow affords the basic mechanism where both positively and negatively charged species can be separated simultaneously by capillary electrophoresis. Normally, μ_e is greater than μ_o, but not always.

High voltage operation under typically high resistance generates relatively low currents and power, which leads to fast separations with high resolution. Excessive Joule heating can occur, and when it does, can lead to reduced resolution primarily due to increased diffusion and possibly to convective mixing [the capillary readily dissipates heat]. Furthermore, nonreproducible migration times can be experienced because of thermal induced viscosity changes [there is about a 2% change in mobility per degree C]. In HPCE, temperature must be actively controlled; presently many of the commercial systems employ regulated forced air convective cooling. Beckman utilizes a circulating liquid bath cooling system and most laboratory constructed systems employ ambient air cooling with a fan. The temperature problem has been studied (10, 11) and it has been recommended that the power levels not exceed 1 W/m in nonactively cooled HPCE systems (11). Loss of separation efficiency caused by heating is related to increased solute diffusion which results in the reduction of the maximum number of theoretical plates (N)

$$N = (\mu_e - \mu_o) \, V / 2 \, D \qquad\qquad [5]$$

where μ_e, μ_o and V have been defined above and D is the diffusion coefficient. N can be more easily calculated from chromatography theory and equations of merit (12). Coincidentally, resolution (R_s) of two ionic species is also dependent upon the number of theoretical plates

$$R_s = ((\mu_{e,1} - \mu_{e,2})/\mu_{av})(1/4 \; N^{0.5}) \qquad\qquad [6]$$

where $\mu_{e,1}$ is the electrophoretic mobility of species one, $\mu_{e,2}$ is the mobility of the second species, μ_{av} is the average mobility of the two species. Resolution is also dependent of column length (4), thus increasing the effective length of the capillary at constant field results in an increased resolution, i.e., by the factor of $(1/L)^{0.5}$.

<u>Sample Injection:</u> The injection of sample into the capillary is critical. Two major modes of sample introduction are employed for capillary electrophoresis which minimize zone broadening: pressure based injection and electrokinetic injection. Pressure systems can take the form of positive pressure injection, simple hydrostatic pressure differences between the inlet and outlet (siphoning) or controlled vacuum injection systems. In any case, the volume of injection can be calculated from the Poiseuille equation

$$V' = (\Delta P \; \pi \; r^4 \; t)/(8 \; \text{\small n} \; L) \qquad\qquad [7]$$

where V' is the volume injected, ΔP the pressure difference between the ends of the capillary, r the internal radius of the capillary, t the injection time, ^{n} the viscosity and L the total length of the capillary. The advantage of pressure based injection systems is that the amount of sample [mass of all sample components] introduced is independent of the electrophoretic mobility of the species. Electrokinetic injection is useful; however, this method may introduce analyte bias due to mobility differences of the species being injected. The volume injected (V') can be determined as follows (13, 14)

$$V' = (\mu_e + \mu_o) \; V_i \; \pi \; r^2 \; t)/L \qquad\qquad [8]$$

where μ_e, μ_o, r, t, and L are defined above and V_i is the injection voltage. This equation is applicable for the open tube when the conductivity of the sample buffer and the separating buffer is about equal. Theoretical and practical considerations of conductivity differences between the sample buffer and the separating buffer have been discussed (9).

<u>Detection</u>: Detection methodology and detection limits have been reviewed (4-6). The two main detection systems employed routinely by the protein chemist are UV/VIS and fluorescence. The limit of detection for UV/VIS has been established in the area of 10^{-13} to 10^{-15} moles. Indirect fluorescence measures to about 10^{-17} moles, and precolumn derivatization fluorescence detection [laser] can further lower the limit to 10^{-17} to 10^{-20} moles (5).

<u>Preparation of the Capillary:</u> Capillary preparation varies from minimal, in the case of open tube, to extensive, in the cases of surface coatings or gel filled capillaries.

Preparation for specific applications will be discussed below in the appropriate sections. However, all applications have several common features: (A) determining the appropriate length and inside diameter tubing to employ [Polymicro Technologies, Phoenix, AZ is the sole supplier of capillary in the United States]; (B) "burning" in the optical window by removing the protective polyimide coating at the detection point [this is accomplished by using an electrically heated wire, a small flame or by scrapping off the coating with a razor]; (C) cleaning the window with methanol, using a wetted Kimwipe; and (D) rinsing the capillary with base [activating] followed by equilibration with buffer or selected derivatization methodologies. Once the capillary is prepared, it is mounted in the optical detector or protective cartridge depending on the HPCE system being employed. Care must be exercised in handling because the window is fragile once the polyimide is removed.

OPEN TUBE HPCE

<u>Peptides</u>: Several examples of peptide separations by HPCE have been presented (15-18). The majority of these efforts employed acidic buffer systems to insure that the peptides were positively charged. The peptides were injected from the positive terminal and migrated toward the negative terminal with the electroosmotic flow. It should be noted that the electroosmotic flow is greatly reduced at low pH values (19), because the ionization of surface silanol groups is reduced. Very slight changes in the pH can result in a large change in the resolution of the peptides (18). Phosphate buffers are particularly good because of the UV/VIS transparency at low UV wavelengths [$\approx$ 200 nm].

Small peptides, below 40 amino acids in length, are well suited for analysis by HPCE in the open tube. The peptide electrophoretic mobility has been shown to be a linear function of the ratio of the natural log of the charge (q) and the amino acid chain length (n) raised to the 0.43 power (17), see Fig 1. The separation mechanism is based primarily on the effective charge [charge/mass] of the peptide (20, 21). The effective charge is a property of the amino acid composition and the pH of the buffer. Selectivity of peptide separations by HPCE [controlling the ionization of the peptide] is thus based primarily on the judicious choice of buffer pH and any organic additives. Systematic studies of the effects of organic additives and/or ionic strength on peptide separations in HPCE are beginning to appear in the literature (22, 23). Organic additives may be helpful in reducing the electroosmotic flow, thus increasing the separation window by allowing the sample to remain in the column longer, or by mediating the effective charge through ion pairing mechanisms. There is no simple set of conditions that will generate optimum separations for all peptides. As in HPLC separation or mapping strategies, HPCE separations will also be tailored for the specific peptides or protein of interest.

HPCE tryptic peptide mapping has been an area of considerable interest with both standard proteins and some recombinant products. Table 1 lists some of the proteolytically digested proteins and buffer conditions that have been investigated by HPCE tryptic mapping. The majority

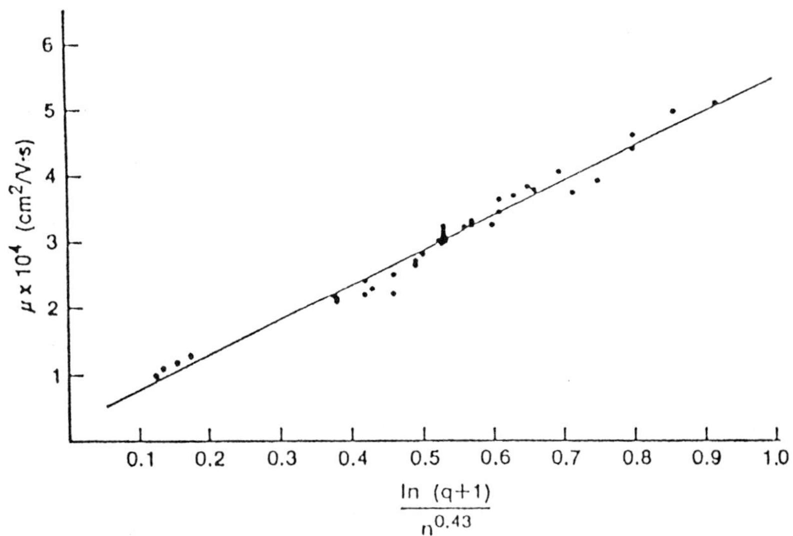

Figure 1: Peptide electrophoretic mobility versus the size (n) and charge (q) for 40 peptides ranging in size from 3 to 40 amino acid residues. (From Ref 17).

of maps were monitored online with UV/VIS detection at 200 or 215 nm. Relative fluorescence (F) or indirect fluorescence (IF) detection schemes have also been employed. Two pH regions are favored: low pH with either phosphate or citrate, and alkaline buffers such as Tricine or Tris. Analysis times are quite comparable, usually complete within 15 to 30 minutes. A representative HPCE peptide map is shown in Fig 2 for denatured, reduced, alkylated and trypsin digested bovine serum albumin. BSA contains 79 trypsin cleavage sites assuming complete reduction, alkylation and digestion. Of these peptides, there are five free amino acids generated and five dipeptides formed; thus to a first approximation, 69 peptide peaks should be observed. 54 peaks, or about 80% of the expected number of peaks, were observed. Reproducibility of the peak migration times was in the area of 0.25 to 0.5 % relative standard deviation. It is known that phosphate complexes with silanol groups (19), and reproducibility was found to be dependent upon the wash cycle employed; furthermore, the wash cycle and equilibration should be optimized for each specific application. In this example, wash and equilibration cycles were as follows: 0.7 min with 0.1N NaOH, 0.7 min with buffer, 1.5 min equilibration. Sample was then pressure injected for 5 sec [0.5 psi] from a stock digest of about 7.5 mg/ml peptides and electrophoresis conducted at the designated voltage. Other investigators have employed similar wash cycle procedures. The same capillary was employed for several weeks without noticeable reduction in peptide separations.

Depending on the complexity of the peptide mixture, it may be necessary to evaluate the efficiency of separations which are largely dependent upon the buffer, washing procedure,

Table 1

Protein	Buffer Conditions	Detection	References
1. ß-Casein	0.38 mM Quinine sulfate/ 0.58 mM H_2SO_4 pH 3.7	IF	24
"	40 mM Tricine/Tris pH 8.1	215 nm	25
2. ß-lactoglobulin	20 mM Citrate pH 2.5	200 nm	16
3. Ovalbumin	12.5 mM Phosphate pH 7	F	26
4. lysozyme	50 mM Phosphate pH 7.0	F	2
5. human Growth Hormone	10 mM Tricine/45 mM morpholine/ 20 mM NaCl, pH 8.0	200 nm	17, 23
	100 mM Phosphate pH 2.6	200 nm	27
6. human tPA	100 mM Phosphate pH 2.5	200 nm	28

temperature, ionic strength and the utilization of organic additives. Considerable work remains to expand the separation window to allow for discriminate identification of the individual peaks from complex digestion mixtures. One organic additive that shows promise in greatly increasing peptide resolution is hexanesulfonic acid [manuscript in preparation].

HPLC peptide mapping on reverse phase columns in conjunction with HPCE provides the opportunity to perform two dimensional analysis. These two techniques are orthogonal, the former separating primarily by hydrophobicity and the latter by effective charge. Several examples of this approach have been presented (17, 27, 29-30). Many well resolved HPLC peaks contain multiple peaks when subjected to HPCE analysis. Alternatively, seemingly well-behaved HPCE peaks have been resolved to multiple peaks by HPLC analysis (29). The importance of orthogonal analysis of these complex separation techniques is emphasized by Bushey and Jorgenson (30) in their construction of an automated HPLC/HPCE instrument capable of online sequential HPCE analysis of HPLC eluted peaks. In many HPLC mapping procedures, the amount of material being analyzed is small, an advantage for HPCE. HPCE analysis of dilute peptide samples has been addressed by Abersold and Morrison (31); dilute peptides are electrophoretically concentrated online prior to separation by HPCE.

HPCE mapping has been employed to identify disulfide reduction, synthetic peptide purity (17), phosphorylated peptides (25) and deamidation states of biosynthetic human insulin (32). Additionally, comparative HPCE peptide mapping of alkylated versus nonalkylated human carbonic anhydrase B tryptic maps has identified a noncleaved tryptic site (33). Titration curves of peptide mobility versus pH can give important information concerning the conformational interactions of the amino acid side chains (34). Overall, HPCE is well suited for rapid screening of digestion patterns since it utilizes very small quantities of material.

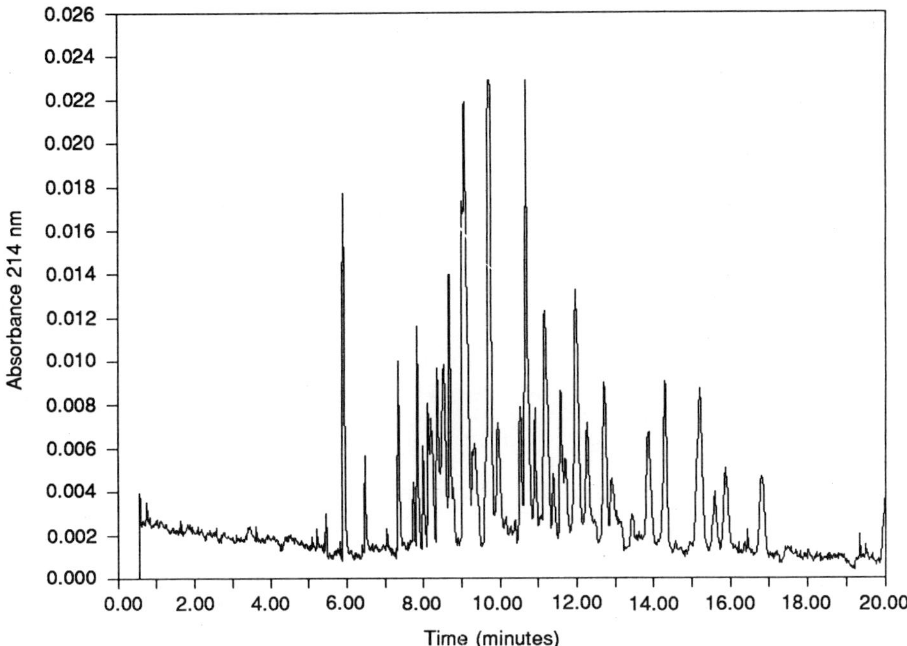

Figure 2: HPCE conditions: 20 mM phosphate, pH 2.5, temperature 20 °C, 20,000 V, I 48 μA, 50 cm effective length, 57 cm total length capillary, 75 μm inside diameter.

<u>Proteins</u>: The major parameters affecting open tube HPCE of proteins are similar to those listed above for peptides, plus the size and shape of the protein, the presence or absence of detergents and the affinity of the protein to adsorb onto the surface of the negatively charged silica wall (2, 4, 35). Like peptides, electrophoretic resolution of proteins is largely, but not totally, dependent upon the differential mobility defined by the effective charge of the protein(s); thus pH is a major selectivity factor. At pH values above the protein's pI, the protein exhibits a net negative charge and is electrostatically repelled from the capillary, while at pH values below the pI, the protein is positively charged and frequently adsorbs onto the capillary electrostatically. Protein interactions with the wall lead to band broadening, reduced separation efficiencies [lower theoretical plates], or loss of signal (2, 30, 36, 37). Two approaches have emerged that partially address the adsorption difficulties: first, pH control to manipulate the protein charge and second, capillary surface coatings to reduce the surface charge.

Adsorption may be minimized by manipulation of the pH such that the protein is negatively charged (19, 30, 35). For basic proteins this approach frequently fails because the pH extremes that are required are not compatable with protein and capillary stability. Additionally, alkaline buffers should be chosen such that the current levels are reasonable, i.e. ($\leq$ 50 μA), so that excessive Joule heating does not occur at normal operating voltages. The use of zwitterionic buffers lowers the conductivity level and the electroosmosis (30), and may allow for the inclusion of salts to aid in the separation by increasing solubility, decreasing wall interactions and

disrupting potential protein protein interactions that can lead
to various states of aggregation. Operating at acidic extremes
[pH ≤ 2], reduces the silica wall charge and thus the
adsorption, but protein stability is questionable at these
extremes. The current is usually very high under these
conditions and can lead to major temperature effects, which may
include conformational alterations, changes in the
oxidation/reduction states of the protein and to degradation or
possible denaturation.

Normally, buffer conditions are chosen such that
adsorption is minimized while maintaining protein stability.
Many proteins appear to separate well in the pH range of 7.5 to
8.5 with Tris or sodium borate buffers. Additives such as
detergents may help, but this is protein dependent. Open tube
separation of proteins still offers the investigator
significant advantages despite the above mentioned
difficulties. Conditions usually can be found that are
applicable to a specific separation.

SURFACE COATED CAPILLARIES

The second approach to reduce protein adsorption is to
employ coated capillaries. Presently, this is an area of
intense research and one involving diverse chemical structures
listed in Table 2. The major criteria for an effective coating
are as follows: (A) chemically stable over wide pH ranges, (B)
nonreactive toward proteins and buffers, and (C) consistant
surfaces (39). The general outcome of a coating meeting these
criteria is improved reproducibility and usually greater
separation ranges. Most of the listed coatings appear not to
be stable above pH 8 over time. Reproducible synthesis of
these coatings is also not trivial and involves multiple
synthetic steps. The polyethyleneimine and the vinyl bound
acrylamide appear to be the most stable coatings tested to date
with respect to pH stability. The two coatings that induce a
charge reversal, i.e., a positive surface charge, will reverse
the direction of the electroosmotic flow. Figure 3 illustrates
the type of results generated from coatings # 4 and # 6 listed
in Table 2.

Open tube HPCE, coated or uncoated capillaries, offers the
investigator more than just a separation technique. This
methodology can be employed to examine physical properties of
protein preparations such as temperature induced conformational
changes, different oxidation states, and the kinetics of
interconversion of various conformational changes (45).
Additionally, several investigators report employing open tube
HPCE to characterize biotechnology derived therapeutic proteins
(17, 23, 27, 32), posttranslational modifications involving
glycosylation (17, 46), acetylation (43), antibody antigen
complexation (17), isoenzyme forms of lactic dehydrogenase
(43), and calcium binding proteins of membrane origin (47).
HPCE has also been employed to evaluate the extent of
polyethylene glycol chemically modified proteins (48), an area
of considerable interest to the pharmaceutical industry. This
methodology should find increased use directed toward
posttranslational modifications of protein structures, as well
as provide a simple, rapid method of characterizing structural
changes induced by point mutation of genetically engineered
proteins. Another area of open tube protein characterization

is that of capillary isoelectric focusing, originally introduced by Hjerten (38).

Preparation of the capillaries involves activation of the silica with base, followed by reaction with a bifunctional reagent [typically Υ-methacryloxypropyltrimethoxysilane] in water at pH 3.5 for an hour [many different bifunctional reagents can be utilized], excess bifunctional reagent rinsed out, and then polymerizing with an acrylamide solution by normal chemistry: i.e., ammonium persulfate and TEMED for 30 min. The excess noncrosslinked acrylamide is then removed by pushing the polyacrylamide out of the tube with a syringe, leaving an acrylamide coating on the surface (38).

Table 2

	Coating	Surface Charge	Reference
1.	10 % Trimethylchlorosilane	neutral	15
2.	polyvinylpyrrolidinone	neutral	19
3.	acrylamide, noncross-linked methylcellulose	neutral	38
4.	vinyl bound acrylamide	neutral	39
5.	pentafluorobenzoyl chloride	neutral	40
6.	polyethyleneimine	positive	41
7.	Maltose, epoxydiol	neutral	42
8.	quaternary amine	positive	43
9.	polyethylene glycol	neutral	44

Two requirements for capillary isoelectric focusing of proteins are as follows: (A) that the capillary be coated such that the electroosmotic flow is reduced to as near zero as possible and (B) that the capillary be actively cooled and the pH gradient formed reproducibly. Basically, capillary IEF experiments are composed of two distinct steps: focusing of the proteins and concurrent generation of the pH gradient with ampholines covering the desired pH range, and mobilization or "elution" of the focused proteins past the detector. Theoretical considerations of mobilization techniques have been presented (49), as well as examples of high resolution capillary IEF (50). Active cooling is desirable because in the initial focusing step, considerable current is generated at the voltages ($\geq$ 350 V/cm) normally employed. The experiment is started by mixing the ampholines with the proteins, completely filling the capillary and applying the voltage. The direction of focusing can be controlled by the choices of electrode polarity and electrode solvents. Typically 50 mM phosphoric acid is used at the positive terminal and 50 mM sodium hydroxide at the negative terminal; the detector is normally on the grounded or negative side. As the sample focuses, the current drops exponentially to near zero, normally within 5 to 6 minutes. At this point mobilization is required to visualize

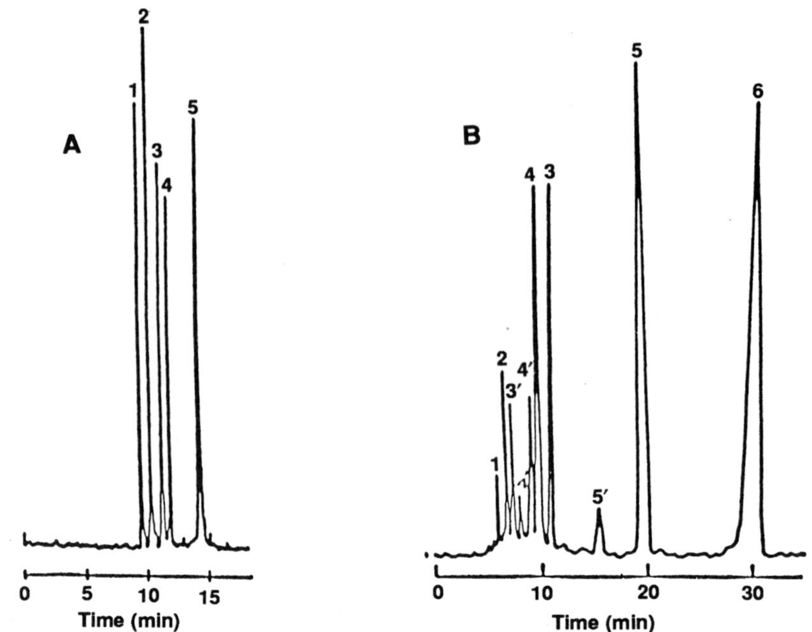

Figure 3A: Examples of coated capillary separation of basic proteins on a vinyl acrylamide coated capillary. **Figure 3B:** PEI-200-EDGE coated capillary From ref 39 and 41 respectively.

the focused proteins. This can be accomplished either by pushing the focused proteins past the detector with low pressure (hydrostatic pressure or low vacuum) or by electrophoretic mobilization (49), which is accomplished by adding 50 mM sodium chloride to the negative electrode buffer and maintaining the voltage. Electrophoretic mobilization is easier and potentially generates higher reproducibility due to the retangular plug flow characterizing HPCE solvent flow; the voltage potential maintains the gradient.

Resolution in IEF is dictated by the steepness of pH gradient employed and has been throughly studied by Righetti (51). An example of capillary IEF with protein standards is shown in Figure 4A for a pH gradient from 3 to 10 (53). The standard proteins employed were as follows: 1 and 2) horse heart myoglobin, pI 7.2 and 6.8 respectively, 3) human erythrocyte carbonic anhydrase B, pI 6.6; 4) bovine erythrocyte carbonic anhydrase B, pI 5.9; and 5) ß-lactoglobulin A, pI 5.1. Figure 4B is a plot of the pI versus the migration time. Linear regression analysis indicates a correlation coefficient of -0.966, nearly linear. Ideally this plot should be absolutely linear, but linearity also depends on the construction of the ampholyte mix, the stability of the coating and the stability of the protein at or near the protein pI. In the present case, the slight nonlinearity is believed to be due primarily to the instability of the acrylamide coating (38, 39). It is obvious that capillary IEF can produce high resolution and fast results, but this is tied to the production of stable coated capillaries. Studies directed toward reproducibility and stability have not been reported for capillary IEF. Solubility difficulties associated with

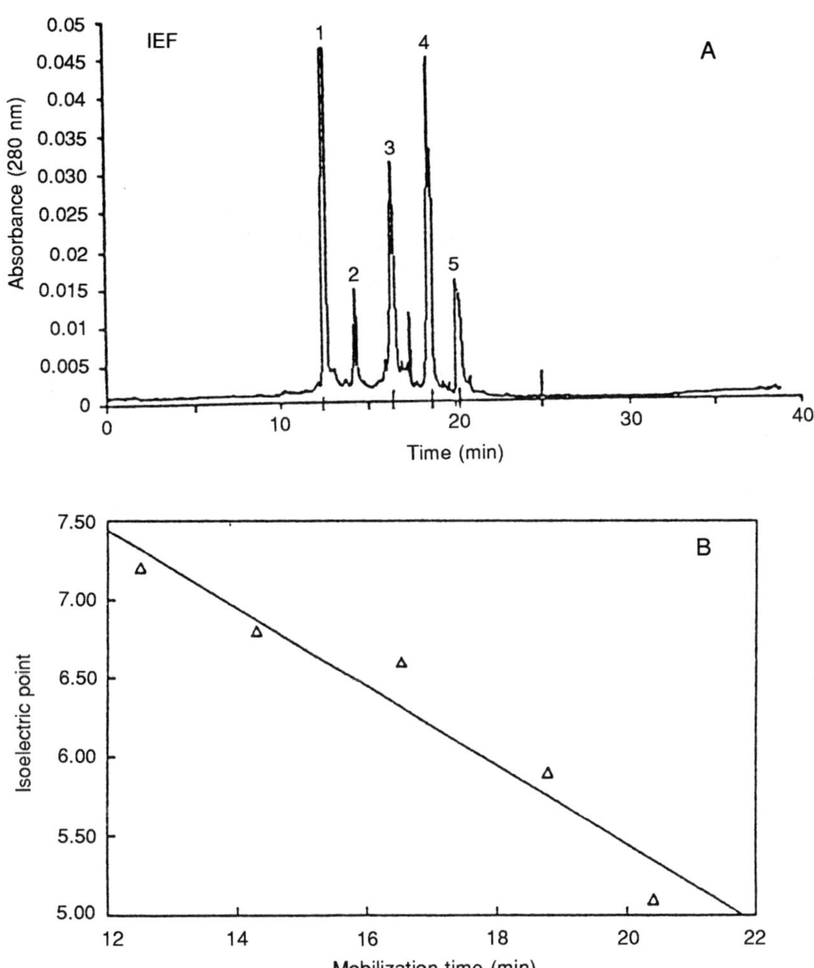

Figure 4A: Separation of a standard protein mixture by capillary IEF (From ref 53). **Figure 4B** is a plot of pI versus mobilization time.

proteins at or near their pI may be partially alleviated by use of nonionic detergents, a subject reviewed by Hjelmeland and Chramback (54). However, utilization of detergents in capillary IEF may result in adsorption of both detergent and sample to the wall, resulting in sample carry-over from injection to injection.

In spite of some of the technical difficulties listed above, capillary IEF is becoming a powerful analytical tool for the protein chemist to elucidate multiple isoelectric forms of proteins (49, 50, 52, 55, 56). Isoelectric points may be approximated by bracketing the protein of interest with the appropriate pI standard proteins, as in classical slab IEF. Continued research on capillary coatings will greatly aid the development of capillary IEF.

POLYACRYLAMIDE GEL FILLED CAPILLARIES

Polyacrylamide gel filled capillary electrophoresis was originally introduced by Cohen and Karger (57) for sodium dodecyl sulfate (SDS) capillary electrophoresis of peptides and proteins. Although the application of this online technique was established, the methodology required experienced operators. A major problem was bubble formation in the gel, which is believed to be due primarily, but not solely, to the following circumstances: (A) inadequate degassing of the prepolymerized acrylamide and buffer solutions; (B) physical changes in the properties of the polymerized gel at high voltages [less bubble formation is observed with low crosslinked gels compared to high crosslinked gels] and (C) movement of the gel due to locally induced electroosmotic flow. With proper degassing, control of the electric field and temperature, bubble formation can be eliminated.

Recently, the molecular weight range of protein SDS-PAGE has been extended to 200,000 daltons (58) in low cross-linked acrylamide filled capillaries. Selection of the buffer conditions was found to be critical in separating proteins of this mass, [detailed results will be presented elsewhere]. A typical example of a medium range SDS molecular weight separation is shown in Figure 5A for lysozyme [14,700], ß-lactoglobulin [17,000], trypsinogen [27,400], ovalbumin [43,000], bovine serum albumin [67,000] and phosphorylase B [92,000] respectively. A 20 cm effective length capillary was employed at 178 V/cm, current 16 μA, monitored online at 280 nm in 0.12 M Tris 0.12 M histidine, pH 8.7 and air cooled by a fan. Electrophoretic injection was employed for 9 sec at 150 V/cm. The number of plates is in the area of hundreds of thousands per meter, depending on the protein [considerably lower than reported for nucleic acid separations, discussed below]. The log molecular weight versus apparent electrophoretic mobility is shown in Figure 5B. Linear regression analysis indicates a correlation coefficient of - 0.992 (see Fig 5B). Thus the capillary can be employed to characterize the molecular weight of most proteins and probably can be employed for any application in which slab gel electrophoresis is routinely used.

The utility of the polyacrylamide filled capillary has been firmly established, especially for nucleic acid separations (59-63). Preparation of the gel filled capillary involves

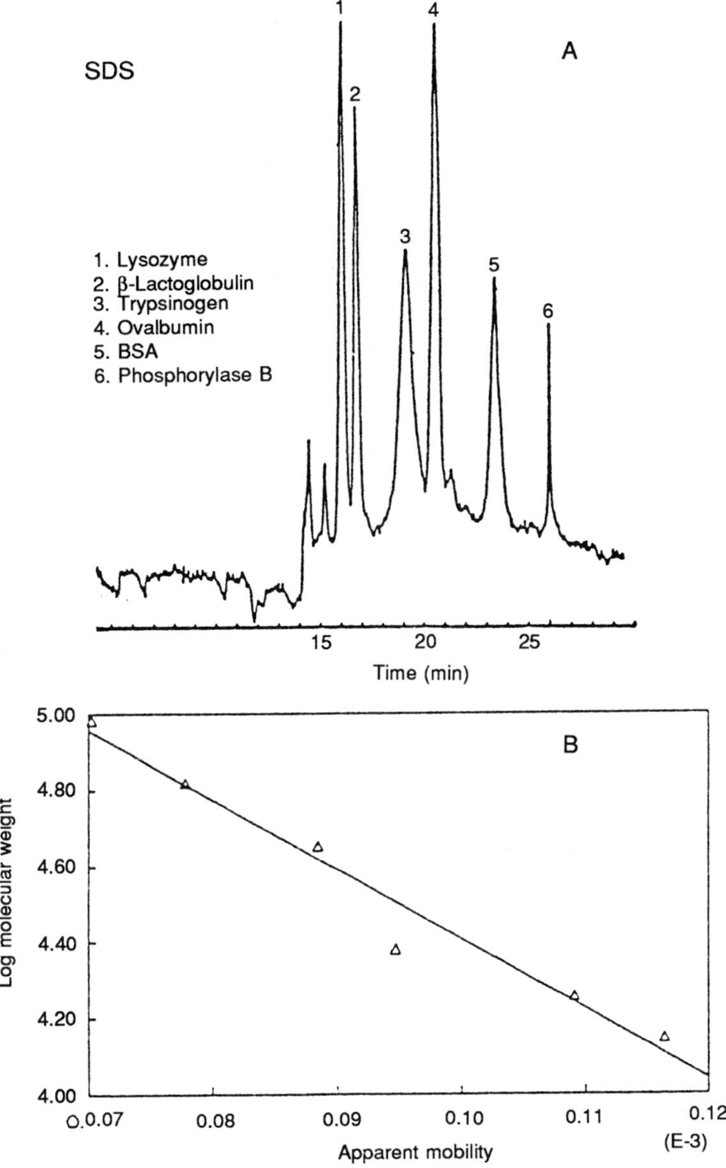

Figure 5A: HPCE SDS-PAGE profile of protein separations on a 5% T 1% C acrylamide filled capillary. Conditions in text (From ref 63). **Figure 5B:** Log molecular weight versus apparent mobility.

several steps with associated precautions. Accordingly, (60-62) the capillary is prepared as follows: base rinse to etch the inner silica wall and expose additional silanol groups, rinse with acid and then water/methanol, and derivatize the inner silica wall with [methylacryloxypropyl)trimethoxysilane (Petrach Systems, Bristol, PA] in methanol. This bifunctional reagent covalently binds the polyacrylamide gel to the wall of the capillary when the polymerization reaction is initiated. Free radical polymerization of the degassed polyacrylamide follows normal polymerization chemistry: ammonium persulfate and TEMED. The gels are allowed to polymerize for various times; although partially complete within an hour, complete polymerization requires overnight. Reproducibility of nucleic acid separations from within a gel, from gel to gel and from batch to batch is excellent and has been discussed (60, 61). Resolution of nucleic acid fragments, even in linear polyacrylamide (noncross-linked), is phenomenally high [plate counts in the range of millions per meter of gel (60-62)].

<u>Collection</u>:   Another area of tremendous importance is post-electrophoretic capillary analysis. Sample collection and subsequent amino acid sequence analysis from a poly-acrylamide filled capillary have been reported (64) for cyanogen bromide cleavage products derived from myoglobin. Similarly, nucleic acid fragments have been collected, evaluated for purity and used to spike known samples for positive identification (58, 59). The procedure of sample collection is based on the concept of field programming (59), which basically involves establishing the separation at high field and collecting the sample at low field, where the velocity is reduced as a result of the voltage reduction. The time window is expanded proportionate to the change in field. Diffusion does not appear to compromise the separation in the gel filled capillaries. Collection can easily be handled under computer control or manually. Samples are collected into 2 or 3 μl of sample buffer or water for subsequent analysis. Depending upon the concentration of the initial sample, several HPCE runs may be required to collect enough material for downstream analysis.

MICELLAR ELECTROKINETIC CHROMATOGRAPHY

Micellar electrokinetic chromatography (MEKC) was originally introduced by Terabe *et al.* (74) for the electrophoretic separation of neutral compounds. MECK is a true chromatographic separation involving buffer additives to effect selectivity and has also been extensively reviewed (3-8). This approach separates by differential partitioning into a mobile phase micelle, which is charged, and therefore migrates in an electric field. It has been shown to be effective for the separation of D- and L- forms of amino acids when cyclodextrins are employed (7), small organic compounds such as polyamines (75), nucleic acid bases (76), precolumn derivatized phenylthiohydantoin-amino acid separations (77) and amino sugars (70). The separation of amino acid enantiomers and peptide isomers by employing Marfey's reagents (78) has also been reported. MEKC has not been extensively employed yet for the characterization of proteins and of protein aggregates, but work in this area is anticipated.

246

SUMMARY

    HPCE is an analytical technique that separates components
based on effective charge and/or sieving. Providing an
orthogonal analysis, this methodology will support data
generated from other more traditional techniques such as
chromatography. HPCE and HPLC are fully complementary
methodologies and when used together greatly augment the
analysis of proteins and peptides. HPCE has been applied to
most current problems commonly associated with protein and
peptide chemistry. The advancement has been very rapid, based
primarily on the experience of methods and equipment
development in high performance liquid chromatography. One
promising area that is currently being developed is that of
interfacing HPCE with mass spectrometry for definitive
structural analysis (65-67). Complex carbohydrate separations
and analysis by HPCE are also under development (68-71) as well
as DNA sequencing by capillary polyacrylamide gel filled
columns (61, 62). Charge coupled devices are now being used
for detection of electropherograms (72), and new approaches to
indirect laser fluorescence are being examined (73). The
future for improved separations of biomolecules utilizing HPCE
technology is very bright indeed. The appearance of several
commercial HPCE systems within the past two years will promote
the development of HPCE methodologies. An absolute flood of
research papers, involving all aspects of modern biochemistry,
is anticipated in the next few years. Commercially prepared
polyacrylamide gel filled and coated capillaries, when proven
effective and when available, will greatly enhance development,
as will the utilization of other gels.

ACKNOWLEDGEMENTS Thanks are extended to my wife, Paula, for
editorial assistance and to Dr. A. S. Cohen and Dr. J.-W, Chen
for helpful discussions and suggestions during the preparation
of this manuscript.

REFERENCES

1. Mikkers, F. E. P., Everaerts, F. M. and Verheggen, P. E.
    M. (1979). *J. Chromatogr.* _169_, 11-20.
2. Jorgenson, J. W. and Lukas, K. D. (1983). *Science* _222_,
    266-272.
3. Gordon, M. J., Huang, X., Pentoney Jr., S. L., Zare, R. N.
    (1988). *Science* _242_, 224-228.
4. Karger, B. L., Cohen, A. S. and Guttman, A. (1989). *J.
    Chromatogr.* _492_, 585-614.
5. Ewing, A. G., Wallingford, R. A., Olefirowicz, T. M.
    (1989). *Anal. Chem.* _61_, 292A-303A.
6. Kuhr, W. G. (1990). *Anal. Chem.* _62_, 403R-414R.
7. Terabe, S. (1989). *Trends in Anal. Chem.* _8_, 129-134.
8. Tsuda, T. (1989). *J. Liq. Chromatogr.* _12_, 2501-2514.
9. Hjerten, S. (1990). *Electrophoresis* _11_, 665-690.
10. Grushka, E., McCormick, R. M. and Kirkland, J. J. (1989).
    Anal. Chem. _61_, 241-246.
11. Nelson, R. J., Paulus, A., Guttman, A., Cohen, A. S. and
    Karger, B. L. (1989). *J. Chromatogr.* _480_, 111-127.
12. Foley, J. P. and Dorsey, J. G. (1983). *Anal. Chem.* _55_,
    730-737.

13. Rose, D.J. and Jorgenson, J. W. (1988). *Anal. Chem.* 60, 642-648.

14. Wallingford, R. A. and Ewing, A. G. (1989). *Adv. Chromatogr.* 29, 1-76.

15. Jorgenson, J. W. and Lucas, K.D. (1981). *Anal. Chem.* 53, 1298-1302.

16. Grossman, P. D,, Wilson, K. J., Petrie, G. and Lauer, H. H. (1988). *Anal. Biochem.* 173, 265-270.

17. Grossman, P. D., Colburn, J. C., Lauer, H. H., Nielsen, R. G., Riggin, R. M., Sittampalam, G. S. and Rickard, E. C. (1989). *Anal. Chem.* 61, 1186-1194.

18. Strickland, M. and Strickland, N. (1990). *Am. Laboratory* Nov, 60-65.

19. McCormick, R. M. (1988). *Anal. Chem.* 60, 2322-2328.

20. Offord, R. E. (1966). *Nature* 211, 591-593.

21. Nyberg, F., Zhu, M.-D., Liao, J.-L. and Hjerten, S. (1988) in *Electrophoresis '88*, (ed. Schafer-Nielen), VCH, Weinheim 141-150.

22. Palmeri, R. Kalberg, S., Osborne, J. (1988). *J. Cell Biol.* 107, 216a.

23. Nielsen, R. G. and Richard, E. C. (1990). *J. Chromatogr.* 516, 99-114.

24. Gross, L. and Yeung, E. S. (1990). *Anal. Chem.* 62, 427-431.

25. Cobb, K. A. and Novotny, M. (1989). *Anal. Chem.* 61, 2226-2231.

26. Green, J. S. and Jorgenson, J. W. (1984). *J. HRC & CC* 7, 529-531.

27. Frenz, J., Wu, S.-L. and Hancock, W. S. (1989). *J. Chromatogr.* 488, 379-391.

28. Palmieri, R., Ohms, J., Field, M., Frenz, J., Hancock, W., Cohen, A. S., and Karger, B. L. (1989). *HPCE '89, Abstr. Intl. Symp. High Perform. Capillary Electrophoresis*. Boston, 65.

29. Nielsen, R. G., Riggin, R. M. and Rickard, E. C. (1989). *J. Chromatogr.* 480, 393-401.

30. Bushey, M. M. and Jorgenson, J. W. (1990). *Anal. Chem.* 62, 978-984.

31. Abersold, R. and Morrison, H. D. (1990). *J. Chromatogr.* 516, 79-88.

32. Nielsen, R. G., Sittampalam, G. S. and Rickard, E. C. (1989). *Anal. Biochem.* 177, 20-26.

33. Rush, R. S., Cohen, A. S. and Karger, B. L. (1991). *Anal. Chem.* (in preparation).

34. Vanorman, B.B. and McIntire, G. L. (1991). *HPCE '91, Abstr. Intl. Symp. High Perform. Capillary Electrophoresis*. San Diego, 51.

35. Deyl, Z., Rohlicek, V. and Struzinsky, R. (1989). *J. Liq. Chromatogr.* 12, 2515-2526.

36. Lauer, H. H. and McManigill, D. (1986). *Anal. Chem.* 58, 166-170.

37. Walbroehl, Y. and Jorgenson, J. W. (1989). *J. Microcol. Separ.* 1, 41-45.

38. Hjerten, S. (1985). *J. Chromatogr.* 347, 191-198.

39. Cobb, K. A., Dolnik, V. and Novotny, M. (1990). *Anal. Chem.* 62, 2478-2483.

40. Swedberg, S. A. (1990). *Anal. Biochem.* 185, 51-56.

41. Towns, J. K. and Regnier, F. E. (1990). *J. Chromatogr.* 516, 69-78.

42. Bruin, G. J. M., Huisden, R., Kraak, J. C. and Poppe, H. (1989). *J. Chromatogr.* <u>480</u>, 339-349.

43. Wiktorowicz, J. E. and Colburn, J. C. (1990). *Electrophoresis* <u>11</u>, 769-773.

44. Bruin, G. J. M., Chang, J. P., Kuhlman, R. H., Zegers, K., Kraak, J. C. and Poppe, H. (1989). *J. Chromatogr.* <u>471</u>, 429-436.

5.   Rush, R. S., Cohen, A. S. and Karger, B. L. (1991). *Anal. Chem.* (in press).

46. Wu, S.-L., Teshima, G., Cacia, J. and Handcock, W. S. (1990). *J. Chromatogr.* <u>516</u>, 115-122.

47. Josic, D., Zeilinger, K., and Reutter, W. (1990). *J. Chromatogr.* <u>516</u>, 89-98.

48. Cunico, R. L., Gruhn, V. A. and Wiktorowicz, J.E. (1991). *HPCE '91, Abstr. Intl. Symp. High Perform. Capillary Electrophoresis*. San Diego, 67.

49. Hjerten, S., Liao, J.-L. and Yao, K. (1987). *J. Chromatogr.* <u>387</u>, 127-138.

50. Hjerten, S., Elenbring, K., Kilar, F., Liao, J., Chen, A. J. C., Siebert, C. J. and Zhu, M.-D. (1987). *J. Chromatogr.* <u>403</u>, 47-61.

51. Righetti, P. G. in *Isoelectric Focusing: Theory, Methodology and Applications*, Elsevier, New York, 1983.

52. Kilar, F. and Hjerten, S. (1989). *Electrophoresis* <u>10</u>, 23-29.

53. Nelson, R., Cohen, A. S., Rush, R. S. and Karger, B. L. (1990). Beckman Application Note DS-750.

54. Hjelmeland, L. M. and Chrambach, A. (1981). *Electrophoresis* <u>2</u>, 1-11.

55. Yim, K. (1991). *HPCE '91, Abstr. Intl. Symp. High Perform. Capillary Electrophoresis*. San Diego, 87.

56. Compton, S. W. and Brownlee, R. G. (1988). *BioTechniques* <u>6</u>, 432-439.

57. Cohen, A. S. and Karger, B. L. (1987). *J. Chromatogr.* <u>492</u>, 409-417.

58. Ganzler, K., Cohen, A. S. and Karger, B. L. (1991). *HPCE '91, Abstr. Intl. Symp. High Perform. Capillary Electrophoresis*. San Diego, 92.

59. Cohen, A. S., Najarian, D. R., Paulus, A., Guttman, A. S., Smith, J. A. and Karger, B. L. (1988). *Proc. Natl. Acad. Sci.* <u>85</u>, 9660-9663.

60. Guttman, A., Cohen, A. S., Heiger, D. N. and Karger, B. L. (1990). *Anal. Chem.* <u>62</u>, 137-141.

61. Heiger, D. N., Cohen, A. S. and Karger, B. L. (1990). *J. Chromatogr.* <u>516</u>, 33-48.

62. Cohen, A. S., Najarian, D. R. and Karger, B. L. (1990). *J. Chromatogr.* <u>516</u>, 49-60.

63. Drossman, H., Luckey, J. A., Kostichka, A. J., D'Cunha, J. and Smith, L. M. (1990). *Anal. Chem.* <u>62</u>, 900-903.

64. Cohen, A. S., Guttman, A., Karger, B. L., Pearson, J. and McCroskey (1989). *HPCE '89, Abstr. Intl. Symp. High Perform. Capillary Electrophoresis*. Boston, 48.

65. Smith, R. D., Loo, J. A., Edmonds, C. G., Barinaga, C. J. and Udseth, H. R. (1990). *J. Chromatogr.* <u>516</u>, 157-165.

66. Moseley, M. A., Deterding, L. J., Tomer, K. B. and Jorgenson, J. W. (1990). *J. Chromatogr.* <u>516</u>, 167-174.

67. Vinther, A., Bjorn, S. E., Sorensen, H. H. and Soeberg, H. (1990). *J. Chromatogr.* <u>516</u>, 175-184.

68. Honda, S., Suzuki, K. and Kakehi, K. (1989). *Anal. Biochem.* **177**, 62–66.

69. Novotny, M., Liu, J., Shirota, O. and Wiesler, D. (1991). *HPCE '91, Abstr. Intl. Symp. High Perform. Capillary Electrophoresis.* San Diego, 50.

70. Liu, J., Hsieh, Y.-Z., Wiesler, D. and Novotny, M. (1991). *Anal. Chem.* **63**, 408–412.

71. Liu, J., Shirota, O. and Novotny, M. (1991). *Anal. Chem.* **63**, 413–417.

72. Sweedler, J. V., Shear, J. B., Fishman, H. A., Zare, R. N. and Scheller, R. H. (1991). *Anal. Chem.* **63**, 496–502.

73. Yeung, E. D. and Kuhr, W. G. (1991). *Anal. Chem.* **63**, 275A–282A.

74. Terabe, S., Otsuka, K., and Ando, T. (1985). *Anal. Chem.* **57**, 834–841.

75. Balchunas, A. T and Sepaniak, M. J. (1988). *Anal. Chem.* **60**, 617–621.

76. Cohen, A. S., Terabe, S., Smith, J. A. and Karger, B. L. (1987). *Anal. Chem.* **59**, 1021–1027.

77. Otsuka, K., Terabe, S. and Ando, T. (1985). *J. Chromatogr.* **332**, 219–226.

78. Tran, A. D., Blanc, T. and Leopold, E. J. (1990). *J. Chromatogr.* **516**, 241–249.

GENETIC ALTERATIONS WHICH FACILITATE PROTEIN PURIFICATION: APPLICATIONS
IN THE BIOPHARMACEUTICAL INDUSTRY

Helmut M. Sassenfeld, Michael Deeley, John Rubero,
Janet C. Shriner and Hassan Madani

Department of Process Development
Immunex Manufacturing Corporation
Seattle, WA

ABSTRACT

A large number of genetic modifications have been employed to facilitate the
purification of numerous recombinant proteins. These techniques have been used for
both commercial and research purposes and in a variety of different expression
systems. This paper will explore the value of this approach as a research tool and
focus primarily on two methods. The first method utilizes a metal-chelate fusion and
the second an immuno-affinity fusion. These two methods can be successfully
employed in virtually any expression system. The resulting purified protein can be
used for a variety of research purposes including refolding studies and pre-clinical
biology.

INTRODUCTION

The use of genetic engineering techniques to aid in the recovery of recombinant
proteins has been the subject of numerous reviews (1-4). In essence, this technique
involves modifying the cDNA of interest such that the expressed protein contains an
additional amino or carboxy terminal peptide. This additional peptide fusion is
designed to facilitate purification. For this reason, it is frequently referred to as a
purification fusion. Other descriptions such as affinity tails, tags, and handles have
also been used. The purification fusion is often designed with a selective cleavage site
to enable production of the native protein.

Currently, there are almost as many purification fusions as there are
purification methods. Undoubtedly, more methods will appear in the future as the
creative potential of genetic engineering is further explored. Many of the researchers
who described these new methods may have been attempting not only to simplify their
separation problems, but also to discover a general purification method for
recombinant proteins. Unfortunately, a single method for purifying any recombinant
protein is difficult to achieve.

General methods have to perform successfully in the numerous expression
systems currently employed in biotechnology. Yeast, bacteria, insect, fungal and
mammalian cells have all been used to express recombinant proteins both inside and
outside of the cell. Collectively, these systems display an enormous range of protein
and non-protein contaminants. Furthermore, certain fusions may be completely
incompatible with the known biochemistry of some host organisms. For example, a
C-terminal polyarginine fusion was designed for ion exchange chromatography and
was employed very effectively on proteins expressed intracellularly in bacteria

Applications of Enzyme Biotechnology, Edited by J.W. Kelly and
T.O. Baldwin, Plenum Press, New York, 1991

(5-7). However, this fusion probably would not be secreted from yeast using the α-factor leader peptide, because this leader sequence is cleaved at a dibasic residue (8). In general, one would not expect a large, charged peptide to be secreted easily.

In an analogous fashion, the unique properties of the recombinant protein may be incompatible with certain fusions. For example, a polycysteine fusion was used for the purification of galactokinase (9). The fusion protein was bound to thiopropyl-Sepharose 6B and eluted with a reducing agent. Obviously, this would be a precarious approach for disulfide containing proteins.

Finally, the removal of some fusions often results in unwanted complications. Although a large variety of chemical and enzymic methods exist, the success of any particular cleavage method depends on the unique nature of the fusion protein. Common difficulties include denaturation, poor efficiency, and undesired cleavages. If the research objective absolutely requires the native protein, then the expression system and cleavage mechanism should be chosen very carefully. A satisfactory compromise may be to use the system described by Suh et al. (10), in which one or two amino acid substitutions created a high affinity metal chelation site. The modified protein was only slightly altered and free from any fusions requiring removal. With this approach, it is important to avoid making substitutions in important structural regions, including those which may affect protein folding.

Usually for research purposes, it is not necessary to remove the purification fusion. Under these circumstances, this technology can be employed with minimal difficulty and a high probability of success. This paper describes two methods which have been successfully employed on numerous occasions in bacterial and yeast expression systems. The purified protein was then used for any of the following purposes: 1)to raise antisera for specific assays and the subsequent purification of the native protein, 2) refolding studies, 3) biology studies, 4) structure/function studies. Examples are given below.

POLYHISTIDINE

Metal chelate chromatography of proteins was first described by Porath in 1975 (11). This purification method primarily depends on the imidazole side chain of histidines coordinating with chelated metals attached to a chromatographic support (12). Other amino acids may also contribute but to a lesser extent (for a recent review see Arnold (13)). Proteins are normally applied at neutral pH values and eluted at low pH values. Alternatively, chelators such as ethylenediamine tetraacetic acid (EDTA) may be used. This type of chromatography has several advantages including commercially available media, high stability, good protein binding capacity and wide solvent compatibility. One drawback of this method is its relatively poor selectivity. Numerous proteins and other contaminants bind to these columns. However, by applying genetic engineering techniques the number of metal interaction sites can be increased, thereby increasing the selectivity. Haymore et al. (14) used site-directed mutagenesis to accomplish this (9). Another approach was described by Hochuli et al. (15). In this case nitrolotriacetic acid (NTA) was used as the chelator for nickel. Recombinant mouse dihydrofolate reductase was produced with a variety of polyhistidine fusions (2-6 residues) and subsequently purified on nickel NTA-Sepharose. With six histidines the affinity was so high that elution at low pH was ineffective unless guanidine HCl was included in the buffer. Conversely, fusions with less than three histidines did not bind in the presence of 6 M guanidine HCl but were able to bind and elute in non-denaturing solvents.

We have taken advantage of these observations to develop a refolding procedure for recombinant human interleukin 7 (IL-7). Our objective was to obtain highly purified, denatured protein and then perform light scattering experiments to determine the aggregation properties of IL-7. Once aggregation-free conditions were identified, the next step was to ensure IL-7 had some secondary structure in these solvents. Ideally, these studies would identify potential refolding conditions, which could then be tested on the native protein.

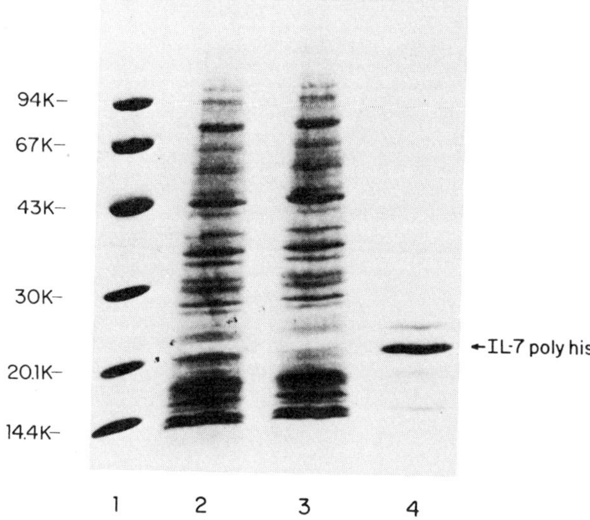

Figure 1. SDS-PAGE analysis of recombinant polyhis IL-7 purified by metal chelate chromatography. Recombinant IL-7 was expressed in *E. coli* with a six residue, C-terminal polyhistidine fusion. Recombinant IL-7 was extracted with 6 M guanidine HCl and purified using NTA-Sepharose essentially by the method of Hochuli et al., 1988. Lane 1, molecular weight standards. Lane 2, applied crude extract. Lane 3, unbound. Lane 4, elution.

Interleukin 7 was expressed in *E. coli* with a six residue, C-terminal polyhistidine fusion and purified under denaturing conditions, essentially by the method of Hochuli et al (15). The purified protein (Fig. 1) was then subjected to laser light scattering in different concentrations of guanidine HCl in order to determine its aggregation properties.

The effect of protein concentration and denaturant concentration on aggregation is shown in Figure 2. This study indicated, that even at low protein concentrations (20 ug/ml), the IL-7 tended to aggregate at concentrations below 2 M guanidine HCl.

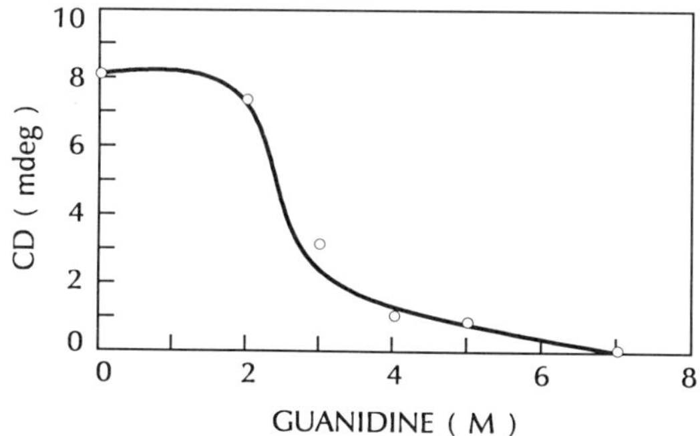

Figure 2. Effects of denaturant and protein concentration on IL-7 aggregation. Light scattering was performed using an argon laser at 488 nm. Scattering intensity was measured at 90 degrees scattering angle. Cells were of 1 cm path length. Intensity measurements were used to follow the aggregation of polyhis IL-7 in gaunidine concentrations up to 4 M. (O) 20 µg/mL IL-7. (●) 100 µg/mL IL-7. (△) 200 µg/mL IL-7.

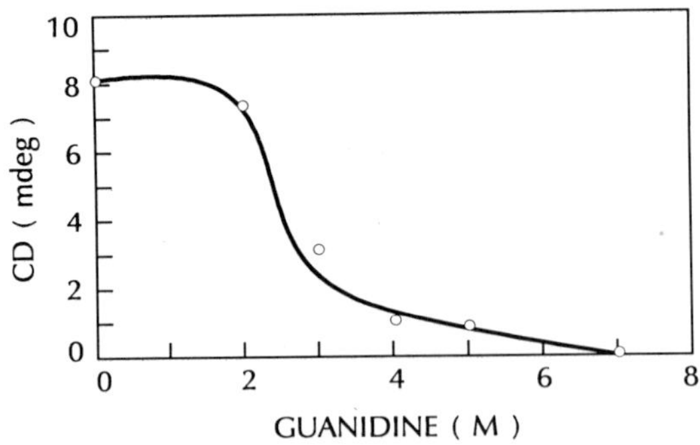

Figure 3.   CD spectra of polyhis IL-7 at 220 nm as a function of guanidine concentration.   CD spectra were determined using a J-600 Spectropolarimeter (Jasco, Inc.) and a .5 mm cell. The mdeg ellipticity at 220 nm as a function of guanidine concentration is shown.

Circular dichroism of IL-7 was performed in 2 M guanidine HCl to ensure that the protein was not completely denatured under these conditions (Fig. 3).   The resulting spectra indicated a substantial amount of secondary structure in this solvent.

Several combinations of reduced and oxidized glutathione were then tested in a neutral pH refolding buffer containing 2 M guanidine HCl.   After air oxidation the samples were dialyzed to a non-denaturing solvent such as phosphate buffered saline. The final specific activity of the refolded fusion protein was comparable to recombinant IL-7 expressed in mammalian cells (Table 1).

This procedure was tested on IL-7 without a polyhis fusion, in a crude, denatured, *E. coli* extract.   After a conventional purification procedure the resulting IL-7 showed comparable activity to both mammalian expressed IL-7 and the polyhis

Table 1

Biological Specific Antibody of *E. coli* and Mammalian Cell
Derived Recombinant Human IL-7

Sample	Specific Activity[1]
IL-7 polyhis (*E. coli*)	32,000 U/µg
IL-7 (*E. coli*)	38,000 U/µg
IL-7 (Mammalian)	34,000 U/µg

[1] Biological activity determined in a murine pre-B cell proliferation assay as described previously (18).

fusion protein (Table 1). The refolding yield was estimated to be about 50% based on the results of phenyl-Sepharose chromatography. This purification step was able to separate active from inactive IL-7 as indicated in Table 2. Approximately, 50% of the IL-7 and essentially all of the biological activity were eluted with a low salt, Tris buffer whereas guanidine was required to elute the remainder of the IL-7. The guanidine eluted IL-7 had a very low specific activity.

Table 2

Phenyl-Sepharose Purification of IL-7

Sample	Units	IL-7 Protein (μg)	Specific Activity U/μg	IL-7 Yield (%) Protein	Units
Applied	1.7×10^7	1,380	13,000	[100]	[100]
Unbound + Wash	ND[1]	ND	-	-	-
Elution (Tris)	1.5×10^7	590	25,500	44	88
Elution (Guanidine)	9.3×10^5	590	1,500	44	5

[1] ND = None detected.

Several features of this approach warrant further discussion. Although refolding experiments can be performed on crude lysates, it is very difficult to determine the success of a renaturation protocol when both the target specific activity and the purification process are unknown. These questions were obviated by using the polyhistidine fusion. Furthermore, light scattering is a very sensitive technique for the detection of aggregates, but it requires extremely pure samples. The polyhis fusion enabled the production of sufficient quantities of IL-7 suitable for these experiments. Other methods of purifying denatured proteins, such as size-exclusion chromatography or repeated extractions of the cell pellet, were considerably less efficient.

Some fusion proteins may require different refolding conditions than their corresponding native proteins and the fusion itself might interfere with proper refolding. These risks are probably offset by the advantages of this approach which include using the purified IL-7 to raise antisera suitable for immunoassays and purification of the native protein.

Removal of polyhis fusions with carboxypeptidase A (CPA) has been reported (15). However, we were unable to remove the polyhis fusion from IL-7 using these published procedures. Since histidine is readily digested by CPA (16), it is possible that the unique tertiary structure of IL-7 may have reduced the enzyme's efficiency.

Heterologous proteins expressed intracellularly in bacteria and yeast often accumulate as denatured aggregates. Developing a purification and refolding procedure for such proteins can be exceptionally time-consuming due to the large number of experiments required. The approach described above can be successfully employed to facilitate the development of a renaturation protocol which may then be employed on the native protein.

Using polyhistidine to purify proteins expressed in a fully functional state may not be as advantageous. If the incorrect number of histidines is used, the affinity may be either too weak or too strong. Furthermore, the secretion of polyhis proteins into a fermentation broth containing trace metals may compromise the chromatography. For secreted proteins, we have found the Flag fusion to be extremely useful.

FLAG

Flag is a small immunogenic fusion designed for the purification of recombinant proteins (17). It consists of a short, 3 residue, immunogenic peptide followed by an enterokinase cleavage site (Fig. 4). We have used this fusion on a large number of recombinant cytokines which were secreted from yeast using the α-factor leader peptide. For the cytokines listed in Table 3, we have not observed any loss of biological activity which could be attributed to the Flag fusion. For most of these cytokines, Flag versions of both the mouse and human protein were expressed. This extensive list indicates the versatility of this purification fusion.

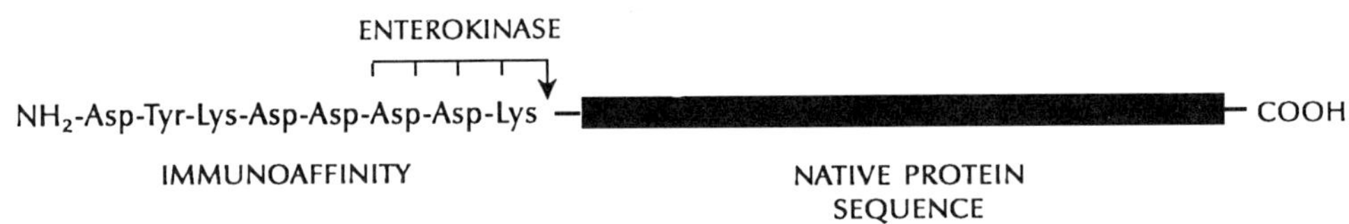

Figure 4. The Flag fusion with enterokinase cleavage site.

Table 3

Flag Proteins Expressed in Yeast

Flag Proteins

Interleukin 2 (IL-2)
Interleukin 3 (IL-3)
Interleukin 4 (IL-4)
Interlerkin 5 (IL-5)
Interleukin 6 (IL-6)
Interleukin 7 (IL-7)
Interleukin 8 (IL-8)
Granulocyte Macrophage Colony Stimulating Factor (GM-CSF)
Granulocyte Colony Stimulating Factor (G-CSF)
Megakaryocyte Growth Factor (MGF)
CSF-1
Tumor Necrosis Factor-α (TNF-α)
Leukemia Inhibitory Factor (LIF)

In most cases, the Flag proteins were purified by first binding the protein at a neutral pH and then eluting at low pH. This is a typical procedure for an immuno-affinity purification. In some cases, the low elution pH may denature the protein. Since the Flag antibody requires calcium for binding, elution can be performed at neutral pH by the addition of a metal chelator such as EDTA.

Another salient advantage of the Flag fusion is that it provides a product specific assay. Flag proteins can be easily detected and quantitated using standard immunochemical techniques. This advantage greatly aids fermentation development.

In theory, the Flag fusion can be removed by virtue of its enterokinase cleavage site. Although frequently successful, this step is somewhat unreliable due to the limited availability of highly purified enterokinase. Furthermore, the secondary structure at the cleavage site may adversely affect the efficiency of cleavage. Nonetheless, this drawback does not prevent the Flag fusion from being employed in a variety of research applications.

For example, this technique was employed to purify a receptor-ligand complex. Tumor necrosis factor -α (TNF-α) was secreted from yeast with a Flag fusion and purified using a low pH elution. The purified protein was then re-applied to an anti-Flag column. After binding the TNF-α, a crude mammalian cell supernatant containing a soluble version of the TNF receptor (18) was applied to the column. After extensively washing the column, the receptor-ligand complex was eluted at low pH. The purified receptor was then separated from TNF-α by cation exchange (Fig. 5).

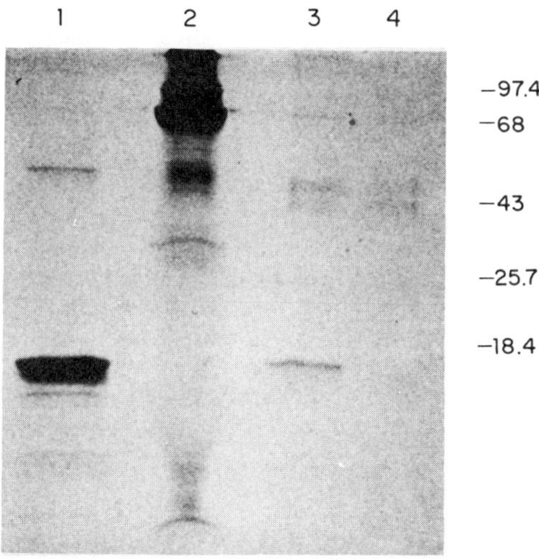

Figure 5. SDS-PAGE analysis of purified Flag TNF-α and TNF-receptor. Flag TNF-α was secreted from yeast and purified on an anti-Flag immunoaffinity column. Lane 1, eluted Flag TNF-α (17 Kd). The purified protein was then used to saturate an anti-Flag column which was subsequently loaded with crude mammalian cell supernatant containing TNF-receptor. The receptor ligand complex was eluted at pH 3 and then separated using cation exchange. Lane 2, crude mammalian cell supernatant. Lane 3, eluted receptor-ligand complex. Lane 4, purified receptor after cation exchange (two bands 43 Kd and 45 Kd). Lane 5, molecular weight standards.

This method also works when the receptor and ligand are mixed in free solution, even if both are present in a crude cell supernatant. In this approach, it is important to ensure that both components are present in approximately the correct molar ratios. In principle, this method can be used to purify new receptors from natural sources.

SUMMARY

The two fusions described above can facilitate research and development of recombinant proteins. Collectively, these two methods can be successfully employed in most expression systems regardless of whether the product accumulates inside or outside the cell. The metal chelate fusion can be used to purify denatured proteins and thereby facilitate renaturation studies. The Flag fusion works extremely well for secreted proteins and also provides an assay.

Both these fusions lack a convenient and reliable method of removal. However, this does not prevent their use as a research tool. Both these methods can be used to produce pure protein virtually as soon as the first fermentation is complete. The resulting protein can be used to raise antisera and for some early biological testing. For these reasons, we find it advantageous to express any newly discovered gene with and without one of these fusions.

REFERENCES

1. S. J. Brewer, B. L. Haymore, T. P. Hopp and H. M. Sassenfeld, Engineering Proteins to Enable Their Isolation in a biologically active form. In: <u>Purification and Analysis of Recombinant Proteins.</u> R. Seetharam, S. K. Sharma eds. Marcel Dekker, Inc., NY (1991) pp. 239-266.

2. M. Uhlen and T. Moks, Gene fusions for purpose of expression: An Introduction. In: <u>Methods in Enzymology.</u> R. Williamson, ed. Academic Press, NY (1981) pp. 129-143.

3. H.M. Sassenfeld, Engineering proteins for purification. TIBTECH 8:88 (1990).

4. R. Sherwood, Protein fusions: Bioseparation and application. TIBTECH 9:1 (1991).

5. H. M. Sassenfeld and S. J. Brewer, A polypeptide fusion designed for the purification of recombinant proteins. Bio/Technology 2:76 (1984).

6. S. J. Brewer and H. M. Sassenfeld, The purification of recombinant proteins using C-terminal polyarginine fusions. Trends in Biotechnol. 5:119 (1985).

7. H. M. Sassenfeld, Engineering proteins for purification: Use of a C-terminal polyarginine fusion. PhD Thesis, University of Liverpool, UK.

8. A. J. Brake, J. P. Merryweather, D. G. Coit, U. A. Heberlein, F. R. Masiarz, G. T. Mullenbach, M. S. Urdea, P. Valenzuela, and P. J. Barr, α-Factor-directed synthesis and secretion of mature foreign proteins in Saccharomyces cerevisiae. Proc. Natl. Acad. Sci. USA 81:4642 (1984).

9. M. Persson, M. G. Bergstrand, L. Bulow, and K. Mosbach, Enzyme purification by genetically attached polycysteine and polyphenylalanine affinity tails. Anal. Biochem. 172:330 (1988).

10. S. S. Suh, B. L. Haymore, and F. H. Arnold, Characterization of His-X$_3$-His sites in α-helices of synthetic metal-binding bovine somatotropin. Proteins in Engineering, in press.

11. J. Porath, J. Carlsson, I. Olsson, and G. Belfrage, Metal chelate affinity chromatography, a new approach to protein fractionation. Nature 258:598 (1975).

12. E. S. Hemdan, Y. Zhao, E. Sulkowski, and J. Porath, Surface topography of histidine residues: A facile probe by immobilized metal ion affinity chromatography. Proc. Natl. Acad. Sci. USA 86:1811-1815.

13. F. H. Arnold, Metal-affinity separations: A new dimension in protein processing. Bio/Technology 9:151 (1991).

14. B. L. Haymore, G. S. Bild, S. S. Abdel-Meguid, M. C. Clare, G. G. Krivi, W. J. Salsgiver, and N. R. Staten, Introducing strong metal-binding sites into somatotropin. Facile and efficient metal-affinity purification, Submitted.

15. E. Hochuli, W. Bannwarth, H. Dobeli, R. Gentz, and D. Stuber, Genetic approach to facilitate purification of recombinant proteins with a novel metal chelate absorbent. Bio/Technology 6:1321 (1988).

16. R. P. Ambler, Enzymatic hydrolyses with carboxypeptidases. In: <u>Methods in Enzymology</u> 25:143 (1972).

17. T. P. Hopp, K. S. Prickett, V. L. Price, R. T. Libby, C. J. March, D. P. Cerretti, D. L. Urdal, and P. J. Conlon, A short polypeptide marker sequence useful for recombinant protein identification and purification. Bio/Technology 6:1204 (1988).

18. A. E. Namen, S. Lupton, K. Hjerrild, J. Wignall, D. Y. Mochizuki, A. Schmierer, B. Mosley, C. J. March, D. Urdal, S. Gillis, D. Cosman, and R. G. Goodwin. Stimulation of B-cell progenitors by cloned murine interleukin-7. Nature 333:571 (1988).

BACILLUS SUBTILIS: A MODEL SYSTEM FOR HETEROLOGOUS GENE EXPRESSION

Roy H. Doi, Xiao-Song He, Paula McCready, and Nouna Bakheit

Department of Biochemistry and Biophysics
University of California
Davis, CA 95616

INTRODUCTION

Bacillus subtilis has several features that make it a favorable system for studying the expression of foreign genes. B. subtilis is a non-pathogenic organism and makes no known endotoxins and therefore poses no health threat. Furthermore, it has been used historically for making food products in the Orient and has been used widely in industry to produce enzymes such as proteases, alpha-amylases and other degradative enzymes in large quantities. The cells grow readily to high densities in relatively simple media. The growth characteristics and physiology of the cell have been studied extensively. Among prokaryotes it is second only to Escherichia coli in the amount of genetic information that is available about a microorganism. Thus genetic manipulations can be accomplished easily and a large number of suitable mutants and plasmid vectors for genetic engineering are available. Finally. a most interesting feature about B. subtilis is that it produces a large number of proteins that are secreted directly into the growth medium. Harnessing the B. subtilis system would facilitate the construction of recombinant genes that are expressed efficiently and whose products are secreted in high quantities. Moreover, if a foreign gene product is synthesized and secreted into the growth medium at high levels, its purification is simplified. Thus from the historical, physiological, genetic, and technical viewpoints, B. subtilis would appear to be a good model system for the expression of foreign genes.

The full potential of B. subtilis as an expression system has not been realized, however, for at least two major reasons: (a) the secretion of several very active extracellular proteases by B. subtilis and (b) the lack of information concerning the secretion mechanism for B. subtilis. The synthesis of at least 5 extracellular proteases by B. subtilis has been reported (Rufo et al., 1990; Sloma et al., 1990abc). A low level of extracellular proteases is found in the growth medium during logrithmic growth. During the early stationary phase a great increase in the synthesis of these proteases occurs and these proteases can accumulate to quite a high level in even under relatively small scale laboratory conditions. The presence of these extracellular proteases causes a rapid degradation of any foreign proteins that might be produced simultaneously.

The second major impediment to the development of the B. subtilis expression system is based on our lack of understanding of the "secretion" mechanism for extracellular proteins. The many questions concerning the secretion process would include not only how are proteins translocated across the membrane, but how the genes for the extracellular proteins are regulated and expressed, how the newly synthesized protein is "processed" (e.g. how is it kept from folding to its mature conformation) before it moves to the membrane, what is the conformation of the protein as it is translocated across the membrane, how is the protein folded to its proper conformation after translocation to the exterior of the cell, and how is the protein processed to its mature form once it is in the exterior medium? These questions arise because most eukaryotic proteins do not appear to be readily translocated by the B. subtilis secretion system. The last statement is based on the observation that eukaryotic proproteins (i.e. signal peptide attached to the mature protein) are often found "stuck" to the membrane and very little of the mature form is found in the medium. However, the absence of extracellular mature forms of the foreign product may be due to the activity of proteases that are secreted along with the foreign protein. Thus it is still uncertain whether eukaryotic proteins can or cannot be readily processed by the B. subtilis secretory system.

Our approach to developing the B. subtilis heterologous gene expression and secretion system has been (a) to undertake a study of the extracellular protease genes with the hope of constructing strains that produce very little or no extracellular protease and (b) to develop model secretion vectors containing foreign genes that can be expressed efficiently and whose protein products could be secreted into the medium. The extracellular protease genes have been studied from the viewpoint of characterizing the number of extracellular proteases that are present, of analyzing the regulation of expression of these genes, and of deleting these genes from the chromosome to yield a B. subtilis strain with reduced levels of extracellular protease. Secretion vectors have been constructed with proper gene expression signals, with suitable ribosome binding sites for translation, with the presence or absence of signal peptides for secretion, and with in-frame heterologous prokaryotic and eukaryotic genes.

In this paper we describe our latest development of a B. subtilis strain that is lacking 5 extracellular and 1 intracellular (Koide et al., 1986) proteases, our current understanding of a regulatory gene for the expression of extracellular protease genes, and a summation of our current status of the efficacy of secretion of foreign proteins from B. subtilis.

CONSTRUCTION OF B. SUBTILIS DB431 LACKING FIVE EXTRACELLULAR PROTEASES

One of the major goals of our program has been to obtain a B. subtilis strain that produced no extracellular proteases. This strain could be used potentially as a host system to secrete foreign gene products which would consequently accumulate to reasonably high levels in the medium. In this regard we began a systematic analysis of extracellular proteases of B. subtilis with the ultimate goal of deleting the genes for these proteases.

Our normal approach for deletion of a particular extracellular protease gene from the chromosome has involved the cloning of the gene of the protease, the in vitro deletion of part of the gene (the deletion usually removed the promoter region and part of the coding region, e.g. the signal peptide, the propeptide, and part of the mature protein sequence) and finally the insertion of the deleted gene into the chromosome by gene conversion techniques (Iglesias and Trautner, 1983; Kawamura and Doi, 1984). We have ordinarily removed the promoter region and the early translated codon regions to prevent the synthesis of truncated proteins that might interfere or compete with the secretion of the proteins of interest. Also deletion of part of the mature coding sequence would absolutely prevent synthesis of an active protease by some read-through transcription process.

Previously we had reported the construction of a _B. subtilis_ mutant strain DB430 which had deletions in the nprE (neutral protease), aprE (alkaline protease or subtilisin), epr, bpf, and ISP1 genes (He et al., submitted, 1991). This mutant still had 0.1% of the extracellular protease activity of the wild type when assayed on gelatin plates and the remaining proteases were capable of rapidly degrading eukaryotic proteins. Since we were deleting the major extracellular protease genes first, we thought that the remaining proteases would be "minor" proteases. However, it has been our experience that when the "major" proteases are removed, the so-called "minor" proteases, which used to be degraded by the major proteases and were present in very small quantities, now are able to accumulate to a higher level and become a significant part of the remaining extracellular protease activity. Furthermore these remaining minor proteases are very active in degrading eukaryotic or heterologous prokaryotic proteins. One of the remaining proteases was coded by the mpr gene (Sloma et al., 1988, 1990a; Rufo et al., 1990).

Since strain DB430 has less than 0.1% extracellular protease activity remaining, it would be difficult to ascertain whether the mpr gene had been deleted by the usual methods. Therefore we devised a modification of the procedures described previously by constructing an antibiotic insertion into the mpr gene which inactivated the mpr gene and facilitated detection of the insertion.

For the purpose of inactivating the mpr gene present in DB430, two ingoing primers were synthesized from the known mpr sequence (Sloma et al., 1990a), and labeled as NBO237 and NBO337. These primers, which included BamHI and SmaI sites to facilitate cloning of the PCR fragment, were used in a polymerase chain reaction (PCR) with PstI cut DB430 chromosomal DNA as the template. An 840 bp partial mpr gene fragment (PCRV) was obtained by BamHI cleavage of the PCR product and cloned into the BamHI site of pUC19 (Fig. 1).

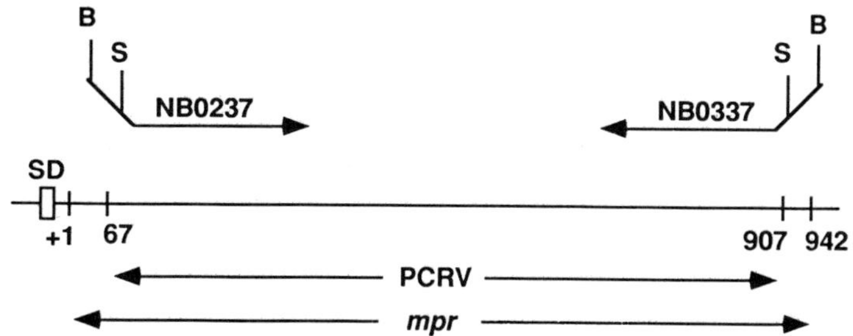

Fig 1.   Construction of PCRV.   NBO237 and NBO337 are the left and right primers used in the PCR reaction with _B. subtilis_ DB430 PstI restricted chromosomal DNA as the template. The NBO237 and NBO337 primers included BamHI and SmaI restriction sites tagged to the mpr 5' sequences from bp 67 to bp 87 and from bp 907 to bp 887. respectively. B, BamHI; S, SmaI; SD, Shine-Dalgarno sequence.

In the next step an antibiotic gene cassette was prepared for insertion into the _mpr_ gene as a marker. The plasmid pC194 carries the chloramphenicol resistance (CmR) gene within a 1.6 kb _ClaI_ fragment. The _mpr_ gene does not have a unique _ClaI_ site, but has a unique _HinpI_ site. _ClaI_ and _HinpI_ upon restriction produce compatible ends that can be ligated but will not regenerate the restriction sequence for either enzyme. However, pUC19 has multiple _HinpI_ sites. Therefore in order to achieve the required _mpr_-CmR-_mpr_ construct the PCRV fragment was cut with _HinpI_ and the enzyme was inactivated by heat treatment. A ligation mixture containing pUC19(_BamHI_), PCRV(_HinpI_, _BamHI_), and pC194(_ClaI_) was prepared. The target construct is shown in Fig. 2.

The ligation mix was used to transform _E. coli_ JM101 competent cells and transformants were selected on medium containing 25, 30, and 50 ug per ml chloramphenicol. The construct of the plasmid carrying the disrupted _mpr_ gene (_mpr_-CmR-_mpr_) was confirmed by restriction analysis and the plasmid was designated as pUC19[_mpr_::Cm].

The pUC19[_mpr_::Cm] was linearized using _ScaI_, which cuts uniquely in the pUC19 vector sequence and used to transform DB430. Transformants were selected for Cm resistance. The insertion of the inactivated _mpr_ gene into the chromosome was confirmed by Southern hybridization where the DNA from DB430(pUC19[_mpr_::Cm]) was cut by _PstI_ and probed separately with the PCRV fragment and pC194. The DNA carrying the chloramphenicol cassette showed a band that ran slower than the wild type and hybridized to both the _mpr_ sequence probe and the pC194 probe. These results confirmed that the _mpr_ gene was inactivated by insertion of the chloramphenicol resistance gene.

REGULATION OF _APRE_ EXPRESSION BY _SENS_

The regulation of the extracellular protein genes of _B. subtilis_ is complex and the expression of these genes is usually related to the stationary phase of the cell. During logarithmic growth very little or no extracellular protein activity is detectable in the growth medium whereas a relatively large amount of extracellular proteases for instance is detected at about 2 hours after the log phase of growth. The complex regulatory system involves a set of negative regulatory factors that apparently repress the expression of the gene during the active growth phase as well as a set of positive regulatory factors that enhance the expression of the gene during the early stationary phase (Valle and Ferrari, 1989; Smith, 1989). To understand more fully the factors that could enhance the expression of extracellular protein genes, we isolated and cloned a _sen_ gene (Wong et al., 1988;

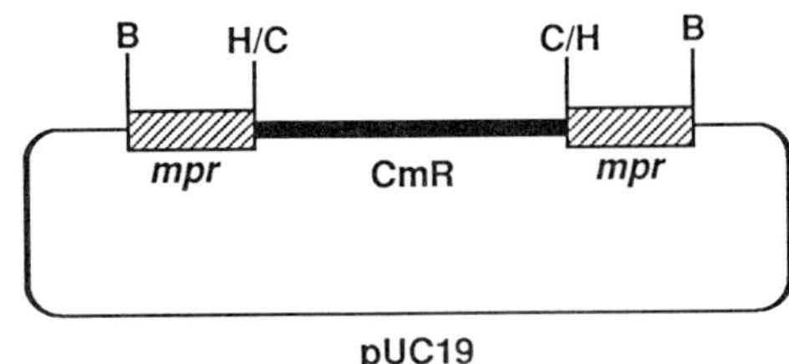

Fig. 2. Disruption of the _mpr_ sequence with a
chloramphenicol cassette. The CmR
(chloramphenicol resistance determinant) is found within a 1.6 kb fragment
from the _Staphylococcus aureus_ plasmid
pC194 (Iordanescu et al., 1978).

Wang and Doi, 1990). When the sen gene was present in high copy number, it significantly stimulated the expression of the subtilisin gene (aprE).

The senS gene of B. subtilis has an interesting sequence organization (Wang and Doi, 1990). senS codes for a small 65 amino acid protein that is highly basic and contains partial homology to B. subtilis RNA polymerase sigma factors and a helix-turn-helix motif found in DNA binding proteins. Preceding the open reading frame of the gene is a nested set of inverted repeats. The leading arm of the first inverted repeat contains a Box A consensus sequence (McCready and Doi, 1990) derived from E. coli which has been shown to be required for some types of antitermination functions (Schauer et al., 1987; Whalen et al., 1988). The first repeat has also been shown to be a strong transcriptional terminator (Wang and Doi, 1990). This type of sequence organization is indicative of a transcriptional attenuator (Yanofsky, 1988).

A transcriptional terminator/attenuator region would cause termination of most transcripts before the senS coding region. Readthrough into the open reading frame may be accomplished through the use of an antitermina tion protein, by preventing formation of the terminator stem-loop and by allowing the formation of an alternate stem-loop. This would cause tran scription readthrough into the downstream open reading frame region of senS.

Another mechanism by which transcription through the attenuator region could be accomplished is by way of the pathway used to antiterminate the ribosomal operons and lambda phage early genes in E. coli (Morgan, 1986; Friedman, 1988). The Box A consensus sequence found within the leading arm of the transcriptional terminator may be recognized by RNA polymerase associated with an antitermination complex. This antitermination complex may be composed of host proteins having similar functions to the E. coli host proteins NusA and NusB, which have been shown to be necessary for this type of regulation (Horwitz et al., 1987).

A further analysis of the senS promoter region has revealed the presence of a sequence immediately 5' of the promoter which is most likely the site of AbrB (Strauch et al., 1989) recognition and binding. Overlapping the promoter recognition sequence and the start of transcription lies a consensus sequence for a catabolite repressor (Weickert and Chambliss, 1990). This senS promoter has been shown to be under the control of both types of regulation (McCready and Doi, unpublished data). Both the AbrB and catabolite repressor are negative regulators and coupled with the downstream attenuator, provide substantial repression of this coding region. Analyses indicate that there are multiple control signals that regulate the expression of this regulatory gene itself. Since senS controls the expression of several extracellular protein genes, it seems likely that conditions in the medium, e.g. low nitrogen and/or low carbon, are transmitted by media sensors to the level of transcription regulation perhaps in a manner similar to that of the two component regulatory systems (Smith, 1989).

To determine how senS may be controlling the activity of extracellular protein genes, we have used the aprE gene as a model system for studying the senS effects. Ordinarily a high copy number of the senS gene stimulates the expression of aprE about 8 fold during the early stationary phase as measur ed by reporter-fusion gene methods. To determine whether the 5' upstream region of the aprE gene is involved in this stimulation by senS, deletion analysis of this region was carried out. If deletion removes some target site for senS action, then the stimulation by high copy numbers of senS would be reduced. This is in fact the case, since the deletion of this up stream region from -180 to -400 bp reduced the stimulation of aprE express ion 4 fold. The fact that expression was still reproducibly stimulated in

this particular deletion mutant suggested to us that there was another down
stream site for senS activation. Further deletions of the upstream region
down to -60 bp indicated that this downstream site was located between -60
and +79, and was responsible for 2 fold stimulation above the control background. These results are summarized in Table 1 as well as in Fig. 3 which
indicates the locations of the regulatory sites for senS in the aprE promoter region as well as for various other regulatory factors for the aprE
gene. It is obvious that the target site for senS may overlap or be coinci-
dent with the target sites of other regulatory factors. These observations
suggest that the product of the senS gene, SenS, either binds directly to
the target site, is controlling the expression of another gene whose product
is in fact the direct DNA binding protein, or is modifying the DNA binding
ability of the regulator at the protein level to allow binding and stimulation of the aprE promoter.

The number of positive and negative trans regulatory factors and
sites in the aprE promoter region is as large as that observed for many
eukaryotic genes. Thus the aprE gene is an excellent model for studying
the complex nature and interaction of regulatory factors in a prokaryotic
system. A full understanding of the control mechanisms will also allow
better control of expression of heterologous genes in B. subtilis and permit maximum expression under suitable growth conditions.

HETEROLOGOUS GENE EXPRESSION IN BACILLUS SUBTILIS

The overall strategy for the expression of a foreign gene in B. subtilis was to develop a host system that produced minimal amounts of extracellular protease (Kawamura and Doi, 1984; Yang et al., 1984; Stahl and Ferrari,
1984; Wang et al., 1989; Wu et al., 1990; Bruckner et al., 1990; Sloma et
al., 1990a) and a vector system containing the proper signals for gene expression, for secretion, and for ease of making a recombinant gene of interest. Basically, the foreign gene would be inserted into a pUB110-based mul-
ticopy plasmid which would be transformed into the appropriate B. subtilis
host resulting in production of the foreign gene protein. The development
of the host system has been described in detail above. The following specific points will be considered in the construction of a suitable vector system.

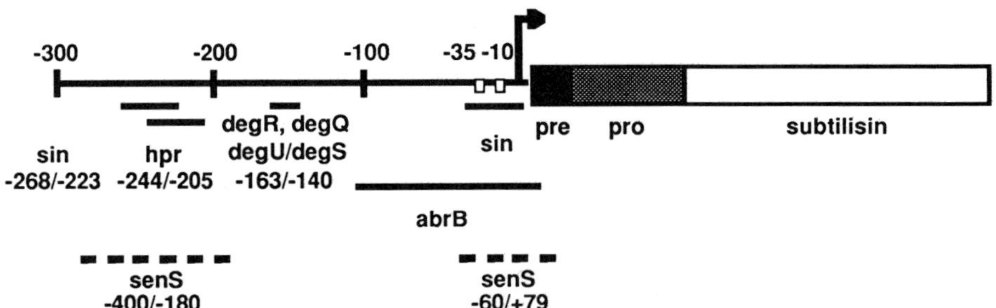

Fig. 3. The upstream regulatory sites for the aprE (subtilisin) gene.
The arrow indicates the site of transcription initiation at
+1. The small open boxes represent the promoter locus at -10
and -35. The dark solid bars indicate the sites for sin, hpr,
degR/degQ/degU/degS, and abrB regulation. The dark broken bars
indicate the regions required for senS regulation.

Table 1

Relative Activity of <u>aprE</u> Promoter as a Function of Its 5'
Flanking Region and <u>senS</u> Copy Number

Promoter Property	Single <u>senS</u> Copy*	Multiple <u>senS</u> Copy*
<u>aprE</u> promoter with wild type 5' flanking region	21	168
<u>aprE</u> promoter with deleted (down to -60 bp) 5' flanking region	42	84

*Miller units (Miller, 1972).

Promoter

Since the <u>B. subtilis</u> transription system is very specific (Moran, 1989), the promoter for transcriptional expression should be a <u>B. subtilis</u> promoter that is transcribed actively during the growth phase of interest. Since most proteases of <u>B. subtilis</u> are produced during the stationary phase of growth, we have tried to avoid this stage and have the foreign gene expressed during the active growth phase when the level of extracellular proteases is very low. Thus we have used the P2 promoter of the <u>B. subtilis</u> <u>rpoD</u> operon (Wang and Doi. 1987) or a cryptic vegetative promoter (Wong and Doi, unpubished data) as the "vegetative" phase promoter of choice. These promoters are not repressed by the presence of glucose and other carbon sources and are expressed at a relatively high rate by the sigma-A form of RNA polymerase.

Ribosome Binding Site (RBS)

<u>B. subtilis</u> has a more stringent requirement for the RBS of the mRNA than <u>E. coli</u> (Hager and Rabinowitz, 1985). Therefore in most cases the RBS of a <u>B. subtilis</u> gene must be used to allow efficient translation. The RBS of the <u>B. subtilis</u> <u>aprE</u> gene has been used routinely to express both prokaryotic and eukaryotic genes. However we have found somewhat surprisingly that the RBS of the <u>E. coli</u> phage T7 gene 10 can be utilized by <u>B. subtilis</u> ribosomes and this RBS has been used as the translation element for expression of human tissue plasminogen activator (t-PA) and for rice alpha-amylase (He et al., submitted for publication, 1991).

Signal Peptide

Part of our goal is not only to express foreign genes in <u>B. subtilis</u>, but also to secrete their products into the medium. The signal peptide is necessary for a secreted protein to be recognized and translocated by the cellular secretion machinery from the inside of the cell to the external medium. The signal peptide of the <u>B. subtilis</u> extracellular enzyme subtilisin (<u>aprE</u>) has been utilized for this purpose (Wong and Doi, 1986; Wong et al.. 1986; Wang et al., 1988).

<u>Expression of E. coli beta-lactamase</u>

The <u>E. coli</u> beta-lactamase gene was placed under the control of a vegatative promoter and the subtilisin RBS and fused to the subtilisin signal peptide. The expression of the beta-lactamase was followed in the newly constructed DB431 strain (<u>nprE aprE epr bpf mpr ispl</u>) and the stability of the beta-lactamase was compared to that produced by strains DB430 (<u>nprE aprE epr bpf ispl</u>) and DB2 (wild type). The results showed a signficant increase in the extracellular beta-lactamase in DB431 relative to that in DB430 and DB2 at the end of the log phase of growth, but during the stationary phase, no great difference in stability of beta-lactamase was noted between DB430 and DB431 (Fig. 4). These results indicate that the remaining proteases are still capable of attacking beta-lactamase. There may be at least two more extracellular proteases produced by B. subtilis (S.-L. Wong, personal commun-

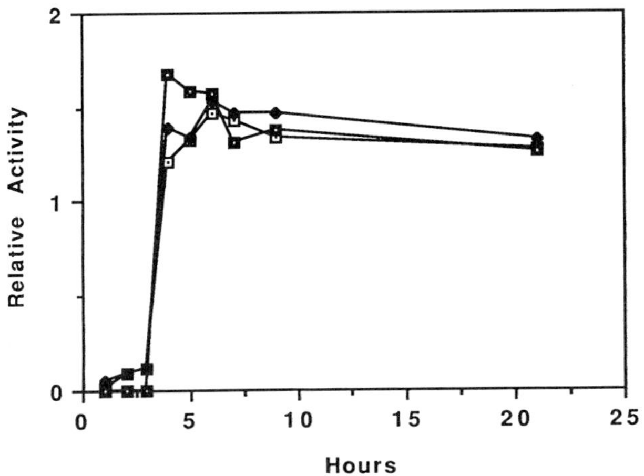

Fig. 4. The relative extracellular protease activities of
<u>B. subtilis</u> DB2 (wild type, open squares), DB430
(closed diamond), and DB431 (closed squares). The
activity was measured by assaying for extracellular
beta-lactamase activity remaining in of the medium.

ication). The amount of total extracellular protease activity of DB431 was reduced to about 0.1 % of that of the wild type cells.

<u>Expression of human atrial natriuretic factor</u> (ANF)

The coding sequence of human atrial natriuretic factor (ANF) was fused with the subtilisin (<u>aprE</u>) RBS and signal peptide sequence under the transcriptional control of the <u>rpoD</u> promoter P2 in plasmid pWL53. In analyzing the product formed after transformation of DB430, the fusion protein containing the signal sequence and mature ANF was detected within the cell, while the processed (mature) ANF was detected in the medium by immunoblotting experiments. About 10 times more ANF was detected within the cell than in the medium. Although the ANF was secreted from <u>B. subtilis</u>

cells, the extracellular yield was extremely low - about 1 mg per liter.
We do not know whether this low yield was due to lack of secretion or whether the remaining extracellular proteases of the host were destroying the
secreted ANF.

Human tissue plasminogen activator (tPA)

The coding sequences for both mature tPA and prepro-tPA were inserted
into the expression vector, pWT19 (Wang and Doi, 1987), under control of
P2 promoter and T7 gene 10 RBS, with or without an in-frame fused subtilisin signal peptide sequence at its 5' end. tPA was found to be expressed
within the cell, with a higher yield when the intracelluar protease defi-

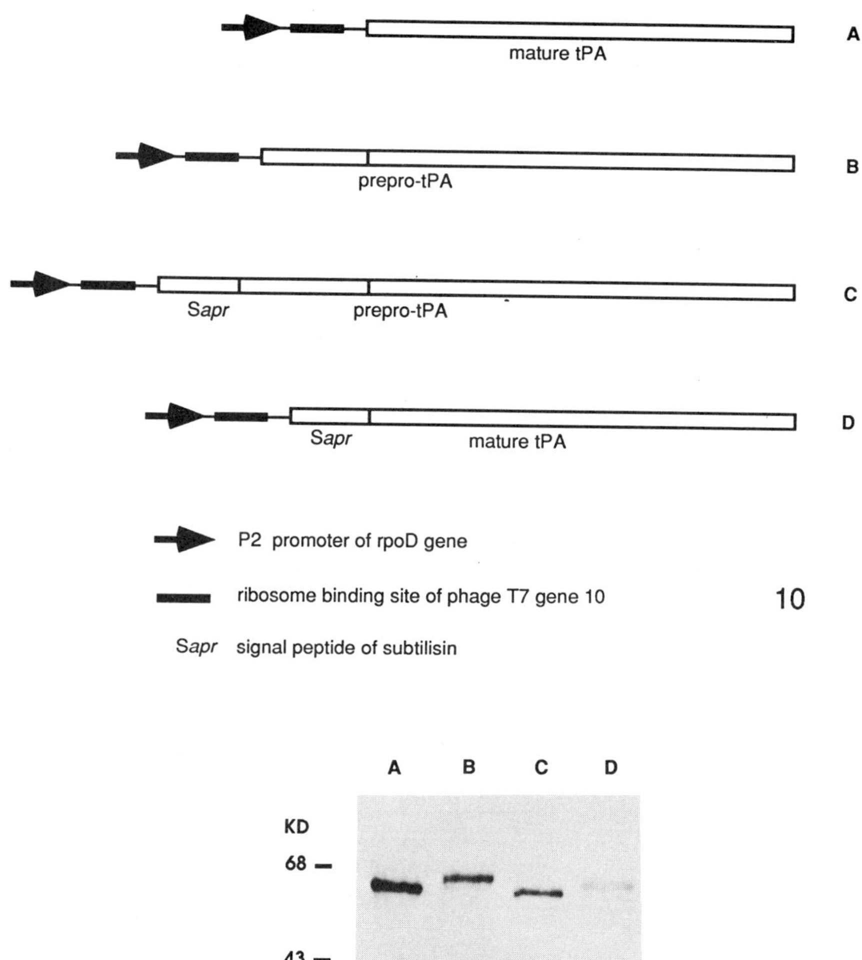

Fig. 5. Structure of various recombinant human tPA genes and their
expression in B. subtilis. The recombinant genes are
labeled A through D. The lower portion of the figure indicates the size of the product obtained after expression of
the genes in B. subtilis and after Western blot analysis.
The 60 kDa band in lane A is typical of the size of control
tPA (Wang et al., 1989).

cient (<u>isp1</u>) strain DB 430 was used instead of the <u>isp+</u> strains. However,
no secreted tPA was ever detected even from DB430. In the case of the fusion product between subtilisin signal peptide and tPA (Sapr-tPA) (Fig. 5),
the expression product was found to be processed by the signal peptidase,
but the product was always found in the insoluble cell debris fraction.
This suggests either that the processed tPA was stuck in the cell membrane, that the processed tPA could not fold correctly into a soluble conformation, or that there was an intermolecular aggregation of tPA into an
insoluble mass.

Rice alpha-amylase

The coding sequence for mature rice alpha-amylase was fused in-frame
with the subtilisin signal peptide and placed under control of P2 promoter
and T7 gene 10 RBS. Expression product was detected by immunoblotting only
within the cell when DB430 was used. This intracellular product was mostly
pre-alpha-amylase, but there was a very low level of processing and the internal amylase did show some activity. It is generally known that unprocess-
ed pre-alpha-amylase does not exhibit enzymatic activity. No secreted amylase activity was detected (He et al., submitted, 1991).

Thus for the few foreign genes that have been tested, only the prokaryotic gene products can be expressed and secreted into the medium. All attempts to show high level expression of eukaryotic genes and secretion of
their products have not been successful. Whether this is due to the incomp-
atability of the <u>B. subtilis</u> secretory mechanism and eukaryotic proteins or
to rapid degradation of the secreted proteins by <u>B. subtilis</u> extracellular
proteases is still in question. However, it appears likely that both of
these factors are playing a role in the low production of eukaryotic gene
products. The elimination of the remaining extracellular protease genes and
a better understanding of the secretion mechanism and its regulation should
lead to a suitable host system for the production of foreign gene products.

REFERENCES

Bruckner, R., Shoseyov, O. and Doi, R.H., 1990. Multiple active forms of a
        novel serine protease from <u>Bacillus subtilis</u>. <u>Mol. Gen. Genet</u>, 221,
        486-490.
Friedman, D.I., 1988, Regulation of phage gene expression by termination
        and antitermination of transcription, in: The Bacteriophages, Vol.
        2, R. Calendar, ed., Plenum Press, New York.
Hager, P.W. and Rabinowitz, J.C., 1985, Translational specificity in <u>Bacill-
        us subtilis</u>, in: "The Molecular Biology of the <u>Bacilli</u>", D.A. Dubnau,
        ed., Academic Press, Inc., Orlando, Florida.
Horwitz, R.J., Li, J., and Greenblatt, J., 1987, An elongation control
        particle containing the N gene transcriptional antitermination pro-
        tein of bacteriophage lambda, <u>Cell</u>, 51, 631-641.
Iglesias, A. & Trautner, T., 1983, Plasmid transformation in <u>Bacillus subti-
        lis</u>: symmetry of gene conversion in transformation with a hybrid
        plasmid containing chromosomal DNA. <u>Mol. Gen. Genet.</u>, 189, 73-76.
Iordanescu, S., Surdeanu, M., Della Latte, P., and Novick, R., 1978, In-
        compatibility and molecular relationships between small <u>Staphylo-
        coccal</u> plasmids carrying the same resistance marker, <u>Plasmid</u>, 1:468
        -479.
Kawamura, F. & Doi, R.H., 1984, Construction of a <u>Bacillus subtilis</u> double
        mutant deficient in extracellular alkaline and neutral proteases. <u>J.
        Bacteriol.</u>, 160, 442-444.
Koide, Y., Nakamura, A., Uozumi, T. & Beppu, T., 1986, Cloning and sequenc-
        ing of the major intracellular serine protease gene of <u>Bacillus sub-
        tilis</u>, <u>J. Bacteriol.</u>, 167, 110-116.

McCready, P. and Doi, R.H., 1990, Possible regulaton of sens by a nusA-related mechanism, in: "Genetics and Biotechnology of Bacilli", Vol. 3, M.M. Zukowski, A.T. Ganesan, and J.A. Hoch, eds., Academic Press, Inc., San Diego, California.

Miller, J.H., 1972, Experiments in Molecular Genetics, Cold Spring Harbor Laboratory, Cold Spring Harbor, New York.

Moran, C. Jr., 1989, Sigma factors and the regulation of transcription, in: "Regulation of Procaryotic Development", I. Smith, R.A. Slepecky, and P. Setlow, eds., American Society for Microbiology, Washington, DC.

Morgan, E.A., 1986, Antitermination mechanisms in rRNA operons of Escherichia coli. J. Bacteriol., 168,1-5.

Rufo Jr., G.A., Sullivan, B.J., Sloma, A. and Pero, J., 1990, Isolation and characterization of a novel extracellular metalloprotease from Bacillus subtilis, J. Bacteriol., 172, 1019-1023.

Schauer, A.T., Carver, D.L., Bigelow, B., Baron, L.S. and Friedman, D.I., 1987, Lambda N antitermination system: functional analysis of phage interaction with the host NusA protein, J. Mol. Biol., 194, 679-690.

Sloma, A., Ally, A., Ally, D. and Pero, J., 1988, Gene encoding a minor extracellular protease in Bacillus subtilis, J.Bacteriol., 170, 5557-5563.

Sloma, A., Rudolph, C.F., Rufo Jr., G.A., Sullivan, B.J., Theriault, K.A., Ally, D. and Pero, J., 1990a, Gene encoding a novel extracellular metalloprotease in Bacillus subtilis, J. Bacteriol., 172, 1024-1029.

Sloma, A., Rufo Jr., G.A., Rudolph, C.F., Sullivan, B.J., Theriault, K.A. and Pero, J., 1990b, Bacillopeptidase F of Bacillus subtilis: purification of the protein and cloning of the gene, J. Bacteriol., 172, 1470-1477.

Sloma, A., Rufo,Jr., G.A., Rudolph,C.F., Sullivan, B.J., Theriault, K.A. and Pero, J., 1990c, Errata:Bacillopeptidase F of Bacillus subtilis: purification of the protein and cloning of the gene, J. Bacteriol., 172, 5520-5521.

Smith, I., 1989, Initiation of sporulation, in: "Regulation of Procaryotic Development", I. Smith, R.A. Slepecky, and P. Setlow, eds., American Society for Microbiology, Washington, DC.

Stahl, M.L. and Ferrari, E., 1984, Replacement of the B. subtilis subtilisin structural gene with an in vitro-derived deletion mutation, J. Bacteriol., 158, 411-418.

Strauch, M.A., Spiegelman, G.B., Perego, M., Johnson, W.C., Burbulys, D., and Hoch, J.A., 1989, The transition state transcription regulator abrB of Bacillus subtilis is a DNA binding protein, EMBO J., 8, 1615-1621.

Valle, R. and Ferrari, E., 1989, Subtilisin: a redundantly temporally regulated gene?, in: "Regulation of Procaryotic Development", I. Smith, R.A. Slepecky, and P. Setlow, eds., American Society for Microbiology, Washington, DC.

Wang, L.-F., Bruckner, R. and Doi, R.H., 1989, Construction of a Bacillus subtilis mutant deficient in three extracellular proteases, J. Gen. Appl. Microbiol., 35, 487-492.

Wang, L.-F. and Doi, R.H., 1987, Promoter switching during development and the termination site of the sigma-43 operon of Bacillus subtilis, Mol. Gen. Genet., 207, 114-119.

Wang, L.-F. and Doi, R.H., 1990, Complex character of sens, a novel gene regulating expression of extracellular protein genes of Bacillus subtilis, J. Bacteriol. 172, 1939-1947.

Wang, L.F., Hum, W.T., Kalyan, N.K., Lee, S.G., Hung, P.P. and Doi, R.H., 1989, Synthesis and refolding of human tissuetype plasminogen activator in Bacillus subtilis, Gene, 84, 127-133.

Wang, L.-F., Wong, S.-L., Lee, S.-G., Kalyan, N.K., Hung, P.P., Hilliker, S., and Doi, R.H., 1988, Expression and secretion of human atrial natriuretic alpha-factor in <u>Bacillus</u> <u>subtilis</u> using the subtilisin signal peptide, <u>Gene</u>, 69, 39-47.

Weichert, M.J. and Chambliss, G.H., 1990, Site-directed mutagenesis of a catabolite repression operator sequence in <u>Bacillus</u> <u>subtilis</u>. <u>Proc.</u> <u>Natl.</u> <u>Acad.</u> <u>Sci.</u> <u>USA</u>. 87, 6238-6242.

Whalen, W., Ghosh, B., and Das, A., 1988, NusA protein is necessary and sufficient <u>in</u> <u>vitro</u> for phage lambda <u>N</u> gene product to suppress a <u>rho</u>-independent terminator placed downstream of <u>nutL</u>, <u>Proc.</u> <u>Natl.</u> <u>Acad.</u> <u>Sci.</u> <u>USA</u>. 85, 2494-2498.

Wong, S.-L. and Doi, R.H., 1986, Determination of the signal peptide cleavage site in the preprosubtilisin of <u>Bacillus</u> <u>subtilis</u>, <u>J.</u> <u>Biol.</u> <u>Chem.</u>, 261, 10176-10181.

Wong, S.-L., Kawamura, F. and Doi, R.H., 1986, Use of the <u>Bacillus</u> <u>subtilis</u> subtilisin signal peptide for efficient secretion of TEM beta-lactamase during growth, <u>J.</u> <u>Bacteriol.</u>, 168, 1005-1009.

Wong, S.-L., Wang, L.-F., and Doi, R.H., 1988, Cloning and nucleotide sequence of <u>senN</u>, a novel 'Bacillus natto' (<u>B.</u> <u>subtilis</u>) gene that regulates expression of extracellular protein genes, <u>J.</u> <u>Gen.</u> <u>Microbiol.</u>, 134, 3269-3276.

Wu, X.-C., Nathoo, S., Pang, A.S.-H., Carne, T. and Wong, S.-L., 1990, Cloning, genetic organization and characterization of a structural gene encoding bacillopeptidase F from <u>Bacillus</u> <u>subtilis</u>, <u>J.</u> <u>Biol.</u> <u>Chem.</u>, 265, 6845-6850.

Yang, M.Y., Ferrari, E. and Henner, D.J., 1984, Cloning of the neutral protease gene of <u>Bacillus</u> <u>subtilis</u> and the use of the cloned gene to create an <u>in</u> <u>vitro</u>-derived deletion mutant, <u>J.</u> <u>Bacteriol.</u>, 160, 15-21.

Yanofsky, C., 1988, Transcription attenuation, <u>J.</u> <u>Biol.</u> <u>Chem.</u>, 263, 609-612.

ASPERGILLUS NIGER VAR. *AWAMORI* AS A HOST FOR

THE EXPRESSION OF HETEROLOGOUS GENES

Randy M. Berka,[1] Frank T. Bayliss,[2] Peggy Bloebaum,[1] Daniel Cullen,[1†]
Nigel S. Dunn-Coleman,[1] Katherine H. Kodama,[1] Kirk J. Hayenga,[1‡]
Ronald A. Hitzeman,[3] Michael H. Lamsa,[1] Melinda M. Przetak,[1,2]
Michael W. Rey,[1] Lori J. Wilson,[1] and Michael Ward[1]

[1]Genencor International Inc., South San Francisco, California
[2]San Francisco State University, San Francisco, California
[3]University of California, Berkeley, California

INTRODUCTION

Among the diversity of cellular systems that have been developed for the expression
of heterologous gene products, certain species of filamentous fungi possess features
which make them exceptionally attractive for this purpose. These include (a) the ability
to produce high levels (>25 grams per liter) of secreted protein in submerged culture,
(b) a long history of safe use in the production of enzymes, antibiotics, and
biochemicals which are used for human consumption, and (c) established fermentation
processes which are inexpensive by comparison with animal cell culture processes done
on a similar scale. These attributes have prompted several biotechnology companies to
explore the use of filamentous fungi as hosts for the expression and secretion of foreign
proteins. Some of the heterologous gene products which have been made using fungal
expression systems are shown in Table 1. Compared to highly refined expression
systems such as *Escherichia coli* or *Saccharomyces cerevisiae*, the evolution of
filamentous fungi as hosts has barely begun, and many fundamental aspects of cell
biology and biochemistry in fungi have not been studied. Fortunately, many of the
molecular details and principles which have been elucidated in yeast and in mammalian
cell systems appear to be applicable to the study of heterologous gene expression and
protein secretion in filamentous fungi as well.

Several years ago, our laboratory began investigating the use of filamentous fungi
for the synthesis and secretion of bovine chymosin (rennin), an aspartyl protease used
in cheesemaking. Chymosin is synthesized in the fourth stomach of unweaned calves
as a zymogen precursor (preprochymosin). An N-terminal signal peptide consisting of
16 amino acids is removed during secretion of the enzyme, and a 42 amino acid
propeptide is autocatalytically removed in the low pH environment of the stomach[1].
Two forms of the enzyme, designated as chymosin A and chymosin B, differing by a
single amino acid substitution, are encoded by alleles of the same gene[2,3]. Chymosin is

---

†Present address: USDA Forest Products Laboratory, Madison, WI

‡Present address: Genentech, Inc., South San Francisco, CA

Applications of Enzyme Biotechnology, Edited by J.W. Kelly and
T.O. Baldwin, Plenum Press, New York, 1991

Table 1. Heterologous Proteins Expressed in Filamentous Fungi

Gene Product	Species of Origin	Fungal Species	Reference
Chymosin	Bovine	*Aspergillus nidulans*	4
Chymosin	Bovine	*Aspergillus niger* var. *awamori*	5
Chymosin	Bovine	*Trichoderma reesei*	6
Chymosin	Bovine	*Aspergillus oryzae*	7
Tissue plasminogen activator (tPA)	Human	*Aspergillus nidulans*	8
Superoxide dismutase	Human	*Aspergillus nidulans*	9
Interferon α2	Human	*Aspergillus nidulans*	10
Endoglucanase	Bacterial	*Aspergillus nidulans*	10
Interferon γ	Human	*Achlya ambisexualis*	11
Thymidine kinase	Herpes virus	*Achlya ambisexualis*	12
Lysozyme	Hen egg	*Aspergillus niger*	13
Epidermal growth factor	Human	*Aspergillus nidulans*	9
Parathyroid hormone	Human	*Aspergillus nidulans*	9
Growth hormone	Human	*Aspergillus nidulans*	9
Corticosteroid binding globulin	Human	*Aspergillus nidulans*	9
Heat labile enterotoxin	Bacterial	Aspergillus nidulans	14
Interleukin 6	Human	*Aspergillus nidulans*	15

the preferred enzyme for milk coagulation during cheesemaking, primarily because it preferentially cleaves the Phe_{105}-Met_{106} peptide bond in κ-casein with high specificity[16]. Chymosin is frequently in short supply and expensive because of market fluctuations and seasonal availability of veal calves. Aspartyl proteases from filamentous fungi such as *Endothia parasitica*, *Rhizomucor miehei* and *Rhizomucor pusillus* have been employed as chymosin substitutes during periodic shortages, however, they lack the specificity of chymosin and are generally considered by the cheese industry to be inferior. Thus, several biotechnology companies have attempted to produce chymosin in microorganisms using recombinant DNA methodology. To date, researchers have expressed bovine chymosin cDNA sequences in bacteria[17, 18], yeasts[19-22], and in filamentous fungi[4-7].

CHYMOSIN EXPRESSION IN *ASPERGILLUS NIDULANS*

We first tested the expression and secretion of chymosin in a laboratory strain of the fungus *Aspergillus nidulans*[4]. Although it is not generally considered to be an industrial production organism, *A. nidulans* is a genetically well-characterized fungus that is taxonomically related to species such as *A. niger* and *A. oryzae* which are used for production of industrial enzymes. We hoped that the observations made in *A. nidulans* would provide information necessary to initiate similar experiments in industrial production strains.

We constructed a series of expression vectors which employed the transcriptional, translational and secretory control elements from the *A. niger glaA* (glucoamylase) gene (Fig. 1). Three of the vectors had prochymosin cDNA joined to the *glaA* sequences following [a] the glucoamylase signal peptide cleavage site (pGRG1), [b] the glucoamylase propeptide cleavage site (pGRG2), or [c] after 11 codons of the mature glucoamylase (pGRG4). A fourth construction involved joining preprochymosin cDNA sequences directly to the *glaA* promoter (pGRG3). All four vectors contained a segment of *Neurospora crassa* genomic DNA which contained the *pyr4* (orotidine-5'-monophosphate decarboxylase) gene which was used as the selectable marker for

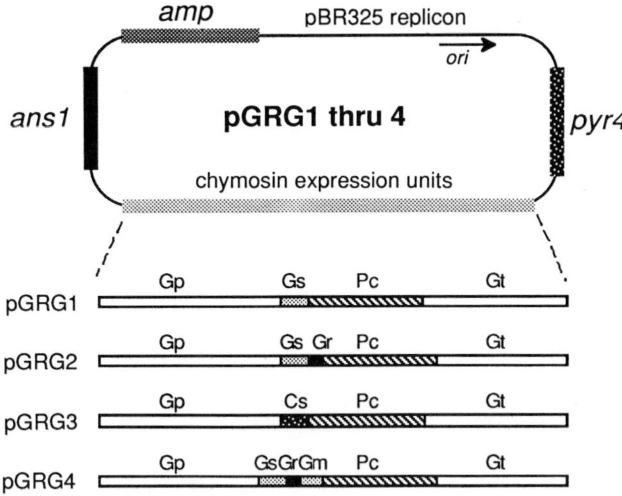

FIGURE 1. Schematic map of chymosin expression vectors pGRG1 thru pGRG4. A pBR325 replicon and ampicillin resistance gene (*amp*) were used for construction and propagation of the vectors in *E. coli*. The selectable *pyr4* gene from *N. crassa* encodes orotidine-5'-monophosphate decarboxylase which relieves an auxotrophic requirement for uridine in *pyrG* mutants of *A. nidulans*[23]. The *ans1* sequence has been shown to increase transformation frequency in *A. nidulans*[24]. The chymosin expression units contained the following components: Gp, *A. niger glaA* promoter; Gs, *A. niger* glucoamylase signal peptide codons; Gr, glucoamylase propeptide codons; Pc, prochymosin B cDNA; Gt, *A. niger glaA* terminator.

transformation of an *A. nidulans pyrG* auxotrophic recipient[23]. In addition, these vectors contained a segment of genomic DNA from *A. nidulans* known as *ans1* which promotes efficient integrative transformation in this species[24].

    A. nidulans transformants generated with each of the four expression vectors secreted active chymosin into the culture medium when grown in the presence of starch as the sole carbon source. When xylose was used as the carbon source, the level of extracellular chymosin was negligible, suggesting that the heterologous *glaA* promoter was starch-inducible in *A. nidulans*, just as it is in the native *A. niger* host. Northern blotting experiments indicated that the expression of chymosin in *A. nidulans* was regulated at the level of transcription (not shown). In contrast to observations made previously in the yeast *S. cerevisiae*[20], chymosin was secreted by *A. nidulans* even when cDNA sequences encoding the native signal peptide were incorporated into the expression vector. This observation hints that filamentous fungi may be more permissive than yeast with respect to heterologous signal peptide function. While the levels of extracellular chymosin varied with the type of expression vector employed, chymosin yields also varied widely among transformants derived from a single vector (Table 2). While integration of the expression vectors into the host chromosomal DNA appeared to involve predominantly non-homologous recombination and varying copy numbers were observed among the transformants, there seemed to be no strict correlation between chymosin yield and the number of integrated gene copies. The extracellular yields of chymosin produced by *A. nidulans* were modest (approximately 20 to 150 µg per gram of dry weight mycelia; 1 to 5 mg/L in a shake flask culture). Nevertheless, the results compelled us to attempt similar experiments in a strain of *A. niger* var. *awamori* which had been selected for its ability to secrete large amounts of glucoamylase.

Table 2. Extracellular Levels of Chymosin Produced by *A. nidulans*

Vector	Extracellular Chymosin (µg/g dry weight mycelia)*
pDJB3 (neg. control)	N.D.
pGRG1	146
pGRG2	22
pGRG3	93
pGRG4	119

*Mean values calculated from enzyme immunoassays of six randomly chosen transformants, grown in medium with starch as the sole carbon source. N.D., not detected.

CHYMOSIN PRODUCTION IN *A. NIGER* VAR. *AWAMORI*

Chymosin expression vectors pGRG1 and pGRG3 were used for transformation of a *pyrG* auxotrophic mutant of *A. niger* var. *awamori* known as strain GC5 (see Fig. 3). Plasmid pGRG1 encoded the glucoamylase signal peptide to direct chymosin secretion while pGRG3 employed codons of the native signal peptide for preprochymosin (see Fig. 1, above). Culture filtrates from a large number of transformants were screened using a milk-clotting assays to measure levels of extracellular chymosin. Those transformants which produced the highest levels were selected for further study. Table 3 displays the yields of chymosin and glucoamylase secreted by these selected transformants. One transformant derived from expression vector pGRG3, designated as strain 107, produced the highest levels of extracellular chymosin (approximately 14 µg/ml in a 50 ml shake flask culture). Several other conclusions can be drawn from these data. First, there was considerably much less chymosin produced than glucoamylase, suggesting that some aspect of chymosin expression (*e.g.*, transcription, mRNA stability, mRNA processing, translation, or secretion) was inefficient. Although the DNA sequence of the *A. niger glaA* promoter used in these experiments was nearly identical to the *A. niger* var. *awamori glaA* promoter[25, 26], the was the possibility that it did not function effectively in *A. niger* var. *awamori*. Therefore, a construction, equivalent to pGRG4, but containing the *A. niger* var. *awamori glaA* promoter, was assembled and used to transform strain GC5. Chymosin expression levels directed from this vector were found to be similar to those obtained with pGRG3 (data not shown), suggesting that both sources of the *glaA* promoter function equivalently in *A. niger* var. *awamori*.

In most of the transformants the presence of extra copies of the *glaA* promoter on the expression vectors had little affect on the production of native glucoamylase. This suggests that there is not a limiting supply of *trans*-acting transcriptional activator(s) which were titrated by the addition of extra copies of the *glaA* promoter. It should be noted, however, that in one transformant (number 62), the level of glucoamylase was exceptionally low. By Southern blotting analysis it was shown that this transformant contained a high number of integrated copies of pGRG1. In contrast, the number of integrated gene copies in strain 107 was low (*i.e.* two or three copies per genome). Thus, it is possible, that in some transformants, the number of copies of the *glaA* promoter may reach a level at which transcriptional regulatory factors may become limiting. Nonetheless, as we observed in *A. nidulans*, there was no apparent correlation between chymosin expression levels in *A. niger* var. *awamori* and the number of integrated copies of the expression vectors. Furthermore, integration of the

Table 3. Extracellular Yields of Chymosin and Glucoamylase in 50 ml Shake Flask Cultures of Selected *A. niger* var. *awamori* Transformants

Strain	Expression Vector	Chymosin (μg/ml)[a]	Glucoamylase (μg/ml)[a]
GC5	None	N.D.	2311
62	pGRG1	4.7	420
82	pGRG1	5.6	3922
83	pGRG1	4.9	3105
13	pGRG3	4.8	1482
18	pGRG3	5.5	3863
20	pGRG3	4.1	2821
107	pGRG3	14.6	3280

N.D., not detectable. [a]Average from two shake flask cultures harvested after three days of growth. Measured by enzyme immunoassay.

vectors occurred at random sites in the genome, and thus, the site of integration may be important for high chymosin yields. Alternatively, some factor other than transcription was limiting chymosin production.

Northern blot experiments were done to compared the steady-state levels of chymosin and glucoamylase mRNA. A combination chymosin-glucoamylase probe fragment was constructed which contained approximately equal amounts of *glaA*- and chymosin-specific DNAs. Thus, a direct comparison of the steady-state levels of the two mRNA species could be made from a single Northern blot hybridized with just one probe. In the same experiments, the efficiency of polyadenylation was examined following oligo-dT cellulose chromatography (Fig. 2). It was apparent from this experiment that the level of chymosin-specific mRNA was slightly higher than the level of glucoamylase mRNA suggesting that transcription of the chymosin cDNA from the *glaA* promoter was reasonably efficient. In addition, analysis of oligo-dT selected RNA indicated that nearly all of the chymosin mRNA was polyadenylated. Therefore, the steps which limit production of bovine chymosin in *Aspergillus* are apparently post-transcriptional, perhaps involving degradation of the enzyme, inefficient translation, secretion or a combination of these.

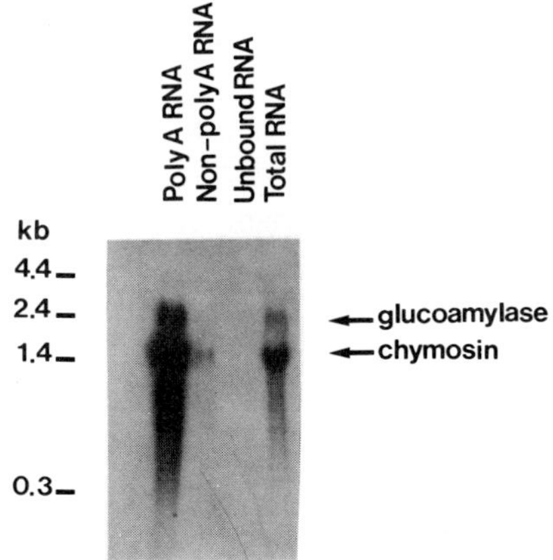

Figure 2. Northern blot hybridization comparing steady-state levels of glucoamylase and chymosin in an *A. niger* var. *awamori* strain which contains several integrated copies of pGRG3.

STRAIN IMPROVEMENT BY MUTAGENESIS AND SCREENING

Repeated rounds of mutagenesis of strain 107 with ultraviolet (UV) light or
nitrosoguanidine (NTG), with subsequent screening for increased chymosin production
resulted in the isolation of improved production strains[27]. Screening for higher
chymosin levels was accomplished by inoculating mutagenized spores onto filter discs
which were laid on the surface on agar growth medium which contained a colony size
restrictor (dichloran). After growth of the colonies, the filters were removed and the
chymosin present in the agar medium was assayed following treatment of the plates
with an aspartyl protease inhibitor, diazoacetyl norleucine methyl ester (DAN). This
was done to inhibit the activity of the host aspartyl protease (chymosin was only
partially inhibited). After each round of mutagenesis and screening, the mutants which
produced the highest levels of chymosin were mutagenized again and screened for
improved productivity. Five serial rounds of mutagenesis and screening produced a
strain which showed a four-fold improvement in chymosin production (Table 4).

Table 4. Chymosin Production by Mutants Derived from an *A. niger* var.
awamori Transformed with pGRG3 (Strain 107)

Strain	Extracellular Chymosin Concentration (μg/ml)[a]	
	Activity Assay	Immunoassay
107	11	21
A96	15	31
1AA84	16	28
2AB26	20	42
2AC109	31	45
2AD73	45	81

[a]Values represent the average from triplicate shake flask cultures.

DEGRADATION OF CHYMOSIN BY ASPERGILLOPEPSIN A

The culture conditions that were used to induce expression of bovine chymosin in
A. niger var. *awamori* also supported secretion of an endogenous host aspartyl
protease known as aspergillopepsin A[28]. This enzyme was responsible for the majority
of the proteolytic activity present in *A. niger* var. *awamori* culture filtrates. We noted
from the results of the mutagenesis and screening program that the levels of chymosin
detected by enzyme activity assays were invariably lower than the levels that were
measured using antibody assays. This observation suggested the presence of inactive
chymosin material in the culture filtrates. Moreover, experiments in which a known
amount of chymosin was added to the culture filtrate showed a gradual loss of
chymosin activity with time, suggesting that an endogenous protease was degrading
chymosin. This degradation was prevented by the aspartyl protease inhibitor pepstatin,
and thus, aspergillopepsin A was implicated as the cause. Additionally, we noted from
the mutagenesis and screening program that several mutants with improved chymosin
productivity showed dramatic decreases in production of aspergillopepsin. Therefore,
we elected to clone the gene encoding aspergillopepsin A and delete it from the genome
of our best production strain.

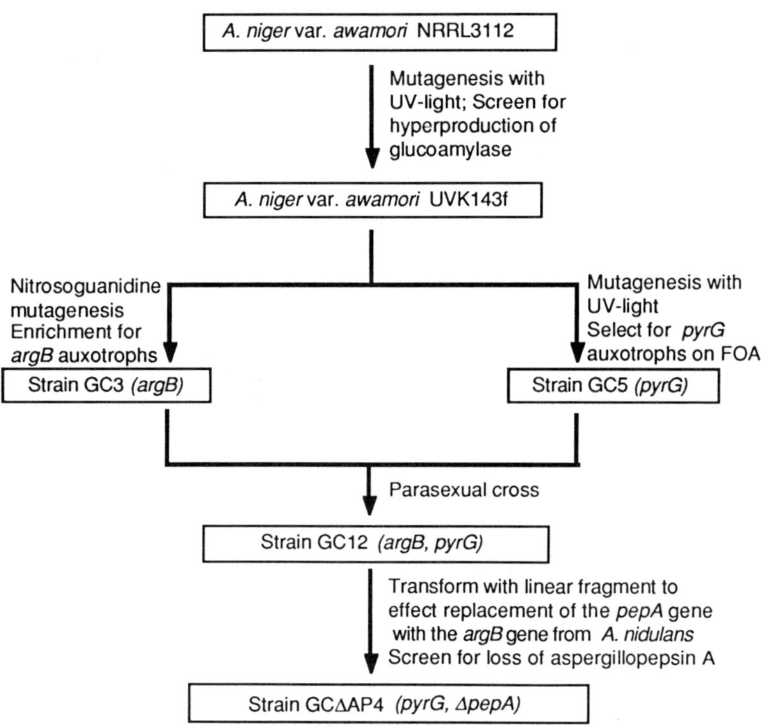

Figure 3. Strategy for deletion of the aspergillopepsin A (*pepA*) gene of *A. niger* var. *awamori*.

The gene encoding aspergillopepsin A (*pepA*) was cloned[29] using a synthetic oligonucleotide probe which corresponded to a portion of the published amino acid sequence[28]. Mutants lacking the *pepA* structural gene were derived[29] using the gene replacement strategy outlined in Figure 3. First, we constructed a plasmid in which a 2.4 kb fragment containing the entire *pepA* coding region was excised from a 9 kb genomic clone and replaced by a synthetic DNA polylinker. Secondly, the selectable *argB* gene from *A. nidulans* was cloned into the polylinker. Lastly, the DNA fragment which contained the *argB* marker, flanked by *pepA* sequences was excised from the plasmid and used to transform *A. niger* var. *awamori* strain GC12 (Fig. 3). A portion of the prototrophic transformants were found to have an aspergillopepsin-deficient phenotype when screened with an immunoassay using aspergillopepsin-specific antibodies. Southern blotting analyses confirmed that these mutants resulted from a gene replacement event at the *pepA* locus. A proposed mechanism for the generation of these aspergillopepsin-deficient (*ΔpepA*) strains is shown in Figure 4.

Although it was clear that the major aspartyl protease activity had been removed in our *ΔpepA* strains, a small amount of residual activity remained, suggesting the presence of one or more additional proteases. The residual proteolytic activity was not inhibited by pepstatin, but was partially inhibited by DAN and completely inhibited at neutral or alkaline pH values. In this regard, a pepstatin-insensitive carboxyl protease has been purified from culture filtrates of *A. niger* var. *macrosporus* and the corresponding gene has been cloned[30]. It is conceivable that *A. niger* var. *awamori* produces a similar enzyme and that the corresponding gene might be cloned by cross-hybridization with the gene from *A. niger* var. *macrosporus*. Subsequently, the gene could be deleted using a strategy similar to that which we have described.

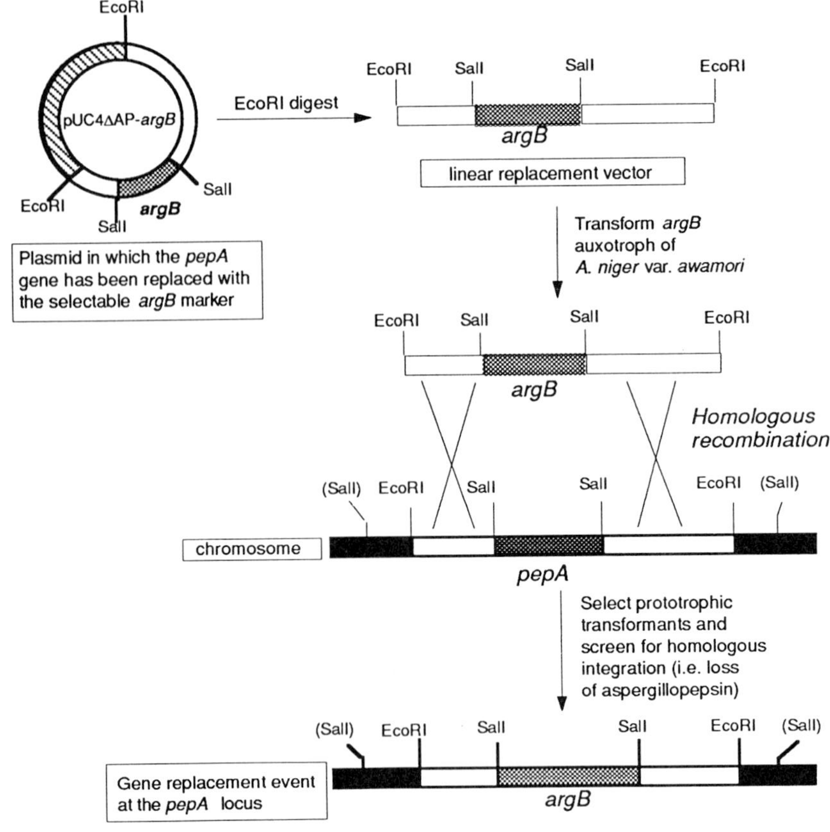

Figure 4. Proposed mechanism for the generation of aspergillopepsin-deficient (*ΔpepA*) strains of *A. niger* var. *awamori* by gene replacement.

Chymosin yields in *ΔpepA* strains were improved approximately two-fold over strains which expressed aspergillopepsin. It was noted that after seven days in a shake flask culture, the level of extracellular chymosin produced by *ΔpepA* strains was still increasing, whereas, the yield from a non-deleted strain was decreasing (not shown).

EFFICIENCY OF CHYMOSIN SECRETION

Experiments to estimate the efficiency of chymosin secretion in *A. niger* var. *awamori* were done by measuring the percentage of chymosin which remained cell-associated. The concentration of intracellular chymosin was measured by extracting the protein from fresh mycelia which were lyophilized and ground to a powder in a mortar and pestle. The powder was resuspended in a buffer which contained pepstatin and phenyl methyl sulfonyl fluoride (PMSF) as protease inhibitors to minimize degradation of chymosin in the sample. Extracts were further treated by adding sodium hydroxide then centrifuged to remove cellular debris. Control experiments showed that the sodium hydroxide treatment reduced the amount of chymosin detected by immunoassays by about 25%, but this treatment was necessary for solubilization of the chymosin from the mycelial extracts. The chymosin concentration was subsequently determined using an enzyme immunoassay. Table 5 shows the percentage of intracellular versus extracellular chymosin produced in pGRG1 and pGRG3 transformants. In a majority of the transformants more than half of the chymosin

Table 5. Comparison of Intracellular versus Extracellular Levels of Chymosin Produced by *A. niger* var. *awamori* pGRG1 and pGRG3 Transformants

	Chymosin Concentration (μg/ml)[a]		
Strain	Intracellular	Extracellular	% Intracellular
GC12[b]	N.D.	N.D.	----
12grg1-1	5.4	1.2	81.8
12grg1-1a	20.8	0.5	97.7
12grg1-3a	0.6	1.1	35.3
12grg1-4a	2.5	1.4	64.1
12grg1-5a	4.6	0.8	85.2
12grg3-3a	2.5	3.7	40.3
12grg3-5a	6.0	0.9	87.0
12grg3-6a	1.7	1.2	58.6

[a]Measured by enzyme immunoassay. [b]Strain GC12 is untransformed. N.D., not detectable.

that was synthesized remained cell-associated, suggesting that secretion of the heterologous chymosin gene product was not efficient.

IMPROVED SECRETION/STABILITY OF CHYMOSIN BY GLYCOSYLATION

We noted that most of the abundant extracellular proteins produced by *A. niger* var. *awamori* are glycoproteins. In contrast, only a small percentage of the chymosin produced in this host is glycosylated. Thus, we attempted to improve the secretion efficiency of chymosin by creating a site for N-linked glycosylation of the enzyme. Using techniques of site-directed mutagenesis, we generated two mutations ($Ser_{74} \rightarrow Asn_{74}$ and $His_{76} \rightarrow Ser_{76}$) in the chymosin coding sequence, thereby creating a consensus N-glycosylation site on the "flap" region of the chymosin molecule[31]. This location is homologous to one of two N-linked glycosylation sites found in the aspartyl protease of the Zygomycete fungus *Rhizomucor miehei*[32].

The expression vector used for these experiments was a derivative of pGRG3 called pUC*pyr*-GLYCHY which was constructed by inserting the mutant chymosin fragment, together with the *glaA* promoter and terminator, into plasmid pUC*pyr* [33]. As a control, a similar vector, pUC*pyr*GRG3, was constructed using the glucoamylase-chymosin expression unit without the mutations which introduce the glycosylation site. Transformants which contained pUC*pyr*-GLYCHY were grown in shake flask culture, and the chymosin polypeptides were analyzed by immunoblotting of culture filtrates. The *glycochymosin* produced by *A. niger* var. *awamori* was heterogeneous in size, migrating as a smear on the gel extending from about 37,000 MW to greater than 55,000 MW. Treatment of the samples with endoglycosidase H prior to electrophoresis removed the attached carbohydrate and gave a single band at the true molecular weight of chymosin (*i.e.*, 37,000 MW).

The yields of extracellular chymosin among transformants derived from this vector increased approximately ten-fold compared to pUC*pyr*-GRG3 transformants, and in most cases, greater than 90% of the enzyme was extracellular (Table 6). These observations suggest that glycochymosin was either more efficiently secreted by *A. niger* var. *awamori* or was more resistant to degradation than wild-type chymosin. We have not yet distinguished between these two possibilities. By comparing the amount

Table 6. Comparison of Intracelluar and Extracellular Levels of Chymosin and Glycochymosin Produced by *A. niger* var. *awamori* GC12

Transformant	Chymosin Type	Chymosin Concentration (μg/ml)[a]		%Secreted
		Intracellular	Extracellular	
12pPyrGRG3-3	Wild-type	0.50	3.30	92
12pPyrGRG3-4		1.39	0.81	36
12pPyrGRG3-5		2.72	4.26	61
12pPyrGRG3-8		0.46	0.31	40
12pPyrGRG3-11		0.49	2.10	81
12GLYCHY8	Glycochymosin	N.D.	14.17	>99
12GLYCHY9		1.66	9.02	84
12GLYCHY17		1.28	20.48	94
12GLYCHY20		1.99	15.32	88
12GLYCHY23		3.15	6.26	67
12GLYCHY25		2.40	33.55	93

[a]Chymosin concentration determined by immunoassay. N.D., not detected.

of chymosin detected in immunoassays with the levels obtained from enzyme activity assays, it was apparent that the specific activity of glycochymosin was reduced four- to five-fold compared to native chymosin, presumably because of the attached oligosaccharide. However, nearly all of the activity could be recovered by treatment with endoglycosidase H. Based on the location of the glycosylation site that is predicted from the known x-ray crystallographic structure of chymosin[31], it is possible that the oligosaccharide present on glycochymosin interferes with either substrate binding or catalytic function of the enzyme.

IMPROVING THE EXTRACELLULAR YIELDS OF GLYCOCHYMOSIN

In the yeast *Saccharomyces cerevisiae*, there is growing evidence that ATPases are involved in the protein secretion pathway. They may play a significant role in the translocation of peptides across membranes[34], the maintenance of cytoplasmic ion concentrations which control vesicular traffic[35], and they are possibly involved in secretion of proteins via a novel pathway which is independent of the classical secretory mechanism[36]. It was recently discovered that calcium ion ATPases located in the rough endoplasmic reticulum (ER) sequester calcium ions in the lumen of the ER[35]. Since all proteins which follow the classical secretory pathway must pass through the ER, where they may receive a core-oligosaccharide, it seems possible that perturbation of calcium ion transport in the ER may affect secretion and/or glycosylation of proteins. In fact, disturbance of the normal concentration of calcium ions can cause rapid secretion of resident ER proteins[37]. Rudolph *et al.*[38] reported that null mutations in the yeast *PMR1* gene, which encodes a P-type calcium ATPase of the ER, can result in a 5- to 50-fold enhancement in the secretion of heterologous proteins in *S. cerevisiae*. Furthermore, analysis of invertase produced by *PMR1* null mutants indicates that they are deficient in the addition of outer chain oligosaccharides (*i.e.*, only the core-oligosaccharide is added). Thus, we reasoned that if a function analogous to the ATPase encoded by the *PMR1* gene existed in *A. niger* var. *awamori*, mutations which vanquished its activity might induce increased yields and elicit less heavily glycosylated chymosin in pUC*pyr*-GLYCHY transformants.

Our first attempts to isolate ATPase mutants in *A. niger* var. *awamori* involved the use of sodium orthovanadate (vanadate) as the selective agent. Vanadate is a toxic analogue of phosphate. Since it acts on phosphorylated cellular enzymes such as sodium, potassium, and calcium ATPases[39, 40], thereby affecting membrane ion pumps, it is conceivable that vanadate might also affect post-translational processing, sorting and secretion of proteins. There are multiple mechanisms by which cells can become resistant to the effects of vanadate. Since entry of vanadate requires a cellular active transport process[41], one mechanism for resistance might involve a change in the transport system which either reduces uptake or enhances excretion of vanadate. Alternatively, cells may detoxify vanadate or the target enzymes may be altered by mutations which render them insensitive. Willsky *et al.*[39] have identified five complementation groups for vanadate resistant mutants of *S. cerevisiae*. Recent experiments by Ballou *et al.*[42] have also demonstrated the existence of five complementation groups in *S. cerevisiae*. Additionally, it was shown that mutants in all five complementation groups exhibited defects in protein glycosylation[42]. These glycosylation defects presumably reflect alterations in glycoprotein trafficking which are caused by perturbations in cellular ion pumps. Thus, we initiated a program to isolate ATPase mutants having a phenotype similar to the *PMR1*-null mutants described by Rudolph *et al.*[38] by screening among spontaneous mutants that were resistant to vanadate.

Vanadate resistant mutants of *A. niger* var. *awamori* were selected by plating germling mycelia on minimal medium agar plates which contained various concentrations of orthovanadate (1.5, 3 and 5 mM) but no added phosphate. Spontaneous mutants were isolated at a frequency of approximately 0.5×10^{-6}. These mutants were purified, and subsequently tested for their ability to produce glycochymosin in shake flask cultures. The results shown in Table 7 indicate that most vanadate resistant mutants produce less glycochymosin than the parental strain. However, a small number of mutants were identified which gave increased yields ranging from 120% to 200% higher than the parental strain (Table 7). Culture filtrates from the two highest chymosin-producing mutants were analyzed by immunoblotting. As shown in Figure 5, there was no apparent difference in glycosylation of glycochymosin in these strains, suggesting that neither has a phenotype similar to *PMR1* null mutants in *S. cerevisiae*. Nevertheless, it seems clear that by selecting for vanadate resistance, strains with improved productivity can be isolated. Whether these strains actually harbor mutations in ATPase genes remains to be demonstrated.

Another approach used to select ATPase mutants in *A. niger* var. *awamori* was to select for multiple drug resistance (*MDR*). Multiple drug resistance has been studied in both yeast and mammalian cells[34, 36, 43]. In both types of cells, a gene conferring the *MDR* phenotype (*STE6* in yeast and *MDR1* in mammalian cells) encodes a P-glycoprotein translocator which appears to be involved in secretion of proteins via a novel pathway which is independent of the classical secretory route. If amplified or overexpressed, the yeast *STE6* gene effects a five-fold increase in the rate of secretion of *a*-mating pheromone[34]. Although *a*-factor is a secreted polypeptide, it does not possess the features which are typical of secretory proteins. For example, it lacks a hydrophobic signal peptide at the amino terminus and it is secreted even when temperature-sensitive secretion deficient (*sec*) mutants are grown at the nonpermissive temperature[44]. Hence, *a*-factor is probably secreted by a non-classical mechanism. *STE6* shows striking homology to its mammalian homologue, *MDR1*, suggesting that a non-classical secretory route might also exist in other eukaryotes. Interestingly, it has

Table 7. Percent Increase in Glycochymosin Production by Vanadate Resistant Mutants of *A. niger* var. *awamori*

Vanadate conc.	Number of mutants identified[a]						
(mM)	<100%	100%	120%	140%	160%	180%	200%
1.5	81	10	3	5	6	1	1
3.0	33	18	8	6	1	0	0
5.0	111	99	4	3	3	2	0

[a]Indicates the number of strains identified (from a total of 395 vanadate resistant mutants) which produced either less chymosin or more than the parental strain (= 100%). Values were determined using a milk-clotting assay to measure enzyme activity.

been postulated that human interleukin-1β, which also lacks a signal sequence, is excreted via a novel secretion pathway[45].

In order to isolate *A. niger* var. *awamori* mutants possessing phenotypes similar to *STE6* or *MDR1*, we selected spontaneous mutants that were resistant to the aminoglycoside antibiotics hygromycin B and G418 (geneticin). It was our hope that these mutants would show enhanced secretion of glycochymosin. In addition, one would expect that diverting glycochymosin from the classical secretion mechanism to a novel pathway might involve bypassing the golgi apparatus, and therefore, alterations in glycosylation might be observed. Briefly, *MDR* mutants of *A. niger* var. *awamori*

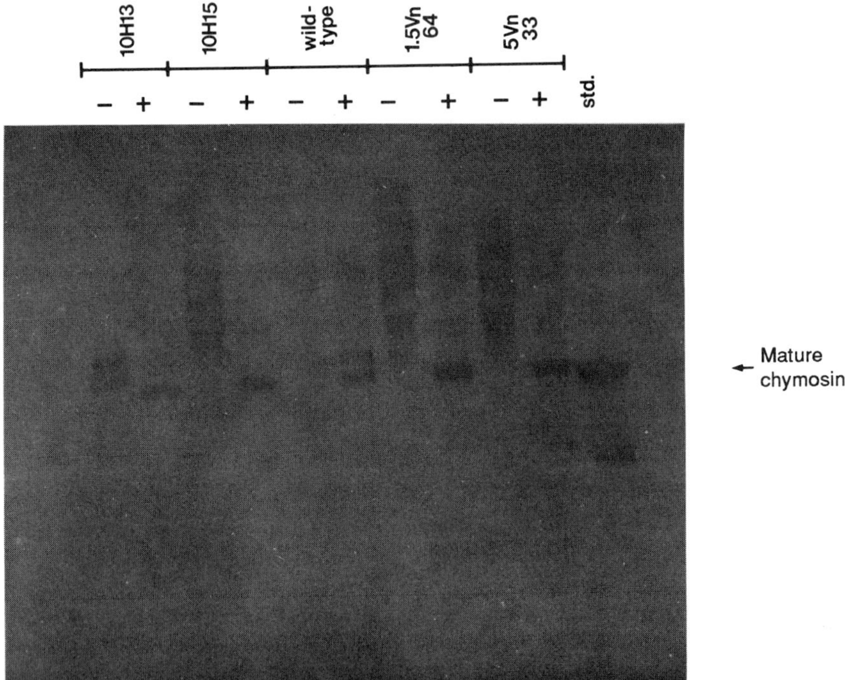

Figure 5. Immunoblotting analysis of glycochymosin polypeptides in culture filtrates of vanadate or hygromycin (*MDR*) mutants of *A. niger* var. *awamori*. Pretreatment of the filtrate proteins with endoglycosidase H to remove the attached N-linked carbohydrate is indicated by a + sign above the lanes. A − sign denotes an untreated sample. Strains 10H13 and 10H15 are hygromycin resistant (*i.e.*, *MDR*) mutants, whereas, strains 1.5Vn64 and 5Vn33 are vanadate resistant mutants. Culture filtrate from a non-mutant (*i.e.*, wild-type) strain and authentic bovine chymosin were included for comparison.

Table 8. Percent Increase in Glycochymosin Production by Multiply Drug Resistant (MDR) Mutants of *A. niger* var. *awamori*

Total number of mutants tested	Number of mutants producing various chymosin levels[a]								
	<100%	100%	120%	140%	160%	180%	200%	250%	350%
70	45	2	6	7	2	6	0	1	1

[a]Indicates the number of strains identified from a total of 140 *MDR* mutants which produce either less chymosin or more than the parental strain (= 100%). Values were determined using a milk-clotting assay to measure enzyme activity.

were selected by plating germling mycelia onto complete medium agar plates (CMA) supplemented with 10 µg/ml of hygromycin B. After six days of incubation, the resistant colonies were picked to fresh CMA plates which contained G418 at 10, 50, or 100 µg/ml to screen for the *MDR* phenotype. Once identified, the *MDR* mutants were grown in shake flask culture and the level of extracellular chymosin was compared to that of the parental strain (Table 8). As seen with the vanadate resistant mutants, most *MDR* mutants produced less glycochymosin than the parental strain. However, increases in chymosin yields by a few *MDR* strains appeared to be much greater than those seen with either wild-type or vanadate resistant mutants. One mutant, designated as 10H13, produced about 3.5-fold greater yields than the wild-type parent strain. Figure 6 shows that the rate of extracellular glycochymosin production in mutant mutant 10H13 was approximately three times greater than the parental (*i.e.*, non-*MDR*) strain.

Two *MDR* mutants which produced the highest levels of chymosin activity (strains 10H13 and 10H15) were analyzed by immunoblotting with chymosin-specific antibodies. As shown in Figure 6, mutant 10H13 secretes a glycochymosin polypeptide which is more homogeneous in size and has greater electrophoretic mobility than the corresponding polypeptides from 10H15 or wild-type strains. Inferred from this observation is that the 10H13-derived material is less extensively glycosylated than that which is produced by strain 10H15 or the parental strain. Treatment with endoglycosidase H increases the electrophoretic mobility of strain 10H13-derived glycochymosin further to that of an authentic bovine chymosin standard. Thus, it appears that mutant 10H13 is not completely deficient in its ability to glycosylate proteins, but it may be altered in some aspect of outer chain oligosaccharide addition. However, more experiments are needed to confirm this hypothesis.

Although the genetics and molecular mechanisms which could explain these results are unknown at the present time, it seems clear that selection of multiply drug resistant strains is a technique which can be successfully employed to develop strains which give enhanced secretion heterologous proteins. For successful commercialization of chymosin produced in *A. niger* var. *awamori*, it was highly desirable to produce an enzyme which was identical in all respects to the authentic bovine enzyme. Therefore, efforts to increase expression of glycochymosin were abandoned in favor of alternative strategies described below.

IMPROVED YIELDS FROM A *glaA*-PROCHYMOSIN GENE FUSION

In an attempt to obtain higher levels of chymosin without compromising the primary structure or activity of the enzyme, we constructed an expression vector called pGAMpR, which could direct the expression of a full-length glucoamylase-

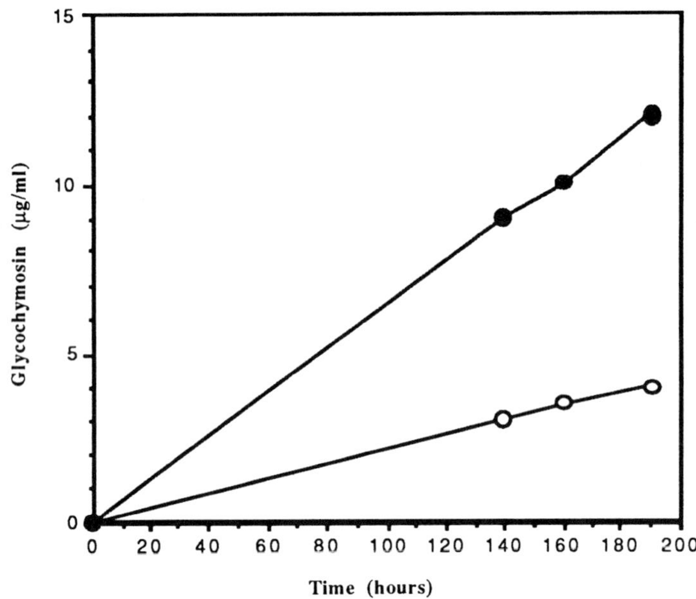

Figure 6. Comparison of glycochymosin production in a wild-type strain (o) and an *MDR* mutant strain (●) of *A. niger* var. *awamori* transformed with pUC*pyr*-GLYCHY. Chymosin concentration was measured by a milk-clotting activity assay. The specific activity of glycochymosin produced in these experiments is about 20% of native chymosin, hence, the values are lower than expected.

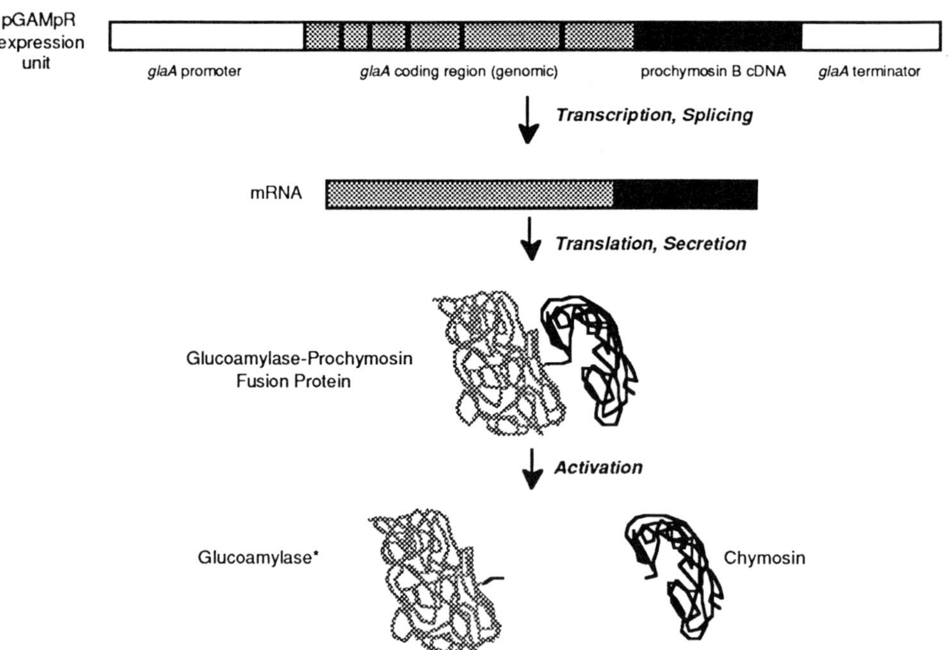

Figure 7. Schematic showing the expected transcription and translation products directed from the expression unit in pGAMpR. Introns in the *glaA* coding region of pGAMpR are indicated by black vertical lines. The final glucoamylase product resulting from proteolytic cleavage of the fusion protein contains a portion of the chymosin propeptide (denoted by glucoamylase*).

286

prochymosin fusion protein[46]. The rationale for this strategy presupposes that if there are sequences contained within the structure of glucoamylase mRNA and/or protein which promote efficient translation and/or secretion, they should also be present in a glucoamylase-prochymosin fusion to provide high levels of secreted product. Subsequently, chymosin may be released from the fusion protein by autocatalytic cleavage of its propeptide. This scheme is illustrated in Figure 7. Analysis of *A. niger* var. *awamori* transformants derived from pGAMpR revealed that active chymosin was produced extracellularly, and yields had improved dramatically. One transformant, designated as strain GC4-1, produced 286 mg per liter of active enzyme. Further analysis of this strain in a time course experiment showed that the glucoamylase-prochymosin fusion product appeared in the culture medium during the first 48 hours of incubation if the medium was strongly buffered at pH 6. At later times in the culture, or at lower pH values, the majority of the chymosin was in the mature active form. Processing of the glucoamylase-prochymosin fusion protein was inhibited by pepstatin, and thus, we suspect that release of chymosin from the fusion occurs via an autocatalytic cleavage mechanism.

IMPROVED STRAINS DERIVED BY MUTAGENESIS AND SELECTION

The chymosin-producing strain GC4-1 which contained pGAMpR was used as the starting point in a yield improvement program which involved six successive rounds of NTG mutagenesis with screening for the best producers after each round[47]. As shown in Table 9, the yield of chymosin increased more than two-fold as a result of this effort.

Table 9. Chymosin Production by Mutants Derived from a pGAMpR Transformant of *A. niger* var. *awamori*

Strain	Chymosin Concentration (mg/L)
Parental strain	
GC4-1	286[a]
NTG mutants	
GC1HA176	273[a]
GC1HB48	280[a]
GC1HC9	377[a]
GC1HD2	475[a]
GC1HE1	510[a]
GC1HF1	646[a]
Deoxyglucose resistant mutants	
GC1HF1-3;*dgr2*	1000[b]
GC1HF1-3;*dgr246*	1200[b]
Transformed with additional copies of pGAMpR	
GC1HF1-3;*dgr246/MC*	1300[b]
Cured then re-transformed with pGAMpR	
GC1HF1-3;dgr246/RT	1300[b]

Chymosin concentration measured by milk-clotting assay.
[a]Exact values obtained from a single experiment. [b]Approximate values obtained from a different experiment.

Additional improvements in the productivity of these strains was made by selecting for spontaneous mutants that were resistant to 2-deoxyglucose. The rationale for this approach is illustrated in the following observations. Fiedurek *et al.*[48] identified mutants of *A. niger*, selected on the basis of deoxyglucose resistance (*dgr*), which showed up to two-fold increases in glucoamylase over the parental strain. Although the molecular basis of deoxyglucose resistance in *A. niger* is unknown, work with *N. crassa* indicates that *dgr* mutants are deficient in glucose transport system I and partially or completely constitutive for the production of glucoamylase, invertase, and the elements of glucose transport system II[49]. Since chymosin production in pGAMpR transformants is dependent upon the ability of the cells to produce high levels of glucoamylase, it seemed reasonable to expect yield improvements from strategies which increased glucoamylase titers.

Deoxyglucose resistant mutants of GC1HF1 were isolated following NTG mutagenesis and selection on minimal medium agar plates which contained fructose as a carbon source and 2% deoxyglucose as the selective agent. Resistant mutants were isolated at a frequency of about 5.0×10^{-6}. By screening 100 resistant mutants for their ability to secrete chymosin, two mutants were identified which showed an approximate two-fold enhancement in yield (Table 9). The chymosin titer from these strains was more than one gram per liter. This represents a total improvement of approximately 80-fold compared to the best pGRG3 transformant (strain 107, see Table 3). Introduction of additional copies of the pGAMpR expression unit on a vector which contained the selectable *amdS* marker had little affect on chymosin productivity (Table 9). Thus, it is possible that barrier(s) to further enhancements are post-transcriptional, possibly involving translation, secretion or degradation.

The mutations introduced during the course of our yield improvement program appear to be host mutations rather than in the integrated vector sequences. Strain GC1HF1-3;*dgr246* was "cured" of its integrated pGAMpR sequences by plating on fluoro orotic acid, which selects for loss of the *pyr4* gene function on the vector (Southern blots confirmed that the integrated sequences had been excised). The resulting strain was subsequently re-transformed with pGAMpR, and transformants were tested for their ability to produce active chymosin. Strains were identified which produced levels equal to those of the original GC1HF1-3;*dgr246*.

CONCLUSIONS

From our experiments it can be concluded that secretion is a major limiting step in the production of heterologous proteins in *A. niger* var. *awamori* despite the fact that copious amounts of endogenous enzymes can be produced in this system. We have developed methods for improving the extracellular yield of a foreign protein, bovine chymosin, which include the use of classical mutagenesis and screening, selection for mutations which confer resistance to toxic metabolites, and molecular biology approaches (*e.g.*, site-directed mutagenesis to introduce a glycosylation site, or construction of glucoamylase-prochymosin gene fusion). It appears likely that the improvements to chymosin production obtained as a result of expressing either glycosylated chymosin or a glucoamylase-prochymosin fusion were the result of an increase in secretion efficiency. Interestingly, a gene fusion was also employed to improve the secretion efficiency of interleukin-6 in *A. nidulans*[15]. Strain improvements made by random mutagenesis or by selection on toxic metabolites apparently result from mutations in the host cell genome. Thus, the strains that are derived from our mutagenesis and selection programs may be useful for the production of other heterologous proteins in *A. niger* var. *awamori*.

REFERENCES

1. V. B. Pedersen, K. A. Christensen, and B. Foltmann, Investigations on the activation of prochymosin, *Eur. J. Biochem.* 94: 573-580 (1979).
2. B. Foltmann, V. B. Pedersen, H. Jacobsen, D. Kauffman, and G Wybrandt, The complete amino acid sequence of prochymosin, *Proc. Nat. Acad. Sci. USA* 74: 2321-2324 (1977).
3. B. Foltmann, V. B. Pedersen, D. Kauffman, and G. Wybrandt, The primary structure of calf chymosin, *J. Biol. Chem.* 254: 8447-8456 (1979).
4. D. Cullen, G. L. Gray, L. J. Wilson, K.J. Hayenga, M. H. Lamsa, M. W. Rey, S. Norton, and R. M. Berka, Controlled expression and secretion of bovine chymosin in *Aspergillus nidulans, Bio/Technol.* 5: 369-376 (1987).
5. M. Ward, Production of calf chymosin by *Aspergillus awamori, in*: "Genetics and Molecular Biology of Industrial Microorganisms," C. L. Hershberger, S. W. Queener, and G. Hegeman, eds., American Society for Microbiology, Washington, DC (1989).
6. A. Harkki, J. Uusitalo, M. Bailey, M. Penttilä, and J. K. C. Knowles, A novel fungal expression system: secretion of active calf chymosin from the filamentous fungus *Trichoderma reesei, Bio/Technol.* 7: 596-603 (1989).
7. E. Boel, T. Christensen, and H. F. Wöldike, Process for the production of protein products in *Aspergillus oryzae* and a promoter for use in *Aspergillus*, European Patent Application 0 238 023 (1987).
8. A. Upshall, A. A. Kumar, M. C. Bailey, M. D. Parker, M. A. Favreau, K. P. Lewison, M. L. Joseph, J. M. Maraganore, and G. L. McKnight, Secretion of active human tissue plasminogen activator from the filamentous fungus *Aspergillus nidulans, Bio/Technol.* 5: 1301-1304 (1987).
9. R. W. Davies, Molecular biology of a high level recombinant protein production system in *Aspergillus*, in: "Molecular Industrial Mycology. Systems and applications for filamentous fungi," S. A. Leong and R. M. Berka, eds., Marcel Dekker, Inc., New York, NY (1991).
10. D. I. Gwynne, F. P. Buxton, S. A. Williams, S. Garven, and R. W. Davies, Genetically engineered secretion of active human interferon and a bacterial endoglucanase from *Aspergillus nidulans, Bio/Technol.* 5: 713-719 (1987).
11. W. -C. Leung, G. Z. Jing, and M. F. K. Leung, Expression and secretion of human interferon gamma in filamentous fungus *Achlya ambisexualis*, Abstracts of the 19th Lunteran Conference on Molecular Genetics, F24(b) (1987).
12. W. -C. Leung, Characterization of herpes simplex virus thymidine kinase activity synthesized in recombinant filamentous fungus *Achlya ambisexualis*, Abstracts of the 19th Lunteran Conference on Molecular Genetics, F24(d) (1987).
13. D. B. Archer, D. J. Jeenes, D. A. MacKenzie, G. Brightwell, N. Lambert, G. Lowe, S. E. Radford, and C. M. Dobson, Hen egg white lysozyme expressed in and secreted from *Aspergillus niger* is correctly processed and folded, *Bio/Technol.* 8: 741-745 (1990).
14. I. F. Turnbull, K. Rand, N. S. Willetts, and M. J. Hynes, Expression of the *Escherichia coli* enterotoxin subunit B gene in *Aspergillus nidulans* directed by the *amdS* promoter, *Bio/Technol.* 7: 169-174 (1989).

15. R. Contreras, D. Carrez, J. R. Kinghorn, C. A. M. J. J. van den Hondel, and W. Fiers, Efficient KEX2-like processing of a glucoamylase-interleukin-6 fusion protein by *Aspergillus nidulans* and secretion of mature interleukin-6, *Bio/Technol.* 9: 378-381 (1991).

16. B. Foltmann, Gastric proteinases. Structure, function, evolution, and mechanism of action, *Essays Biochem.* 17: 53-84 (1981).

17. K. Nishimori, N. Shimizu, Y Kawazuchi, M. Hidaka, T. Uozumi, and T. Beppu, Expression of cloned calf prochymosin gene sequences in *Escherichia coli, Gene* 19: 337-344 (1982).

18. J. S. Emtage, S. Angal, M. T. Doel, T. J. R. Harris, B. Jenkins, G. Lilley, and P. A. Lowe, Synthesis of calf prochymosin (prorennin) in *Escherichia coli*, Proc. Nat. Acad. Sci. USA 80: 3671-3675 (1983).

19. J. Mellor, M. J. Dobson, N. A. Roberts, M. F. Tuite, J. S. Emtage, S. White, P. A. Lowe, T. Patel, A. J. Kingsman, and S. M. Kingsman, Efficient synthesis of enzymatically active calf chymosin in *Saccharomyces cerevisiae, Gene* 24: 1-14 (1983).

20. D. T. Moir, J. Mao, M. J. Duncan, R. A. Smith, and T. Kohno, Production of calf chymosin by the yeast *Saccharomyces cerevisiae, in*: "Developments in Industrial Microbiology," Vol. 26, L. Underkofler, ed., Society for Industrial Microbiology, Arlington, VA (1985).

21. C. G. Goff, D. T. Moir, T. Kohno, T. C. Gravius, R. A. Smith, E. Yamasaki, and A. Taunton-Rigby, Expression of calf prochymosin in *Saccharomyces cerevisiae, Gene* 27: 35-46 (1984).

22. J. A. van den Berg, K. J. van der Laken, A. J. J. van Ooyen, T. C. H. M. Renniers, K. Rietveld, A. Schaap, A. J. Brake, R. J. Bishop, K. Schultz, D. Moyer, M. Richman, and J. R. Schuster, *Kluyveromyces* as a host for heterologous gene expression: expression and secretion of prochymosin, *Bio/Technol.* 8: 135-139 (1990).

23. D. J. Ballance, F. P. Buxton, and G. Turner, Transformation of *Aspergillus nidulans* by the orotidine-5'-phosphate decarboxylase gene of *Neurospora crassa, Biochem. Biophys. Res. Commun.* 112: 284-289 (1983).

24. D. J. Ballance and G. Turner, Development of a high-frequency transforming vector for *Aspergillus nidulans, Gene* 36: 321-331 (1985).

25. E. Boel, M. T. Hansen, I. Hjort, I. Høegh, and N. P. Fiil, Two different types of intervening sequences in the glucoamylase gene from *Aspergillus niger, EMBO J.* 3: 1581-1585 (1984).

26. J. H. Nunberg, J. H. Meade, G. Cole, F. C. Lawyer, P. McCabe, V. Schweickart, R. Tal, V. P. Wittman, J. E. Flatgaard, and M. A. Innis, Molecular cloning and characterization of the glucoamylase gene of *Aspergillus awamori, Mol. Cell. Biol.* 4: 2306-2315 (1984).

27. M. Lamsa and P. Bloebaum, Mutation and screening to increase chymosin yield in a genetically-engineered strain of *Aspergillus awamori, J. Industr. Microbiol.* 5: 229-238 (1990).

28. V. I. Ostoslavskaya, L. P. Revina, E. K. Kotlova, I. A. Surova, E. D. Levin, E. A. Timokhina, and V. M. Stepanov, Primary structure of aspergillopepsin A – an aspartyl proteinase from *Aspergillus awamori, Bioorg. Khim.* 8: 1030-1047 (1986).

29. R. M. Berka, M. Ward, L. J. Wilson, K. J. Hayenga, K. H. Kodama, L. P. Carlomagno, and S. A. Thompson, Molecular cloning and deletion of the gene encoding aspergillopepsin A from *Aspergillus awamori, Gene* 86: 153-162 (1990).

30. K. Takahashi, M. Tanokura, H. Inoue, M. Kojima, Y. Muto, M. Yamasaki, O. Makabe, T. Kimura, T. Takizawa, T. Hamaya, E. Suzuki, and H. Miyano, Structure and function of a pepstatin-insensitive acid proteinase from *Aspergillus niger* var. *macrosporus*, Abstracts of the Aspartic Proteinase Conference, Sonoma, CA (1990).

31. G. L. Gilliland, E. L. Winborne, J. Nachman, and A. Wlodawer, The three-dimensional structure of recombinant bovine chymosin at 2.3 Å resolution, *Proteins* 8: 82-101 (1990).

32. A. M. Bech and B. Foltmann, Partial primary structure of *Mucor miehei* protease, *Neth. Milk Dairy J.* 35: 275-280 (1981).

33. M. W. Rey, R. M. Berka, L. J. Wilson, and M. Ward, Cloning vectors for filamentous fungi: construction of a novel pUC19-derivative containing the *Neurospora crassa pyr4* gene, Abstracts of the 14th Fungal Biology Conference, Asilomar, (1987).

34. K. Kuchler, R. E. Sterne, and J. Thorner, *Saccharomyces cerevisiae STE6* gene product: a novel pathway for protein export in eukaryotic cells, *EMBO J.* 8: 3973-3984 (1989).

35. J. F. Sambrook, The involvement of calcium in transport of secretory proteins from the endoplasmic reticulum, *Cell* 61: 197-199 (1990).

36. J. P. McGrath and A. Varshavsky, The yeast *STE6* gene encodes a homologue of the mammalian multidrug resistance P-glycoprotein, *Nature* 340: 400-404 (1989).

37. C. Booth and G. L. E. Koch, Perturbation of cellular calcium induces secretion of luminal ER proteins, *Cell* 59: 729-737 (1989).

38. H. K. Rudolph, A. Antebi, G. R. Fink, C. M. Buckley, T. E. Dorman, J. LeVitre, L. S. Davidow, J. Mao, and D. T. Moir, The secretory pathway is perturbed by mutations in *PMR1*, a member of a Ca^{2+}-ATPase family, *Cell* 58: 133-145 (1989).

39. G. R. Willsky, J. O. Leung, P. V. Offermann, Jr., E. K. Plotnick, and S. F. Dosch, Isolation and characterization of vanadate-resistant mutants of *Saccharomyces cerevisiae*, *J. Bacteriol.* 164: 611-617 (1985).

40. C. Kanik-Ennulat and N. Neff, Vanadate-resistant mutants of *Saccharomyces cerevisiae* show alterations in protein phosphorylation and growth control, *Mol. Cell. Biol.* 10: 898-909 (1990).

41. S. M. Penningroth, Erythro-9-[3-(2-hydroxynonyl)]adenine and vanadate as probes for microtubule-based cytoskeletal mechanochemistry, *Methods Enzymol.* 134: 477-487 (1986).

42. L. Ballou, R. A. Hitzeman, M. S. Lewis, and C. E. Ballou, Vanadate-resistant yeast mutants are defective in protein glycosylation, *Proc. Nat. Acad. Sci. USA* 88: 3209-3213 (1991).

43. G. Bradley, P. F. Juranka, and V. Ling, Mechanism of multidrug resistance, *Biochim. Biophys Acta* 948: 87-128 (1988).

44. A. J. Brake, C. Brenner, R. Najariam, P. Laybourn, and J. Merryweather, *in:* "Protein Transport and Secretion," M. J. Gething, ed., Cold Spring Harbor Laboratory Press, Cold Spring Harbor, NY (1985).

45. A. Rubartelli, F. Cozzolino, M. Talio, and R. Sitia, A novel secretory pathway for interleukin-1β, a protein lacking a signal sequence, *EMBO J.* 9: 1503-1510 (1990).

46. M. Ward, L. J. Wilson, K. H. Kodama, M. W. Rey, and R. M. Berka, Improved production of chymosin in *Aspergillus* by expression as a glucoamylase-chymosin fusion, *Bio/Technol.* 8: 435-440 (1990).

47. N. S. Dunn-Coleman, P. Bloebaum, R. M. Berka, E. Bodie, N. Robinson, G. Armstrong, M. Ward, M. Przetak, G. L. Carter, R. LaCost, L. J. Wilson, K. H.

Kodama, E. F. Baliu, B. Bower, M. Lamsa, and H. Heinsohn, Commercially viable levels of chymosin production by *Aspergillus*, submitted for publication.

48. J. Fiedurek, A. Paszcznski, G. Ginalska, and Z. Ilczuk, Selection of amylolytically active *Aspergillus niger* mutants to 2-deoxy-D-glucose, *Zentralbl. Mikrobiol.* 142: 407-412 (1987).

49. K. E. Allen, M. T. McNally, H. S. Lowendorf, C. W. Slayman, and S. J. Free, Deoxyglucose-resistant mutants of *Neurospora crassa*: isolation, mapping, and biochemical characterization, *J. Bacteriol.* 171: 53-58 (1989).

POXVIRUS VECTORS: MAMMALIAN CYTOPLASMIC-BASED EXPRESSION SYSTEMS

Bernard Moss

Laboratory of Viral Diseases
National Institute of Allergy and Infectious Diseases
National Institutes of Health
Bethesda, Maryland 20892

INTRODUCTION

The unique biological characteristics of poxviruses have been
exploited for the development of vaccinia virus into a powerful and
versatile expression vector with applicability to many areas of research
and biotechnology (Moss, 1991). The poxviruses comprise a group of
genetically related DNA viruses that are unusual in their ability to
propagate in the cytoplasm, rather than in the nucleus, of infected cells.
Electron microscopic and autoradiographic examinations have revealed
cytoplasmic factories of viral DNA replication and particle assembly. The
enzymes and factors required for transcription and replication are encoded
within the poxvirus genome, which may contain 200 or more genes. The
previous use of vaccinia virus as a live vaccine for the eradication of
smallpox, and the potential for recombinant vaccinia viruses containing
genes from other microorganisms to serve as live vaccines against current
diseases, contributed to the interest in this vector. For these reasons,
recombinant vaccinia viruses have been extensively used by immunologists
and virologists to determine the targets of humoral and cell mediated
immune responses to microbial infections and candidate human and veterinary
live recombinant vaccines are currently being tested. Poxviruses encode
their own DNA-dependent RNA polymerase and high levels of expression of
foreign genes have been obtained by optimizing poxvirus promoter function.
Thus, recombinant vaccinia viruses are particularly useful for protein
production in cultured mammalian cells. A modified system has been made by
importing the bacteriophage T7 RNA polymerase gene into vaccinia virus and
using bacteriophage promoters for high level expression. With both systems,
the recombinant proteins appear to be properly processed and transported
within the mammalian cell.

MOLECULAR BIOLOGY OF VACCINIA VIRUS

Vaccinia virus is the best studied member of the poxvirus family (Moss,
1990a). The infectious particles, called virions, are brick-shaped and
approximately 300 nm in diameter. A lipoprotein membrane surrounds the
core structure which contains a linear 200,000 base pair DNA molecule with
hairpin loops connecting the two strands at each end. Also within the core
are virus-encoded enzymes, including a multisubunit DNA-dependent RNA
polymerase, a transcription factor, capping and methylating enzymes, and a
poly(A) polymerase, for the production of mRNA. Virus attachment to the
cell is followed by the fusion of viral and cellular membranes leading to
the entry of the core into the cytoplasm. The viral RNA polymerase

Applications of Enzyme Biotechnology, Edited by J.W. Kelly and
T.O. Baldwin, Plenum Press, New York, 1991

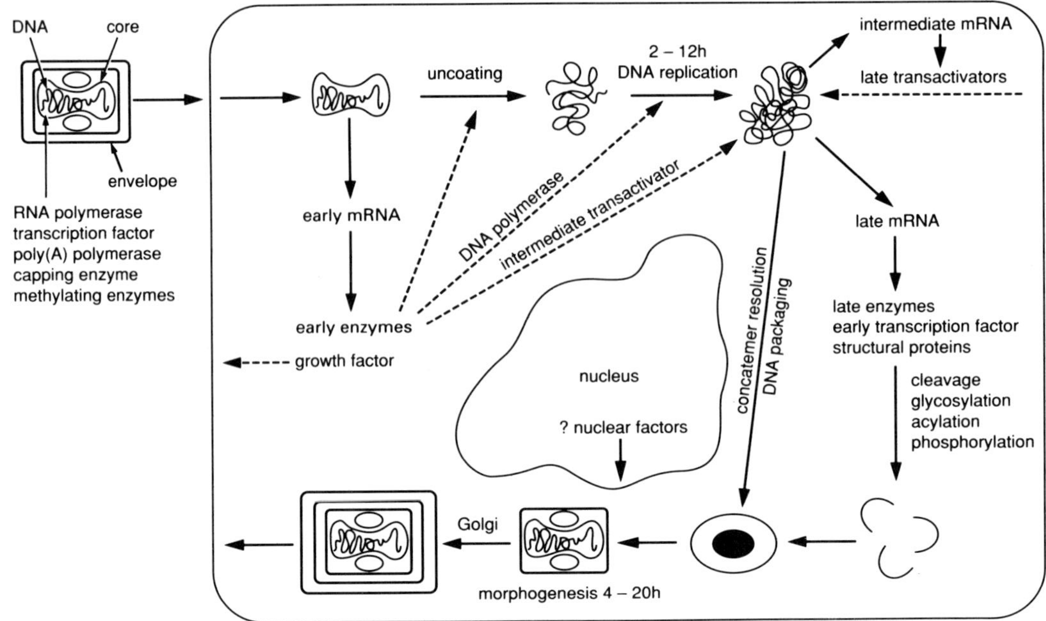

Fig.1. Vaccinia virus replication cycle. Reproduced from Moss, 1991.

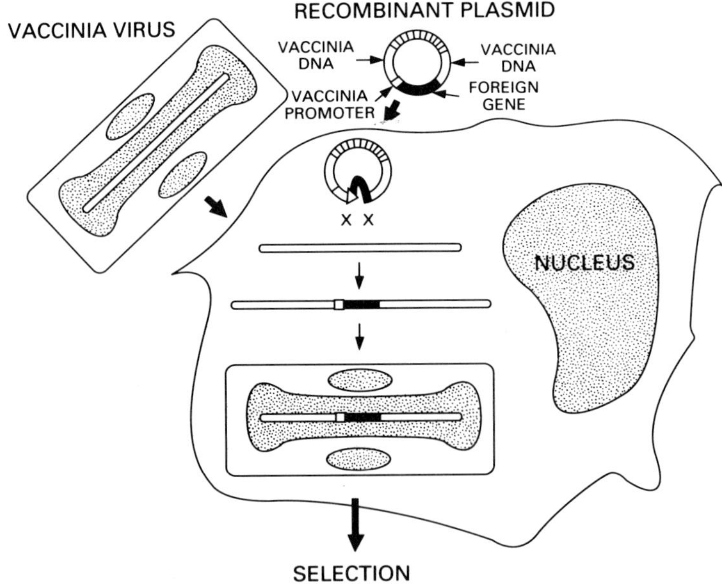

Fig.2. Formation of recombinant vaccinia virus. Reproduced from Moss, 1991.

transcribes the early genes,which number about 100 and have a variety of
functions leading to the replication of the viral genome (Fig. 1). The
replicated DNA molecules provide templates for the expression of
intermediate and late genes which have their own characteristic
transcriptional promoter sequences. The sequential synthesis of specific
viral proteins that recognize the different classes of promoters results in
a programmed cascade mechanism of gene regulation (Moss, 1990b; Moss et
al., 1991). The DNA and structural proteins are assembled into virus
particles and some are enveloped by an additional membrane and released
from the cell.

CONSTRUCTION OF RECOMBINANT VIRUSES

     The large size of the vaccinia virus genome has made homologous DNA
recombination a practical way of inserting new genes (Mackett et al., 1982;
Panicali and Paoletti, 1982). As a first step in the insertion process,
standard "cutting and ligating" methods of manipulating DNA are used to
flank a foreign gene with vaccinia virus DNA sequences and the resulting
plasmid is transfected into a cell that has been infected with vaccinia
virus (Fig. 2). Recombination, between homologous genomic and plasmid DNA
sequences, leads to the insertion of the foreign DNA. In this way 25,000 or
more base pairs of DNA can be added to the vacinia virus genome, without
interfering with its packaging (Smith and Moss, 1983). By directing the
insertion into a non-essential region of the genome, the recombinant virus
will still be capable of independent replication.

     Expression of the foreign gene is dependent on poxvirus
transcriptional regulatory signals; hence it is desirable to place a strong
vaccinia virus promoter adjacent to the foreign gene. Expression directed
by early promoters occurs during the first few hours but then usually
diminishes, whereas expression directed by intermediate or late promoters
occurs after the onset of DNA replication. The greatest amounts of protein
have been obtained with strong natural (Falkner and Moss, 1988; Patel et
al., 1988) or synthetic (Davison and Moss, 1990) late promoters.

     Some special considerations must be observed with regard to the
properties of the poxvirus transcription system. For instance, cDNA copies
of mRNAs or genes without introns must be used since splicing does not
occur in this system. With early promoters, it is advisable to alter any
TTTTTNT sequences that may be present in the non-coding strand of the gene
to be inserted (Earl et al., 1990) since transcriptional termination occurs
when that sequence is present in early (but not late) vaccinia virus genes
(Shuman and Moss, 1988; Yuen and Moss, 1987).

     A variety of available plasmid insertion vectors were made to
simplify the construction and isolation of recombinant viruses (Chakrabarti
et al., 1985; Davison and Moss, 1990; Falkner et al., 1987; Mackett et al.,
1984). These vectors contain a segment of vaccinia virus DNA with an
expression cassette consisting of a vaccinia virus promoter and unique
restriction endonuclease sites for insertion of the foreign gene.
Recombinants comprise approximately 0.1% of the progeny virus making it
possible to screen virus plaques by DNA hybridization or for expression of
the desired foreign gene product. However, since such screening is not
convenient, several different general selection methods have been devised.
These include: selection for the thymidine kinase (TK) negative phenotype
in TK deficient cells (Mackett et al., 1984), antibiotic selection (Boyle
and Coupar, 1988; Falkner and Moss, 1988; Franke et al., 1985; Isaacs et
al., 1990) and screening for β-galactosidase expression (Chakrabarti et
al., 1985; Panicali et al., 1986), host range (Perkus et al., 1989) or
plaque size (Rodriguez and Esteban, 1989). Some selection and screening
procedures, such as TK⁻ selection and β-galactosidase expression, can be
combined.

HYBRID T7 BACTERIOPHAGE/VACCINIA VIRUS EXPRESSION SYSTEM

The high efficiency and stringent promoter specificity of the single subunit DNA-dependent RNA polymerase encoded by T7 bacteriophage has been extremely useful for prokaryotic expression vectors (Studier et al., 1990). To adapt this technology to eukaryotic systems, we have placed the T7 RNA polymerase gene under the control of a vaccinia virus promoter and inserted it into the vaccinia virus genome (Fuerst et al., 1986). When mammalian cells are infected with this recombinant virus, functional T7 RNA polymerase is made. For analytical purposes, a plasmid containing a target gene with a T7 promoter can be transfected into the cell. For larger scale production, the T7 promoter regulated gene can be inserted into the genome of a second vaccinia virus so that expression occurs when cells are doubly infected (Fuerst et al., 1987).

The hybrid expression system was shown to make large amounts of RNA, but since only a small percentage was capped at the 5' end, translation was inefficient (Fuerst et al., 1989). To overcome this problem, cap-independence was conferred by incorporating a cDNA of the encephalomyocarditis virus untranslated leader sequence downstream of the T7 promoter (Elroy-Stein et al., 1989). With this modification, the recombinant chloramphenicol acetyltransferase protein made up more than 10% of the total cell protein in 24 hours.

The use of two separate recombinant viruses, one to produce T7 RNA polymerase and the other to provide the T7 promoter regulated foreign gene template, prevents expression until the cells are coinfected. This is advantageous when expressing proteins that are toxic or that would diminish the yield of virus when growing stocks. Under some circumstances, it would be simpler to have the T7 RNA polymerase gene and the T7 promoter-regulated gene in the same vaccinia virus genome. However, initial attempts to construct such a double recombinant virus failed apparently because they were not viable (Fuerst et al., 1987). Recent studies suggest that this problem can be circumvented by using the *Escherichia coli lac* operator-repressor system to limit expression of the T7 RNA polymerase gene until addition of inducer (Alexander et al., 1991). As an alternative solution, a cell line was stably transfected with the bacteriophage T7 RNA polymerase gene (Elroy-Stein and Moss, 1990). Expression was then obtained when this cell line was infected with a single recombinant vaccinia virus containing the bacteriophage promoter-regulated foreign gene.

VECTOR APPLICATIONS

Characterization of proteins

Vaccinia virus replicates in many mammalian and avian cell lines and can synthesize proteins in some amphibian cells (Massung and Moyer, 1990) providing a wide choice for expression studies. Fortunately, recombinant proteins retain their ability to be posttranslationally modified and N- and O-glycosylation, myristylation, phosphorylation, and specific proteolytic cleavage all have been demonstrated (Moss and Flexner, 1987). In addition, polarized membrane and nuclear transport and secretion occur depending on the cell line used.

For analytical purposes, transient infection protocols are convenient because there is no need to construct recombinant viruses. Thus, the gene of interest can be cloned into a plasmid containing a late vaccinia promoter and then used to transfect cells that have been infected with ordinary vaccinia virus (Cochran et al., 1985). Similarly, genes cloned into plasmids with bacteriophage T7 promoters can be expressed when transfected into cells that have been infected with a recombinant vaccinia virus containing the T7 RNA polymerase gene. As mentioned earlier, use of the EMCV untranslated leader region provides maximum expression, but the amounts of protein made with genes cloned into plasmids with T7 promoters (designed primarily for in vitro transcription) may be adequate. CaPO4 (Fuerst et al., 1986) and cationic lipid (Whitt et al., 1989) transfection methods work well, with the latter resulting in 90% of the cells expressing the gene (Whitt et al., 1991). Although cell lines that stably express T7 RNA polymerase can be used for transient expression, vaccinia virus infection is still needed for reasonable efficiency (Elroy-Stein and Moss, 1990). There have been many applications of the vaccinia/T7 transfection expression system. These include analysis of: CD4 binding to the HIV envelope glycoprotein (Buonocore and Rose, 1990; Mizukami et al., 1988), rescue of conditionally lethal virus mutants (Li et al., 1988), epitope mapping of antibodies (Keil and Wagner, 1989), virus-induced cell fusion (Ashorn et al., 1990), virion assembly (Whitt et al., 1989), expression of voltage gated ion channels (Yang et al., 1991) and the cystic fibrosis transmembrane conductance regulator (Rich et al., 1990). Expression can be obtained from several plasmids transfected simultaneously (Pattnaik and Wertz, 1991). In addition, expression has been obtained following injection of recombinant virus and plasmids containing T7 promoter regulated genes into amphibian oocytes (Yang et al., 1991).

For large scale work, it is necessary to construct recombinant viruses containing the foreign gene. With present vectors, similar amounts of protein are made using a strong vaccinia promoter or a T7 promoter for gene expression. If the recombinant protein is inhibitory to virus replication or toxic to cells, then the bacteriophage T7/vaccinia hybrid system is best. Typically, the vaccinia expression system has been used for relatively small scale laboratory research investigations. In one report, however, Vero cell microcarrier cultures of 40 l were used to synthesize the HIV-1 envelope protein using the T7/vaccinia hybrid system (without the EMCV untranslated leader) and a yield of 11.2 mg/liter was obtained (Barrett et al., 1989).

Immunological studies

Vaccinia virus vectors have been used for a variety of immunological studies (Moss, 1991). The ability of recombinant vaccinia viruses to infect cells or animals and produce proteins with native conformation has led to their use as a tool for identifying both the type and target of protective immune responses. Thus, a set of recombinant viruses expressing individual proteins of a complex pathogen can be constructed. After infection of animals, the antibody and cell mediated immune responses can be measured and the ability of the vaccinated animal to resist disease determined. The list of viruses for which protection has been demonstrated includes: murine cytomegalovirus, Epstein-Barr, Friend leukemia, hepatitis B, herpes simplex, influenza, Japanese encephalitis, lymphocytic choriomeningitis virus, measles, parainfluenza, pseudorabies, rabies, respiratory syncytial, rinderpest, Venezuelan equine encephalitis, vesicular stomatitis. In the majority of cases, immunity was correlated with antibody but in some cases it appeared to be due to cell mediated immunity.

There have been examples of the use of recombinant vaccinia virus for experimental tumor prevention and therapy. Recombinants that express the envelope glycoprotein gene of Friend leukemia virus, the T antigen genes of polyoma virus, the rat *neu* oncogene and human melanoma-associated antigen gene provided protection that was probably due to cell mediated immunity.

LIVE RECOMBINANT VACCINES

The characteristics of vaccinia virus that contributed to its effectiveness as a smallpox vaccine have suggested its use as a recombinant vaccine against other diseases. Nevertheless, considerably more work needs to be done to improve safety and efficacy before such vaccines will be widely used in humans. A recombinant vaccinia virus that expresses the rabies virus glycoprotein is currently being field tested as an oral bait vaccine for raccoons and foxes (Blancou et al., 1986; Rupprecht et al., 1986; Wiktor et al., 1984). Initial results with a recombinant vaccinia

virus that expresses the envelope glycoprotein genes of rinderpest, an
endemic cattle disease in Africa, also seems promising (Yilma et al.,
1988). A human phase I trial for safety and immunological response to a
live recombinant vaccinia virus vaccine expressing HIV-1 envelope
glycoprotein is in progress (Cooney et al., 1991). In addition, a phase 1
trial was recently started with a subunit vaccine made from purified HIV-1
envelope (gp160) protein made with a bacteriophage T7/vaccinia virus
expression vector (Barrett et al., 1989).

CONCLUSIONS

Recombinant vaccinia virus vectors provide a novel way of expressing
proteins in the cytoplasm of eukaryotic cells derived from mammals, birds,
and amphibians. Their usefulness as a tool for laboratory investigations
has been well established. Further studies will determine the practicality
of vaccinia vectors for commercial production of proteins and as live
recombinant vaccines for medical and veterinary purposes.

REFERENCES

Alexander, W. A., Moss, B. and Fuerst, T. R., 1991. Novel system for
        regulated high level expression in vaccinia virus. Vaccines 90:221.
Ashorn, P., Berger, E. A. and Moss, B., 1990. Human immunodeficiency virus
        envelope glycoprotein/CD4-mediated fusion of non-primate cells with
        human cells. J. Virol. 64:2149.
Barrett, N., Mitterer, A., Mundt, W., Eibl, J., Eibl, M., Gallo, R. C.,
        Moss, B. and Dorner, F., 1989. Large-scale production and
        purification of a vaccinia recombinant-derived HIV-1 gp160 and
        analysis of its immunogenicity. AIDS Res. Human Retroviruses. 5:
        159.
Blancou, J., Kieny, M. P., Lathe, R., Lecocq, J. P., Pastoret, P. P.,
        Soulebot, J. P. and Desmettre, P., 1986. Oral vaccination of the
        fox against rabies using a live recombinant vaccinia virus. Nature.
        322:373.
Boyle, D. B. and Coupar, B. E. H., 1988. A dominant selectable marker for
        the construction of recombinant poxviruses. Gene. 65:123.
Buonocore, L. and Rose, J. K., 1990. Prevention of HIV-1 glycoprotein
        transport by soluble CD4 retained in the endoplasmic reticulum.
        Nature. 345:625.
Chakrabarti, S., Brechling, K. and Moss, B., 1985. Vaccinia virus
        expression vector: Coexpression of β-galactosidase provides visual
        screening of recombinant virus plaques. Mol. Cell. Biol. 5:3403.
Cochran, M. A., Mackett, M. and Moss, B., 1985. Eukaryotic transient
        expression system dependent on transcription factors and regulatory
        DNA sequences of vaccinia virus. Proc. Natl. Acad. Sci. USA. 82:19.
Cooney, E. L., Collier, A. C., Greenberg, P. D., Coombs, R. W., Zarling,
        J., Arditti, D. E., Hoffman, M. C., Hu, S. L. and Corey, L., 1991.
        Safety of and immunological response to a recombinant vaccinia
        virus vaccine expressing HIV envelope glycoprotein. Lancet.
        337:567.
Davison, A. J. and Moss, B., 1990. New vaccinia virus recombination
        plasmids incorporating a synthetic late promoter for high level
        expression of foreign proteins. Nucleic Acids Research. 18:4285.
Earl, P. L., Hügen, A. W. and Moss, B., 1990. Removal of cryptic poxvirus
        transcription termination signals from the human immunodeficiency
        virus type 1 envelope gene enhances expression and immunogenicity
        of a recombinant vaccinia virus. J. Virol. 64:2448.
Elroy-Stein, O., Fuerst, T. R. and Moss, B., 1989. Cap-independent
        translation of mRNA conferred by encephalomyocarditis virus 5'
        sequence improves the performance of the vaccinia
        virus/bacteriophage T7 hybrid expression system. Proc. Natl. Acad.
        Sci. USA. 86:6126.

Elroy-Stein, O. and Moss, B., 1990. Cytoplasmic expression system based on constitutive synthesis of bacteriophage T7 RNA polymerase in mammalian cells. Proc. Natl. Acad. Sci. USA. 87:6743.

Falkner, F. G. and Moss, B., 1988. Escherichia coli gpt gene provides dominant selection for vaccinia virus open reading frame expression vectors. J. Virol. 62:1849.

Falkner, G. G., Chakrabarti, S. and Moss, B., 1987. pUV I: a new vaccinia virus insertion and expression vector. Nucl. Acids. Res. 15:7192.

Franke, C. A., Rice, C. M., Strauss, J. H. and Hruby, D. E., 1985. Neomycin resistance as a dominant selectable marker for selection and isolation of vaccinia virus recombinants. Mol. Cell. Biol. 5:1918.

Fuerst, T. R., Earl, P. L. and Moss, B., 1987. Use of a hybrid vaccinia virus T7 RNA polymerase system for expression of target genes. Mol. Cell. Biol. 7:2538.

Fuerst, T. R.and Moss, B., 1989. Structure and stability of mRNA synthesized by vaccinia virus-encoded bacteriophage T7 RNA polymerase in mammalian cells. Importance of the 5' untranslated leader. J. Mol. Biol. 206:333.

Fuerst, T. R., Niles, E. G., Studier, F. W. and Moss, B., 1986. Eukaryotic transient-expression system based on recombinant vaccinia virus that synthesizes bacteriophage T7 RNA polymerase. Proc. Natl. Acad. Sci. USA. 83:8122.

Isaacs, S. N., Kotwal, G. J. and Moss, B., 1990. Reverse guanine phosphoribosyltransferase selection of recombinant vaccinia viruses. Virology. 178:626.

Keil, W. and Wagner, R. R., 1989. Epitope mapping by deletion mutants and two chimeras of two vesicular stomatitis virus glycoprotein genes expressed by a vaccinia virus vector. Virol. 170:392.

Li, Y., Luo, L., Snyder, R. M. and Wagner, R. R., 1988. Expression of the M gene of vesicular stomatitis virus cloned in various vaccinia virus vectors. J. Virol. 62:776.

Mackett, M., Smith, G. L. and Moss, B., 1982. Vaccinia virus: a selectable eukaryotic cloning and expression vector. Proc. Natl. Acad. Sci. USA. 79:7415.

Mackett, M., Smith, G. L. and Moss, B., 1984. General method for production and selection of infectious vaccinia virus recombinants expressing foreign genes. J. Virol. 49:857.

Massung, R. F. and Moyer, R. W., 1990. Orthopoxvirus gene expression in Xenopus laevis oocytes: a component of the virion is needed for late gene expression. J. Virol. 64:2280.

Mizukami, T., Fuerst, T. R., Berger, E. A. and Moss, B., 1988. Binding region for human immunodeficiency virus (HIV) and epitopes for HIV-blocking monoclonal antibodies of the CD4 molecule defined by site-directed mutagenesis. Proc. Natl. Acad. Sci. USA. 85:9273.

Moss, B., 1990a. Poxviridae and their replication. in: Virology., B. N. Fields, D. M. Knipe, R. M. Chanock, M. S. Hirsch, J. Melnick, T. P. Monath and B. Roizman, ed., Raven Press, New York.

Moss, B., 1990b. Regulation of vaccinia virus transcription. Ann. Rev. Biochem. 59:661.

Moss, B., 1991. Vaccinia virus: a tool for research and vaccine development. Science. 252:1662.

Moss, B., Ahn, B.-Y., Amegadzie, B., Gershon, P. D. and Keck, J. G., 1991. Cytoplasmic transcription system encoded by vaccinia virus. J. Biol. Chem. 266:1355.

Moss, B. and Flexner, C., 1987. Vaccinia virus expression vectors. Ann. Rev. Immunol. 5:305.

Panicali, D., Grzelecki, A. and Huang, C., 1986. Vaccinia virus vectors utilizing the β-galactosidase assay for rapid selection of recombinant viruses and measurement of gene expression. Gene. 47:193.

Panicali, D. and Paoletti, E., 1982. Construction of poxviruses as cloning vectors: insertion of the thymidine kinase gene from herpes simplex virus into the DNA of infectious vaccinia virus. Proc. Natl. Acad. Sci. USA. 79:4927.

Patel, D. D., Ray, C. A., Drucker, R. P. and Pickup, D. J., 1988. A poxvirus-derived vector that directs high levels of expression of cloned genes in mammalian cells. Proc. Natl. Acad. Sci. USA. 85:9431.

Pattnaik, A. K. and Wertz, G. W., 1991. Cells that express all five proteins of vesicular stomatitis virus from cloned cDNAs support replication, assembly, and budding of defective interfering particles. Proc. Natl. Acad. Sci. USA. 88:1379.

Perkus, M. E., Limbach, K. and Paoletti, E., 1989. Cloning and expression of foreign genes in vaccinia virus, using a host range selection system. J. Virol. 63:3829.

Rich, D. P., Anderson, M. P., Gregory, R. J., Cheng, S. H., Paul, S., Jefferson, D. M., McCann, J. D., Klinger, K. W., Smith, A. E. and Welsh, M. J., 1990. Expression of cystic fibrosis transmembrane conductance regulator corrects defective chloride channel regulation in cystic fibrosis airway epithelial cells. Nature. 347:358.

Rodriguez, J. F. and Esteban, M., 1989. Plaque size recombinants as a selectable marker to generate vaccinia virus recombinants. J. Virol. 63:997.

Rupprecht, C. E., Wiktor, T. J. A., Johnston, D. H., Hamir, A. N., Dietzschold, B., Wunner, W. H., Glickman, L. T. and Koprowski, H., 1986. Oral immunization andprotection of raccooons (Procyon lotor) with a vaccinia-rabies glycoprotein recombinant virus vaccine. Proc. Natl. Acad. Sci. USA. 83:7947.

Shuman, S. and Moss, B., 1988. Factor-dependent transcription termination by vaccinia virus RNA polymerase: Evidence that the cis-acting termination signal is in nascent RNA. J. Biol. Chem. 263:6220.

Smith, G. L. and Moss, B., 1983. Infectious poxvirus vectors have capacity for at least 25,000 base pairs of foreign DNA. Gene. 25:21.

Studier, F. W., Rosenberg, A. H., Dunn, J. J. and Dubendorff, J. W., 1990. Use of T7 RNA polymerase to direct the expression of cloned genes. Meth. Enzymol. 185:60.

Whitt, M. A., Buonocore, L., Rose, J. K., Ciccarone, V., Chytil, A. and Gebeyehu, G., 1991. TransfectACE[TM] Reagent: Transient transfection frequencies >90%. Focus (Life Technologies, Inc.). 13:8.

Whitt, M. A., Chong, L. and Rose, J. K., 1989. Glycoprotein cytoplasmic domain sequences required for rescue of a vesicular stomatitis virus glycoprotein mutant. J. Virol. 63:3569.

Wiktor, T. J., Macfarlan, R. I., Reagan, K. J., Dietzschold, B., Curtis, P., Wunner, W., H., Kieny, M. P., Lathe, R., Lecocq, J. P., Mackett, M., Moss, B. and Koprowski, H., 1984. Protection from rabies by a vaccinia virus recombinant containing the rabies virus glycoprotein gene. Proc. Natl. Acad. Sci. USA. 81:7194.

Yang, X.-C., Karschin, A., Labarca, C., Elroy-Stein, O., Moss, B., Davidson, N., Lester, H.A., 1991. Expression of ion channels and receptors in Xenopus oocytes using vaccinia virus. FASEB J. 5:2209.

Yilma, T., Hsu, D., Jones, L., Owens, S., Grubman, M., Mebus, C., Yamanaka, M. and Dale, B., 1988. Protection of cattle against rinderpest with vaccinia virus recombinants exprssing the HA or F gene. Science. 242:1058.

Yuen, L. and Moss, B., 1987. Oligonuclotide sequence signaling transcriptional termination of vaccinia virus early genes. Proc. Natl. Acad. Sci. USA. 84:6417.

APPENDIX

Alun G. Jones and Max D. Summers presented lectures at the Ninth IUCCP Symposium but did not submit manuscripts for inclusion in this volume. Abstracts of their talks are attached below. If desired, reader can correspond directly with the lecturer of interest for more information.

BACULOVIRUS EXPRESSION VECTORS

M.D. Summers

Texas A & M University
The Center for Advanced Invertebrate Molecular Sciences
and the Department of Entomology
College Station, Texas

The baculovirus expression vector is a new eucaryotic DNA viral vector for the cloning and expression of genes in cultured lepidopteran insect cells and insects. It is now an important tool for studies of the molecular biology of invertebrate cells and insect systems. The baculovirus expression vector is a helper-independent recombinant baculovirus which, during infection, can express abundant quantities of recombinant mRNA and proteins from foreign gene constructs inserted under the transcriptional regulation of the strong baculovirus polyhedrin gene promoter. Chimeric genes for potentially any protein can now be expressed under the transcriptional regulation of baculovirus promoters representing genes of immediate early, delayed early, late and the very late promoter classes in order to introduce recombinant products into infected cells or the organism at selected times during viral infection and replication. Most of the recombinant proteins produced in the baculovirus system are functionally authentic. The characteristics of this cloning and expression system, the protein processing systems which are known to function in insect cells and applications to medicine and agriculture will be reviewed.

METHODS OF LABELING ANTIBODIES WITH RADIONUCLIDES OF TECHNETIUM

Alun G. Jones

Department of Radiology
Harvard Medical School
Boston, Massachusetts

Since the introduction of monoclonal antibodies (MoAbs), many efforts have been made to exploit their specificity. In nuclear medicine, this has taken the form of targeting radionuclides for both diagnostic and therapeutic purposes in cancer, cardiovascular disease and other forms of pathology. Most of the investigative experience in this country thus far has been obtained with conjugated proteins incorporating indium-111, an isotope with a half-life of three days, or with molecules labeled with iodine radionuclides.

From the standpoint of diagnostic imaging, the most favorable nuclide is Tc-99M, a 140 KeV gamma emitter with a half-life of 6 hours that is readily available in hospitals from a generator system. The nuclide is produced in the form of $[TcO_4]^-$, a complex in the highest oxidation state of the element. At present, more than 85% of the 7.5 million nuclear medicine studies performed each year in the US involve the administration of this nuclide in some chemical form. Because of its central importance in the field, there have been numerous methods devised for labeling MoAbs or their fragments with [99m]Tc. These fall broadly into two categories: direct labeling and indirect labeling.

Direct labeling involves methods that take advantage of functional groups naturally present on the protein to coordinate technetium in some reduced chemical form. These groups, such as $-NH_2$, -OH, -COOH, or -SH, may be free on the native MoAb or exposed through prior treatment of the protein. Reduction of the pertechnetate precursor has been effected in several ways including the use of stannous tin and by the production of reactive intermediates such as nitrido species.

The second method is indirect labeling, more commonly known as the bifunctional chelate approach. This generally involves the covalent attachment of a strong chelating agent on the antibody at $-NH_2$ or -SH groups, or on the carbohydrate moiety. These chelators then provide a number of high affinity binding sites for the reduced technetium. Variations on the technique depend upon the type of bifunctional chelate used, the point of attachment to the protein, and also whether the chelating agent is complexed with the metal radionuclide prior to or subsequent to conjugation to the antibody. In principle, similar approaches may be taken with the congener transition metal rhenium, since both [186]Re and [188]Re are beta emitters that may prove useful in radionuclide therapy. The experience here, however, is much less than with technetium since the radiopharmaceutical chemistry of rhenium has been less intensively studied.

This talk will discuss these methods and their evolution through the innovative application of a steadily increasing body of knowledge in technetium chemistry.

INDEX

models for the diiron center in the
    hydroxylase, 76
modular protein assembly, 140
molecular chaperonins recognition site, 134
molecular weight determination by
    capillary electrophoresis, 244
molten globule, 141
monoclonal antibodies, 4, 5, 15, 29
Mössbauer spectra, 81
Mössbauer spectroscopy of MMO, 41
mpr gene, 263
mRNA, 277
multiple drug resistance, 283
mutagenesis, 278, 287

2D NMR, 142
N,N'-bis(2-hydroxybenzyl)-1-(4-
    bromoacetamido-benzyl)-1,2-
    ethylene-diamine-N,N'-dicetic acid
    (BøHBED), 10
N,N-bis(2-hydroxybenzyl)
    ethylenediamine-N,N',diacetic
    acid (HBED), 10
N-(2-hydroxy-3,5-dimethylbenzyl)-N-(2-
    hydroxy-polyclonal antibody, 3
N-linked glycosylation, 281
N-succinimidyl-3-[At-211]astatobenzoic
    acid, 24
N-succinimidyl 2,4-dimethoxy-3-
    iodobenzoate (SDMIB), 22
N-succinimidyl esters, 18
N-succinimidyl 8-{(4'-[F-
    18]fluorobenzyl)amino}suberate
    (SFBS), 26
N-succinimidyl-3-iodobenzoate, 18, 19, 21
N-succinimidyl 4-iodobenzoate, 20
N-succinimidyl 5-iodo-3-
    pyridinecarboxylate (SIPC), 22
N-succinimidyl 3-trialkyl-stannylbenzoates,
    21
native state, 140
6-(p-nitrobenzyl)-1,4,8,11-
    tetracyclotetradecane-N,N',N'',N'''-
    tetraacetic acid (p-nitrobenzyl
    TETA), 10
nitrogenase, 58
nitrosoguanidine (NTG), 278
NTG mutagenesis,287
nucleic acid separation by filled capillary
    electrophoresis, 244
NusA, 265

O_2, 70
O_2 activation, 70
octopine dehydrogenase, 145
operon engineering, 184
optical spectrum of MMO, 82
organocobalt, 198
organohalide, 191

organophosphorus acid (OPA) anhydrase,
    171
ovarian carcinomas, 23
oxene, 46
oxidation of saturated hydrocarbons, 87
oxidation of trichloroethylene
    by dioxygenases, 196
    by monooxygenases, 196
oxidation to ketones, 89
oxidize methane anaerobically, 57
oxygenase, 39
oxygenase dehalogenation, 195
oxygenic photosynthesis, 57

parathion hydrolase, 171, 187
    from *Bacillus stearothermophilus*, 172
    from *Flavobacterium*, 171
    from *Pseudomonas diminuta*, 171
    from *Streptomyces lividands*, 172
partially folded intermediates, 129
PCR, 263
pediatric intracranial tumors, 29
penicillin, 89
pepstatin-insensitive carboxyl protease, 279
peptide and protein separations, 233
peptidyl-proyl cis-trans isomerase (PPI),
    147
peroxidase, 48
peroxide bridged intermediate, 77
peroxide shunt, 47, 50
phospholipase A2, 205
phospholipase D, 206
picolinic acid as a ligand, 89
pineal tumor, 35
pMMo, 57
PMRI-null mutants, 283
poly-His IL-7, 253
polyacrylamide gel filled capillary
    electrophoresis
polyadenylated, 277
polyarginine, 251
polycysteine fusion, 252
polyhistidine, 252
polysaccharides as supports for whole cells,
    157
position emission tomograph (PET), 1
positron emission tomography, 15
pre-steady state kinetics of MMO reaction,
    60
preprochymosin, 273
product distribution of MMO oxidations, 46
proline hypothesis, 118
promoter, 267
promoter sequences, 295
proteases in *B. subtilis*, 263
protein aggregation, 129
protein assembly, 145
protein B, 59
protein disulfide isomerase (PDI),147
protein folding, 137
protein folding and cyclophilins, 118

waste management using immobilized
  bacterial cells, 154
wrong protein aggregation, 145

X-ray crystal structure of the B2 protein of
  ribonucleotide reductase, 65

yeast PMRI gene, 282
yttrium-90, 7

zirconium-89, 7